Benchmark Papers in Genetics

Series Editor: David L. Jameson
University of Houston

Volume

RELATED TITLES IN OTHER BENCHMARK SERIES

Benchmark Papers
in Genetics / 10

A BENCHMARK ® Books Series

HUMAN GENETICS
A Selection of Insights

Edited by

WILLIAM J. SCHULL

University of Texas at Houston

and

RANAJIT CHAKRABORTY

University of Texas at Houston

Dowden, Hutchinson
& Ross, Inc.
STROUDSBURG, PENNSYLVANIA

LIBRARY OF CONGRESS CATALOGING IN PUBLICATION DATA

Main entry under title:
Human genetics.
    (Benchmark papers in genetics ; 10)
    Includes index.
    1. Human genetics—Addresses, essays, lectures.
I. Schull, William J. II. Chakraborty, Ranajit,
1946-   joint author.
QH431.H836   573.2'1   78-13701
ISBN 0-87933-321-9

Distributed world wide by Academic Press,
a subsidiary of Harcourt Brace Jovanovich,
Publishers.

SERIES EDITOR'S FOREWORD

The study of any discipline assumes the mastery of the literature of the subject. In many branches of science, even one as new as genetics, the expansion of knowledge has been so rapid that there is little hope of learning of the development of all phases of the subject. The student has difficulty mastering the textbook, the young scholar must tend to the literature near his own research, the young instructor barely finds time to expand his horizons to meet his class-preparation requirements, the monographer copes with wider literature but usually from a specialized viewpoint, and the textbook author is forced to cover much the same material as previous and competing texts to respond to the user's needs and abilities.

Few publishers have the dedication to scholarship to serve primarily the limited market of advanced studies. The opportunity to assist professionals at all stages of their careers has been recognized by Dowden, Hutchinson & Ross and by a distinguished group of editors knowledgeable in specific portions of the genetic literature. These editors have selected papers and portions of papers that demonstrate both the development of knowledge and the atmosphere in which that knowledge was developed. There is no substitute for reading great papers. Here you can learn how questions are asked, how they are approached, and how difficult and essential it is to obtain definitive answers and clear writing.

The editors of this volume have presented us with a selection of papers demonstrating the interlocking of human genetic studies with the main historical threads of basic genetic studies. The first section is largely historical and traces the development of genetics from the turn of the century through six decades. The second section relates the genetic and chromosomal nature of many human diseases while the third section examines the nature and maintenance of genetic variability in man. The authors close with an especially timely essay on the social and legal problems which accompany the knowledge now available to mankind. The development of new social responsibility structures are clearly called for but the nature of these await extensive interdisciplinary analysis that must include philosophical and humanitarian and social disciplinary efforts.

DAVID L. JAMESON

PREFACE

It is an intrepid act, if not a presumptuous one, to attempt to identify a series of "benchmark" contributions in a discipline of as many parts as human genetics. A selection solely of clinical contributions will not satisfy the views of the nonclinician; similarly, a preponderance of papers of a biochemical nature will disappoint the nonbiochemist. One must accept, therefore, that no selection will achieve universal approbation. We have, as a consequence, adopted a generalist position and chosen a series of papers that addresses three broad areas: the formal genetics of man; genes, chromosomes, and disease; and man's genetic variability. Within each, the choice of a particular paper reflects our perception of its importance, its accessibility, and the eductational value it may have aside from the specific scientific contribution it makes.

These principles are, of course, more readily stated than rigorously adhered to. We have, for example, incorporated only one publication in a language other than English. Parochialism was certainly not our intent; rather we merely acknowledge the lessened emphasis on foreign languages in the scientific curricula of American universities and the loss in the sense of immediacy that accompanies most translations. This approach has led to omissions for which we apologize. Wilhelm Weinberg, for example, is not represented among the authors of any of these selections, nor are those members of the Maxim Gorky Institute who published primarily in Russian. We have attempted to acknowledge our indebtedness to our colleagues abroad in the prefatory remarks that accompany each section. Hopefully the more dedicated reader will seek out these publications and immerse herself or himself, or at least attempt to, in the intellectual milieu that prompted these creative efforts.

Historians of science often allude to *zeitgeist* in the characterization of the import of a particular paper or scientific event. Often such characterization is, we feel, pejorative, but most sciences, including human genetics, have their fads or at least transitory emphases. We have tried both in our selection of papers and in our remarks to distinguish between those creative acts with no apparent, immediate antecedents and those, which like the specification of the genetic code, may have been creative from the laboratory viewpoint, but represent the culmination of an intellectual ferment so pervasive as to dominate the activities of numerous laboratories. We are reluctant to place the latter contributions, despite the elegance of the laboratory technique, on a par with the former. We grant our views to be moot, but surely a contribution of the former kind does not lie outside the prospects of any gifted scientist whereas the latter too frequently depends on accoutrements rather than insight.

We are indebted to many friends and colleagues who have responded unselfishly to our requests for opinions on the merits of our selections and to our solicitations for alternative suggestions. A similar sense of unselfishness tempts us to share with them responsibility for the bad choices as well as the good, but patently this would be unfair. They have contributed importantly to the recognition of the latter, but the former can only represent our lack of insight, appreciation, or sensitiveness of the cross-currents that have characterized, and still do, our discipline. We expecially wish to acknowledge our colleagues here at the Center, Drs. Masatoshi Nei, Wen-Hsiung Li, Kenneth Weiss, and Robert Ferrell, and in addition, Drs. Margery Shaw and T. Edward Reed who have tried to protect us from our own exuberances. Once imbued with the spirit of John—*Apud verbo erat*—it is difficult to stay one's visions of the order of things. But *"Cria cuervos y te sacaran los ojos."* * Finally, to Ms. Sara Barton and Jeryl Silverman without whose patience and forebearance and gentle insistence this volume would never have been assembled—many, many thanks.

WILLIAM J. SCHULL
RANAJIT CHAKRABORTY

*"Raise crows and they will pick out your eyes" [an old Spanish adage].

CONTENTS

PART I: FORMAL GENETICS OF MAN

Contents

CONTENTS BY AUTHOR

HUMAN GENETICS

Part I

FORMAL GENETICS OF MAN

Editors' Comments
on Papers 1 Through 10

Long before 1900, even before Mendel conducted his experiments with garden peas, physicians encountered patients with obvious but unusual abnormalities, and many of these patients found places in their memoirs. Perusal of the descriptions led more inquisitive minds to believe that at least some of the abnormalities were inherited. Thus, in 1750 Maupertuis (1689–1759) in Berlin described the familial occurrence of polydactyly; in 1820 Nasse apprehended the salient features of X-linked recessive inheritance of hemophilia; and in 1876 Horner, a Swiss ophthalmologist, recognized the X-linked recessive pattern of color blindness. Means to alleviate the impact of these disorders did not generally exist, although exceptions did occur. For example, the Talmud includes a provision to dispense with circumcision in newborn males whose older brothers or maternal uncles have displayed a bleeding tendency.

Although the descriptions and reasonings were not in terms of "genes" and "loci," the impact of the works of the early, genetically oriented clinicians is immeasurable. Mayer (1972), for example, has written of Prosper Lucas, one of these clinicians, as follows:

> This was the stage and set for the pioneer work of a French psychiatrist, Prosper Lucas (1808–1885) whose "Philosophical and Physiological Treatise on Natural Heredity" (Paris, 1847–1850, 2nd vol.) served as the basic inspiration of many mid-XIXth century students of human genetics and eugenics, so much so that Prosper Lucas ought to be resurrected and placed upon a pedestal as the true founder of human genetics.

While Lucas' book had its effect on scientific arguments of eugenics, the publication of "A Treatise on the Supposed Hereditary Properties of Disease" by Joseph Adams in 1814 is claimed by some to be the first documentation of the use of the principles of human genetics as a guide for heredity counseling (see Motulsky 1959). In fact, Motulsky suggests that this London apothecary-physician is the "forgotten founder" of medical genetics.

Genetic defects have long been given social value. In India, for example, Manu's law* asserts that a person of uncertain parentage will transmit the bad quality of his or her father or mother to his or her children. Indeed, this notion allegedly resulted in the hereditary stratification of occupation and classes seen in India. A proper understanding of inherited traits and their social import emerged, however, only recently. The late nineteenth century and the first decade of this century saw a grim battle of scientific views being fought in and around the Evolution Committee of the Royal Society. These caustic and very

---

*According to Hindu mythology, Manu is one of the fourteen progenitors of the human race; he codified the laws that are still observed by Hindus.

often vicious exchanges between the biometricians—led by Galton, Pearson, and Weldon—and the Mendelians—led by Bateson—eventually resulted in a unified theory explaining genetic variation causing evolutionary changes in all organisms. These controversies contributed much to the recognition of genetics as the science of natural variation. Indeed, few definitions of genetics are more apt than Dobzhansky's description of it as the "physiology of variation"—that is, as the study of the function of variation.

Our first few selections are drawn from the period soon after Hugo de Vries, Carl Correns, and Erich von Tschermark rediscovered Mendel's theory of heredity. Although Archibald Garrod, for example, was not directly involved in the controversy, his discovery (Paper 1) of the first instance of recessive inheritance in describing the metabolic disorder called "alkaptonuria" (in which a substance called alkapton is excreted in the urine, which turns black on standing) was immediately taken up by Bateson. Bateson used it in a rebuttal in the debate before the Evolution Committee to show the impact of consanguineous marriages on the manifestation of rare recessive genes in the light of Mendel's theory. The general biological consequences of consanguinity had been described much earlier in a report of the American Medical Association in 1859 published by Bemiss. Garrod's paper marks the beginning of human biochemical genetics. It inspired his later concept (Paper 16) of "inborn errors of metabolism" (Garrod 1908; see also other papers reproduced in Part II of this volume), and ultimately led to George Beadle's and Edward Tatum's Nobel Prize winning work on the one-gene-one-enzyme hypothesis (see Knox 1972 for a recent review that has some bearing on this subject).

Bateson's note (Paper 2), which appeared as a footnote on his report to the Evolution Committee, was of "extraordinary" importance to him in justifying the claim that "a recessive allelomorph may even persist as a gamete without the corresponding homozygote having ever reached maturity in the history of the species." Bateson's recognition of the importance of consanguinity saw further attempts to delineate the effects of inbreeding on the genetic composition of a population (see, e.g., Pearl 1913, 1914; Fish 1914; Jennings 1914, 1916, 1917; Robbins 1917, 1918; East and Jones 1919; Wright 1921). In fact, as recent studies have shown (see, e.g., Schull and Neel 1965), Bateson's discovery has bearing not only on simple Mendelian recessive traits but also on others with much more complex genetic etiologies.

Farabee's contribution (Paper 3), on the other hand, indicates that abnormal human traits can be autosomal dominant in nature. The abnormality he studied, brachydactyly or short fingers, was physically conspicuous. The chemical basis of the action of such genes is obscure.

When Ottenberg presented his and Epstein's findings (Paper 4) on a simplified method of typing blood groups at a meeting of the New York Pathology Society in 1908, it was a boon to pathologists since Landsteiner's method requires a greater volume of blood, which is difficult to obtain from infants or in cases of great urgency. The similarity of the blood groups of brothers and sisters through agglutination tests performed by Ottenberg and Epstein led them to believe that if "the sharply opposed nature of blood characteristics" are inherited, "they will form a very good example of the Mendelian law of heredity." The next two decades underscored their insight. First, there were suggestions that the ABO blood groups might be a multilocus system. However, through the statistical efforts of Bernstein (1924) and the population studies of Hirschfeld and Hirschfeld (1919), it became clear that the ABO system is in fact a single locus with multiple alleles. The inheritance patterns of other blood groups, with which Landsteiner's name is closely associated, went through a similar series of controversies, with the Rh-system being the most noteworthy.

Landsteiner and Levine's lucid presentation (Paper 5) is important not only for the discovery of the MN-system but also for the methodology it exhibits. First presented in 1928, this methodology is routinely followed even today for elucidating the Mendelian inheritance of any new trait. Family data collected and analyzed by Landsteiner and Levine also indicate that Mendelian traits can be of immense help in forensic and medico-legal practices and thus contribute to the resolution of a number of practical problems. It would be appropriate to mention Alexander S. Weiner's name here in view of the numerous contributions to this field that he made in later years. The reader may also refer to a recent paper (Chakraborty et al. 1974) to judge the immense foresight Landsteiner and Levine had in writing: "These findings offer the prospect of forensic application to cases of disputed paternity and, in our opinion, a correct decision could already be given, at least with great probability, provided the reagents are available and the method properly applied."

Levit (Paper 6) opened another avenue of research in human genetics in 1936. He was concerned with genes of pathological importance. Fisher (1929, 1931) had already advocated that the majority of mutant genes in man as well as in most plants and animals that had been studied are recessive. Through a critical assessment of the literature, Levit came to the conclusion in Paper 6: that "the majority of mutant pathological genes in man are conditionally dominant"—that is, the phenotype of the heterozygote may not be the same as that of an individual homozygous for the rare allele and thus true dominance does not obtain. Studies of brachydactyly (see Mohr and Wreidt 1919) and other dominant pathological characteristics (Snyder and Doan 1944) are pertinent in this respect. Levit also dealt with questions such as when can a gene be recognized as dominant and on what grounds one can establish the conditional dominance of a particular gene. Some of Levit's comments on the incidences of a trait among relatives and the use of these figures to ascertain the mode of inheritance were also timely since the theory of polygenic inheritance in man was only then being developed (Fisher 1918; Wright 1921). Levit went on to argue that the poor penetrance of conditionally dominant genes can be explained from the viewpoint of the evolution of dominance.

Genetics and disease is also the theme of the paper by Macklin (Paper 7). One of a number of important but unsung early contributors to human genetics, she carefully documented the relationship between the age of onset of a disease, its severity, and its mode of inheritance. In Paper 7 Macklin emphasizes the distinction between familial and hereditary forms of a disease—a distinction that has been made by others but rarely as clearly. The importance of the location of mutations in governing the transmission and severity of disease is also discussed.

A critical examination of the manifestation and severity of diseases also led Morton and Chung (Paper 8) to believe several genetic classes exist within a single form of disorder, muscular dystrophy. In this paper and in a companion one (Chung and Morton 1959), the authors used one of the major techniques of statistical analysis, discriminant analysis, for the first time to solve questions in human medical genetics. Although Fisher (1938) was the first to define a discriminant function, the idea that such a function can be constructed is implicit in Mahalanobis' D^2-statistic (Mahalanobis 1936). For several forms in which such a function can be applied, see Rao's treatises (Rao 1952, 1973). Discriminant analysis and maximum likelihood scores helped Morton and Chung to delineate at least four different genetic categories of muscular dystrophy, in each of which genetic and empirical risks were predicted from data on relationship, age, and the number of normal relatives. As an approach to the formal genetics of man, their treatment is relatively sophisticated, but they were not concerned with the dynamics of such genes in populations. In this sense the approach may be viewed as a static one.

Edwards (Paper 9) brings out the operational distinction between quasi-continuous variation and the effects of major genes through the notion of continuous liability as the basis for discrete phenotypes (see also Edwards 1969). Although Penrose's (1953) findings (Paper 20, Part II) have some bearing on this treatment, Falconer developed the notion more fully (Falconer 1965). Falconer's textbook also presents some allied techniques based on his animal experiments (Falconer 1960). The etiologies of complex human disorders, however, are more difficult to unravel, for the interaction of nature and nurture is far more obscure in the "real world." More recently, the statistical aspects of model building have been considered at various depths by a number of investigators (see, e.g., Smith 1971; Mendell and Elston 1974; Morton et al. 1970; Morton 1974).

These nine papers, individually considered, are of sufficient historical merit to warrant inclusion in this volume, we believe, but in no one is the logic of the formal genetics of man explicitly elaborated. The Croonian lecture by Professor Haldane (Paper 10) addresses this aspect and, in our opinion, affords a fitting conclusion to this section. Here he considers problems arising from ascertainment biases for simple Mendelian traits, human partial linkage studies related to sex chromosomes, and finally the difficulties inherent in the evaluation of mutation rates. The theme is that "breeding behavior is the ultimate test" in the study of the formal genetics of man.

REFERENCES

Bernstein, F. 1924. Ergebnisse einer biostatischen zusammenfassenden betractung uber die erblichem blutstrukturen des menschen. *Clin. Wsch.* **3**:1495–97.

Chakraborty, R., M. Shaw, and W. J. Schull. 1974. Exclusion of paternity: The current state of the art. *Am. J. Hum. Genet.* **26**:477–88.

Chung, C. S., and N. E. Morton. 1959. Discrimination of genetic entities in Muscular Dystrophy. *Am. J. Hum. Genet.* **11**:339–359.

Dobzhansky, T. 1937. *Genetics and the origin of species.* Columbia Univ. Press, New York.

East, E. M., and D. F. Jones. 1919. *Inbreeding and Outbreeding.* J. B. Lippincott, Philadelphia.

Edwards, J. H. 1969. Familial predisposition in man. *Brit. Med. Bull.* **25**:58–64.

Falconer, D. S. 1960. *Introduction to Quantitative Genetics.* Ronald Press, New York.

Falconer, D. S. 1965. The inheritance of liability to certain diseases, estimated from the incidence among relatives. *Ann. Hum. Genet.* **29**:51–76.

Fish, H. D. 1914. On the progressive increase of homozygosis in brother-sister matings. *Am. Nat.* **48**:759–61.

Fisher, R. A. 1918. The correlation between relatives on the supposition of Mendelian inheritance. *Trans. Roy. Soc. Edinburgh* **52**:399–433.

Fisher, R. A. 1929. The evolution of dominance: Reply to Professor Sewall Wright. *Am. Nat.* **63**:553–56.

Fisher, R. A. 1931. The evolution of dominance. *Biol. Rev.* **6**:345–68.

Fisher, R. A. 1938. The statistical utilization of multiple measurements. *Ann. Eugen.* **8**:376–86.

Garrod, A. E. 1908. Inborn errors of metabolism. *Lancet* ii:1–7.

Hirschfeld, L., and H. Hirschfeld. 1919. Serological differences between the blood of different races. The result of researches on the Macedonian front. *Lancet* ii: 675–79.

Jennings, H. S. 1914. Formulae for the results of inbreeding. *Am. Nat.* **48**:693–96.

Jennings, H. S. 1916. The numerical results of diverse systems of breeding. *Genetics* **1**:53–89.

Jennings, H. S. 1917. The numerical results of diverse systems of breeding with respect to two pairs of characters, linked or independent, with special relation to the effects of linkage. *Genetics* 2:97–154.

Knox, W. E. 1972. Phenylketonuria. In J. B. Stanbury, J. B. Wyngaarden, and D. S. Fredrickson (eds.), *The Metabolic Basis of Inherited Disease,* 3rd ed. McGraw-Hill, New York, pp. 266–95.

Mahalanobis, P. C. 1936. On the generalized distance in statistics. *Proc. Nat. Inst. Sci. India* 12:49–55.

Mayer, C. F. 1972. From Plato to Pope Paul. In J. de Grouchy, F. J. G. Ebling, and I. W. Henderson (eds.), *Human Genetics.* Excerpta Medica, Amsterdam, pp. 471–80.

Mendell, N. R., and R. C. Elston. 1974. Multifactorial qualitative traits: Genetic analysis and prediction of recurrence risks. *Biometrics* 30:41–57.

Mohr, O. L., and C. Wreidt. 1919. A new type of hereditary brachyphalangy in man. Carnegie Institute (publication 295), Washington, D.C., pp. 5–64.

Morton, N. E. 1974. Analysis of family resemblance. I: Introduction. *Am. J. Hum. Genet.* 26:318–30.

Morton, N. E., S. Yee, R. C. Elston, and R. Lew. 1970. Discontinuity and quasi-continuity: Alternative hypotheses of multifactorial inheritance. *Clin. Genet.* 1: 81–94.

Motulsky, A. G. 1959. Joseph Adams (1756–1818): A forgotten founder of medical genetics. *AMA Arch. Int. Med.* 104:490–96.

Pearl, R. 1913. A contribution towards an analysis of the problem of inbreeding. *Am. Nat.* 47:577–614.

Pearl, R. 1914. On the results of inbreeding in a Mendelian population: A correction and extension of previous conclusions. *Am. Nat.* 48:57–62.

Penrose, L. S. 1953. The genetical background of common diseases. *Acta Genet.* 4: 257–65.

Rao, C. R. 1952. *Advanced Statistical Methods in Biometric Research.* Wiley and Sons, New York.

Rao, C. R. 1973. *Linear Statistical Inference and its Applications.* Wiley and Sons, New York.

Robbins, R. B. 1917. Some applications of mathematics to breeding problems. *Genetics* 2:489–504.

Robbins, R. B. 1918. Applications of mathematics to inbreeding problems. II, III: *Genetics* 3:73–92, 375–89.

Schull, W. J., and J. V. Neel. 1965. *The Effects of Inbreeding on Japanese Children.* Harper & Row, New York:

Smith, C. 1971. Discrimination between different modes of inheritance in genetic disease. *Clin. Genet.* 2:303–14.

Snyder, L. H., and C. A. Doan. 1944. Clinical and experimental studies in human inheritance: Is the homozygous form of multiple telangiectasia lethal? *J. Lab. Clin. Med.* 29:1211–16.

Wright, S. 1921. Systems of mating. I–V: *Genetics:*111–78.

1

Copyright © 1902 by Lancet, Ltd.

Reprinted from *Lancet* II:1616–1620 (1902)

THE INCIDENCE OF ALKAPTONURIA: A STUDY IN CHEMICAL INDIVIDUALITY

Archibald E. Garrod

ALL the more recent work on alkaptonuria has tended to show that the constant feature of that condition is the excretion of homogentisic acid, to the presence of which substance the special properties of alkapton urine, the darkening with alkalies and on exposure to air, the power of staining fabrics deeply, and that of reducing metallic salts, are alike due. In every case which has been fully investigated since Wolkow and Baumann [1] first isolated and described this acid its presence has been demonstrated and re-examination of the material from some of the earlier cases also has led to its detection. The second allied alkapton acid, uroleucic, has hitherto only been found in the cases investigated by Kirk and in them in association with larger amounts of homogentisic acid. [2] By the kindness of Dr. R. Kirk I have recently been enabled to examine fresh specimens of the urines of his patients who have now reached manhood and was able to satisfy myself that at the present time even they are no longer excreting uroleucic acid. After as much of the homogentisic acid as possible had been allowed to separate out as the lead salt the small residue of the alkapton acid was converted into the ethyl ester by a method recently described by Erich Meyer [3] and the crystalline product obtained had the melting-point of ethyl homogentisate (120° C.). Further observations, and especially those of Mittelbach, [4] have also strengthened the belief that the homogentisic acid excreted is derived from tyrosin, but why alkaptonuric individuals pass the benzene ring of their tyrosin unbroken and how and where the peculiar chemical change from tyrosin to homogentisic acid is brought about, remain unsolved problems.

There are good reasons for thinking that alkaptonuria is not the manifestation of a disease but is rather of the nature of an alternative course of metabolism, harmless and usually congenital and lifelong. Witness is borne to its harmlessness by those who have manifested the peculiarity without any apparent detriment to health from infancy on into adult and even advanced life, as also by the observations of Erich Meyer who has shown that in the quantities ordinarily excreted by such persons homogentisic acid neither acts as an aromatic poison nor causes acid intoxication, for it is not excreted as an aromatic sulphate as aromatic poisons are, nor is its presence in the urine attended by any excessive output of ammonia. However, regarded as an alternative course of metabolism the alkaptonuric must be looked upon as somewhat inferior to the ordinary plan, inasmuch as the excretion of homogentisic acid in place of the ordinary end products involves a certain slight waste of potential energy. In this connexion it is also interesting to note that, as far as our knowledge goes, an individual is either frankly alkaptonuric or conforms to the normal type, that is to say, excretes several grammes of homogentisic acid per diem or none at all. Its appearance in traces, or in gradually increasing or diminishing quantities, has never yet been observed, even in the few recorded temporary or intermittent cases. In cases in which estimations have been carried out the daily output has been found to lie within limits which, considering the great influence of proteid food upon the excretion of homogentisic acid and allowing for differences of sex and age, may be described as narrow. This is well illustrated by Table I., in which the cases are arranged in order of age:—

TABLE I.—*Showing the Average Excretion of Homogentisic Acid.*

No.	Sex.	Age.	Average excretion of homogentisic acid per 24 hours on ordinary mixed diet.	Names of observers.
1	M.	2½ years.	3·2 grammes.	Erich Meyer.
2	M.	3½ ,,	2·6 ,,	A. E. Garrod.
3	M.	8 ,,	2·7 ,,	Ewald Stier.
4	M.	18 ,,	5·9 ,,	P. Stange.
5	M.	44 ,,	4·6 ,,	Mittelbach.
6	M.	45 ,,	4·7 ,,	H. Ogdén.
7	M.	60 ,,	5·3 ,,	Hammarsten.
8	F.	60 ,,	3·2 ,,	H. Embden.
9	M.	68 ,,	4·8 ,,	Wolkow and Baumann.

The information available as to the incidence of alkapton uria is of great interest in connexion with the above view of its nature. That the peculiarity is in the great majority of instances congenital cannot be doubted. The staining property of the urine allows of its being readily traced back to early infancy. This has been repeatedly done and in one of my cases [5] the staining of the napkins was conspicuous 57 hours after the birth of the child. The abnormality is apt to make its appearance in two or more brothers and sisters whose parents are normal and among whose forefathers there is no record of its having occurred, a peculiar mode of incidence which is well known in connexion with some other conditions. Thus of 32 known examples, which were presumably congenital, no less than 19 have occurred in seven families. One family contained four alkaptonurics, three others contained three, and the remaining three two each. The proportion of alkaptonuric to normal members is of some interest and Table II. embodies such definite knowledge upon this point as is at present available regarding congenital cases.

[1] Wolkow and Baumann : Zeitschrift für physiologische Chemie, 1891, Band xv., p. 228.
[2] R. Kirk : Journal of Anatomy and Physiology, 1889, vol. xxiii., p. 69. Huppert : Zeitschrift für Physiologische Chemie, 1897, Band xxiii., p. 412.
[3] E. Meyer : Deutsches Archiv für Klinische Medicin, 1901, Band lxx., p. 443.
[4] Mittelbach : Ibid., 1901, Band lxxi., p. 50.

[5] A. E. Garrod : THE LANCET, Nov. 30th, 1901, p. 1484 ; Transactions of the Royal Medical and Chirurgical Society, 1902, vol. lxxxv., p. 69.

TABLE II.—*Showing the Proportion of Alkaptonuric Members to Normal Members in 9 Families.*

No.	Total number of family (brothers and sisters).	Number of alkaptonuric members.	Number of normal members.	Observers.
1	14	4	10	Pavy.
2	4	3	1	Kirk.
3	7	3	4	Winternitz.
4	2	1	1	Ewald Stier.
5	2	2	0	Baumann, Embden.
6	1	1	0	Erich Meyer.
7	10	1	9	Noccioli and Domenici.
8	5	2	3	A. E. Garrod.
9	3	2	1	W. Smith, Garrod.
Totals	48	19	29	—

The preponderance of males is very conspicuous. Thus, of the 40 subjects whose cases have hitherto been recorded 29 have been males and only 11 females.

In a paper read before the Royal Medical and Chirurgical Society in 1901 the present writer pointed out that of four British families in which were 11 congenitally alkaptonuric members no less than three were the offspring of marriages of first cousins who did not themselves exhibit this anomaly. This fact has such interesting bearings upon the etiology of alkaptonuria that it seemed desirable to obtain further information about as many as possible of the other recorded cases and especially of those which were presumably congenital. My inquiries of a number of investigators who have recorded such cases met with a most kindly response, and although the number of examples about which information could still be obtained proved to be very limited, some valuable facts previously unknown have been brought to light and indications are afforded of points which may be inquired into with advantage regarding cases which may come under observation in the future. In a number of instances the patients have been lost sight of, or for various reasons information can no longer be obtained concerning them. To those who have tried to help me with regard to such cases, and have in some instances been at great trouble in vain, my hearty thanks are no less due than to those who have been able to furnish fresh information.[6]

The following is a brief summary of the fresh information collected. Dr. Erich Meyer,[7] who mentioned in his paper that the parents of his patient were related, informs me that as a matter of fact they are first cousins. Dr. H. Ogden[8] states that his patient is the seventh of a family of eight members and that his parents were first cousins. The three eldest children died in infancy ; the fifth, a female, has three children, but neither is she nor are they alkaptonuric. There is no record of any other examples in the family. The patient, whose wife is not a blood relation, has three children none of whom are alkaptonuric. Professor Hammarsten[9] states that the parents of an alkaptonuric man, whose case he recently described, were first cousins. The patient, aged 61 years, has three brothers and the only brother whose urine has been seen is not alkaptonuric. I have learned from Professor Noccioli[10] that the parents of the woman whose case he investigated with Dr. Domenici were not blood relations. The patient, a twin, who is one of two survivors of a family of ten, states that none of her relations have exhibited the condition. Dr. Ewald Stier[11] informs me that the parents of his patient were not related and it is

mentioned in his paper that they were not alkaptonuric. Professor Ebstein[12] states that the parents of the child with "pyrocatechinuria" whose case was investigated by him in conjunction with Dr. Müller in 1875 were not related, but I gather that he would not regard this as an ordinary case of alkaptonuria, the abnormal substance in the urine having been identified as pyrocatechin. Lastly, Professor Osler supplies the very interesting information that of two sons of the alkaptonuric man previously described by Dr. Futcher[13] one is alkaptonuric. This is the first known instance of direct transmission of the peculiarity. The parents of the father, who has an alkaptonuric brother whose case was recorded by Marshall,[14] were not blood relations. The above particulars are embodied with those of the congenital British cases previously recorded in the following tabular epitome (Table III.).

TABLE III.—*Showing the large Proportion of Alkaptonurics who are the Offspring of Marriage of First Cousins.*

A.

Families the offspring of marriages of first cousins.

No.	Total number of family.	Number of known alkaptonuric members.	Observers.
1	14	4	Pavy.
2	4	3	R. Kirk.
3	5	2	A. E. Garrod.
4	1	1	Erich Meyer.
5	8	1	H. Ogden.
6	4	1	Hammarsten.
Total ...	36	12	—

B.

Families whose parents were not related and not alkaptonuric.

No.	Total number of family.	Number of known alkaptonuric members.	Observers.
1	3	2	Armstrong, Walter Smith, and Garrod.
2	2	1	Ewald Stier.
3	10	1	Noccioli and Domenici.
4*	?	2	Marshall and Futcher.
Total ...		6	—

C.

Family in which alkaptonuria was directly inherited from a parent.

No.	Total number of family.	Number of known alkaptonuric members.	Observers.
1*	?	1	Osler and Futcher.
Total ...	—	1	—

* B 4 and C 1 refer to two generations of one family. No information is forthcoming as to the absence of alkaptonuria in previous generations. Ebstein and Müller's case, which is not included in the table for reasons given above, would raise the number of families in list B to 5.

It will be seen that the results of further inquiries on the continent of Europe and in America confirm the impression derived from the British cases that of alkaptonuric individuals a very large proportion are children of first cousins. The above table includes 19 cases in all out of a total of 40 recorded examples of the condition, and there is

6 To Hofrath Professor Huppert and to Professor Osler my very special thanks are due for invaluable aid in collecting information, and I would also express my most sincere gratitude to Professor Hammarsten, Geheimrath Professor Ebstein, Geheimrath Professor Fürbringer, Geheimrath Professor Erb, Professor Noccioli, and Professor Denigès, as also to Dr. F. W. Pavy, Dr. Kirk, Dr. Maguire, Dr. Futcher, Dr. Erich Meyer, Dr. H. Ogden, Dr. H. Embden, Dr. Mittelbach, Dr. Ewald Stier, Dr. Grassi, Dr. Carl Hirsch, and Dr. Winternitz, all of whom have been kind enough to help the inquiry in various ways.
7 E. Meyer, loc. cit.
8 H. Ogden : Zeitschrift für physiologische Chemie, 1895, Band xx., p. 280.
9 Hammarsten : Upsala Läkareförenings Förhandlingar, 1901, vol. vii., p. 26.
10 Noccioli e Domenici : Gazetta degli Ospedale, 1898. vol. xix., p. 303.
11 Ewald Stier : Berliner Klinische Wochenschrift, 1898, Band xxxv., p. 185.
12 Ebstein and Müller : Virchow's Archiv, 1875, Band lxii., p. 554.
13 Futcher : New York Medical Journal, 1898. vol. lxvii., p. 69.
14 Marshall ; Medical News, Philadelphia, 1887, vol. l., p. 35,

little chance of obtaining any further information on the point until fresh cases shall come under observation. It will be noticed that among the families of parents who do not themselves exhibit the anomaly a proportion corresponding to 60 per cent. are the offspring of marriages of first cousins. In order to appreciate how high this proportion is it is necessary to form some idea of the total proportion of the children of such unions to the community at large. Professor G. Darwin,[15] as the outcome of an elaborate statistical investigation, arrived at the conclusion that in England some 4 per cent. of all marriages among the aristocracy and gentry are between first cousins ; that in the country and smaller towns the proportion is between 2 and 3 per cent., whereas in London it is perhaps as low as 1·5 per cent. He suggests 3 per cent. as a probable superior limit for the whole population. Assuming, although this is, perhaps, not the case, that the same proportion of these as of all marriages are fruitful, similar percentages will hold good for families, and assuming further that the average number of children results from such marriages they will hold good for individuals also. A very limited number of observations which I have made among hospital patients in London gave results which are quite compatible with the above figures. Thus, among 50 patients simultaneously inmates of St. Bartholomew's Hospital there was one whose parents were first cousins. On another occasion one such was found among 100 patients, and there was one child of first cousins among 100 children admitted to my ward at the Hospital for Sick Children. It is evident, on the one hand, that the proportion of alkaptonuric families and individuals who are the offspring of first cousins is remarkably high, and, on the other hand, it is equally clear that only a minute proportion of the children of such unions are alkaptonuric. Even if such persons form only 1 per cent. of the community their numbers in London should exceed 50,000, and of this multitude only six are known to be alkaptonuric. Doubtless there are others, but that the peculiarity is extremely rare is hardly open to question. A careful look-out maintained for several years at two large hospitals has convinced me of this, and although the subject has recently attracted much more attention than formerly the roll of recorded examples increases but slowly.

The question of the liability of children of consanguineous marriages to exhibit certain abnormalities or to develop certain diseases has been much discussed, but seldom in a strictly scientific spirit. Those who have written on the subject have too often aimed at demonstrating the deleterious results of such unions on the one hand, or their harmlessness on the other, questions which do not here concern us at all. There is no reason to suppose that mere consanguinity of parents can originate such a condition as alkaptonuria in their offspring, and we must rather seek an explanation in some peculiarity of the parents, which may remain latent for generations, but which has the best chance of asserting itself in the offspring of the union of two members of a family in which it is transmitted. This applies equally to other examples of that peculiar form of heredity which has long been a puzzle to investigators of such subjects, which results in the appearance in several collateral members of a family of a peculiarity which has not been manifested at least in recent preceding generations.

It has recently been pointed out by Bateson[16] that the law of heredity discovered by Mendel offers a reasonable account of such phenomena. It asserts that as regards two mutually exclusive characters, one of which tends to be dominant and the other recessive, cross-bred organisms will produce germinal cells (gametes) each of which, as regards the characters in question, conforms to one or other of the pure ancestral types and is therefore incapable of transmitting the opposite character. When a recessive gamete meets one of the dominant type the resulting organism (the zygote) will usually exhibit the dominant character, whereas when two recessive gametes meet the recessive character will necessarily be manifested in the zygote. In the case of a rare recessive characteristic we may easily imagine that many generations may pass before the union of two recessive gametes takes place. The application of this to the case in question is further pointed out by Bateson, who, commenting upon the above observations on the incidence of alkaptonuria, writes as follows :[17] "Now there may be other

accounts possible, but we note that the mating of first cousins gives exactly the conditions most likely to enable a rare, and usually recessive, character to show itself. If the bearer of such a gamete mate with individuals not bearing it the character will hardly ever be seen ; but first cousins will frequently be the bearers of similar gametes, which may in such unions meet each other and thus lead to the manifestation of the peculiar recessive characters in the zygote." Such an explanation removes the question altogether out of the range of prejudice, for if it be the true account of the matter it is not the mating of first cousins in general but of those who come of particular stocks that tends to induce the development of alkaptonuria in the offspring. For example, if a man inherit the tendency on his father's side his union with one of his maternal first cousins will be no more liable to result in alkaptonuric offspring than his marriage with one who is in no way related to him by blood. On the other hand, if members of two families who both inherit the strain should intermarry the liability to alkaptonuria in the offspring will be as great as from the union of two members of either family, and it is only to be expected that the peculiarity will also manifest itself in the children of parents who are not related. Whether the Mendelian explanation be the true one or no there seems to be little room for doubt that the peculiarities of the incidence of alkaptonuria and of conditions which appear in a similar way are best explained by supposing that, leaving aside exceptional cases in which the character, usually recessive, assumes dominance, a peculiarity of the gametes of *both* parents is necessary for its production.

Hitherto nothing has been recorded about the children of alkaptonuric parents, and the information supplied by Professor Osler and Dr. Ogden on this point has therefore a very special interest. Whereas Professor Osler's case shows that the condition may be directly inherited from a parent Dr. Ogden's case demonstrates that none of the children of such a parent need share his peculiarity. As the matter now stands, of five children of two alkaptonuric fathers whose condition is known only one is himself alkaptonuric. It will be interesting to learn whether this low proportion is maintained when larger numbers of cases shall be available. That it will be so is rendered highly probable by the undoubted fact that a very small proportion of alkaptonurics are the offspring of parents either of whom exhibits the anomaly. It would also be extremely interesting to have further examples of second marriages of the parents of alkaptonurics In the case of the family observed by Dr. Kirk the only child of the second marriage of the father, not consanguineous, is a girl who does not exhibit the abnormality. The only other available example is recorded by Embden. The two alkaptonurics studied by Professor Baumann and himself were a brother and sister born out of wedlock, and as far as could be ascertained the condition was not present in the children of the subsequent marriages which both parents contracted. The patient of Noccioli and Domenici was a twin, and I gather from Professor Noccioli's kind letter that the other twin was also a female, did not survive, and was not alkaptonuric. Further particulars are wanting, and the information was derived from the patient herself, who is described as a woman of limited intelligence but who was aware that in her own case the condition had existed from infancy. It is difficult to imagine that of twins developed from a single ovum one should be alkaptonuric and the other normal, but this does not necessarily apply to twins developed from separate ova.

It may be objected to the view that alkaptonuria is merely an alternative mode of metabolism and not a morbid condition, that in a few instances, not included in the above tables, it appears not to have been congenital and continuous but temporary or intermittent. In some of the cases referred to the evidence available is not altogether conclusive, and it is obvious that for the proof of a point of so much importance to the theory of alkaptonuria nothing can be regarded as wholly satisfactory which falls short of a complete demonstration of the presence of homogentisic acid in the urine at one time and its absence at another. The degree and rate of darkening of the urine vary at different periods apart from any conspicuous fluctuations in the quantity of homogentisic acid which it contains. The staining of linen in infancy is a much more reliable indication, especially if the mother of the child has had previous experience of alkaptonuric staining. In Geyger's case[18] of a diabetic man the

[15] G. Darwin : Journal of the Statistical Society, 1875, vol. xxxviii., p. 153.
[16] W. Bateson : Mendel's Principles of Heredity, Cambridge, 1902.
[17] W. Bateson and Miss E. R. Saunders : Report to the Evolution Committee of the Royal Society, No. 1, 1902, p. 133, note.
[18] A. Geyger : Pharmaceutische Zeitung, 1892, p. 488.

11

intermittent' appearance in the urine of an acid which he identified with the glycosuric acid of Marshall was established beyond all doubt, and the melting-point and proportion of lead in the lead salt render it almost certain that he was dealing with homogentisic acid. In Carl Hirsch's case[19] a girl, aged 17 years, with febrile gastro-intestinal catarrh, passed dark urine which gave the indican reaction for three days. Professor Siegfried extracted by shaking with ether an acid which gave the reactions of homogentisic acid and formed a sparingly soluble lead salt. Neither the melting-point of the acid nor any analytical figures are given. After three days the urine resumed its natural colour and reactions.

Von Moraczewski[20] also records a case of a woman, aged 43 years, who shortly before her death passed increasingly dark urine, rich in indican, from which he extracted an acid which had the melting-point and reactions of homogentisic acid. Such increasing darkening of the urine as was here observed not infrequently occurs with urines rich in indoxyl-sulphate, as Baumann and Brieger first pointed out, and this was probably a contributory factor in the production of the colour first called attention to the condition. Stange[21] has described a case in which the presence of homogentisic acid was very fully established, but he clearly does not regard the mother's evidence as to the intermittent character of the condition as conclusive. Zimnicki's[22] case of intermittent excretion of homogentisic acid by a man with hypertrophic biliary cirrhosis is published in a Russian journal which is inaccessible to me, and having only seen abstracts of his paper I am unacquainted with the details. Of hearsay evidence the most convincing is afforded in Winternitz's cases.[23] The mother of seven children, three of whom are alkaptonuric, was convinced that whereas two of her children had been alkaptonuric from the earliest days of life this had not been so with the youngest child in whom she had only noticed the peculiarity from the age of five years. This is specially interesting as supplying a link between the temporary and congenital cases. In a somewhat similar case described by Maguire[24] the evidence of a late onset is not so conclusive. Slosse's case[25] in which, as in von Moraczewski's, the condition apparently developed in the last stages of a fatal illness, completes the list of those falling into the temporary class. Evidently we have still much to learn about temporary or intermittent alkaptonuria, but it appears reasonable to suppose that those who exhibit the phenomenon are in a state of unstable equilibrium in this respect, and that they excrete homogentisic acid under the influence of causes which do not bring about this result in normal individuals. There is reason to believe that a similar instability plays a not unimportant part in determining the incidence of certain forms of disease in which derangements of metabolism are the most conspicuous features. Thus von Noorden,[26] after mentioning that diabetes occasionally develops at an early age in brothers and sisters and comparatively seldom occurs in the children of diabetic parents, adds that in three instances he has met with this disease in the offspring of marriages of first cousins. In one such family two out of six children, in another two out of three, and in the third the only two children became diabetic at ages between one and four years.

The view that alkaptonuria is a "sport" or an alternative mode of metabolism will obviously gain considerably in weight if it can be shown that it is not an isolated example of such a chemical abnormality, but that there are other conditions which may reasonably be placed in the same category. In the phenomenon of albinism we have an abnormality which may be looked upon as chemical in its basis, being due rather to a failure to produce the pigments of the melanin group which play so conspicuous a part in animal colouration than to any defect of development of the parts in which in normal individuals such pigments are laid down. When we study the incidence of albinism in man we find that it shows a striking resemblance to that of alkaptonuria. It, too, is commoner in males than in females, and tends to occur in brothers and sisters of families in which it has not previously appeared, at least in recent generations. Moreover, there is reason to believe that an undue proportion of albinos are the offspring of marriages of first cousins. Albinism is mentioned by most authors who have discussed the effects of such marriages and Arcoleo,[27] who gives some statistics of albinism in Sicily, states that of 24 families in which there were 62 albino members five were the offspring of parents related to each other in the second canonical degree. On the other hand, Bemiss[28] found that of 191 children of 34 marriages of first or second cousins five were albinos. In a remarkable instance recorded by Devay[29] two brothers married two sisters, their first cousins. There were no known instances of albinism in their families, but the two children of the one marriage and the five children of the other were all albinos. After the death of his wife the father of the second family married again and none of the four children of his second marriage were albinos. Again, albinism is occasionally directly inherited from a parent, as in one instance quoted by Arcoleo, but this appears to be an exceptional occurrence. The resemblance between the modes of incidence of the two conditions is so striking that it is hardly possible to doubt that whatever laws control the incidence of the one control that of the other also.

A third condition which suggests itself as being probably another chemical "sport" is cystinuria. Our knowledge of its incidence is far more incomplete and at first sight direct inheritance appears to play here a more prominent part. However, when more information is forthcoming it may turn out that it is controlled by similar laws. In this connexion a most interesting family described by Pfeiffer[30] is very suggestive. Both parents were normal, but all their four children, two daughters and two sons, were cystinuric. The elder daughter had two children neither of whom was cystinuric. A number of other examples of cystinuria in brothers and sisters are recorded, but information about the parents is wanting, except in the cases of direct transmission. In some of the earlier cases such transmission through three generations was thought to be probable, but the presence of cystin in the urine of parent and child has only been actually demonstrated in two instances. In Joel's[31] often quoted case it was only shown that the mother's urine contained excess of neutral sulphur. E. Pfeiffer[32] found cystin in the urine of a father and son and in a family observed by Cohn[33] the mother and six of her children shared the peculiarity. As more than 100 cases are on record the proportion of cases of direct inheritance has not hitherto been shown to be at all high and Pfeiffer's first case shows that, as with alkaptonuria, the children of a cystinuric parent may escape. A large majority of the recorded cystinurics have been males. There is as yet no evidence of any influence of consanguinity of parents and in the only two cases about which I have information the parents were not related. Neither has it yet been shown that cystinuria is a congenital anomaly, although in one case, at any rate, it has been traced back to the first year of life. Observations upon children of cystinuric parents from their earliest infancy or upon newly-born brothers or sisters of cystinurics would be of great interest and should in time settle this question. Lastly, it seems certain that, like alkaptonuria, this peculiarity of metabolism is occasionally temporary or intermittent. The so frequent association with cystinuria of the excretion of cadaverine and putrescine adds to the difficulty of the problem of its nature and upon it is based the infective theory of its causation. However, it is possible that, as C. E. Simon[34] has suggested, these diamines may themselves be products of abnormal metabolism. Unlike alkaptonuria and albinism cystinuria is a distinctly harmful condition, but its ill effects are secondary to its deposition in crystalline form and the readiness with

19 C. Hirsch : Berliner Klinische Wochenschrift, 1897, Band xxxiv., p. 866.
20 W. von Moraczewski : Centralblatt für die Innere Medicin, 1896, Band xvii., p. 177.
21 P. Stange : Virchow's Archiv, 1896, Band cxlvi., p. 86.
22 Zimnicki : Jeschenedelnik, 1900, No. 4. Abstract, Centralblatt für Stoffwechsel und Verdauungs-Krankheiten, 1900, Band i., p. 348.
23 Winternitz : Münchener Medicinische Wochenschrift, 1899, Band xlvi., p. 749.
24 R. Maguire : Brit. Med. Jour., 1884, vol. ii., p. 808.
25 A. Slosse : Annales de la Société Royale des Sciences Médicales et Naturelles, Bruxelles, 1895, tome iv., p. 89.
26 Von Noorden : Die Zuckerkrankheit, 3. Aufgabe, 1901, p. 47.
27 G. Arcoleo : Sull' Albinismo in Sicilia. See notice in Archivio per l'Anthropologia, 1871, vol. i., p. 367.
28 Bemiss : Journal of Psychological Medicine, 1857, vol. x. 368.
29 Devay : Du Danger des Mariages Consanguins, &c., Paris, 1857.
30 E. Pfeiffer : Centralblatt für Krankheiten der Harn- und Sexual-Organe, 1894, Band v., p. 187.
31 Joel : Annalen der Chemie und Pharmacologie, 1855, p. 247.
32 E. Pfeiffer : Centralblatt für Krankheiten der Harn- und Sexual-Organe, 1897, Band viii., p. 173.
33 J. Cohn : Berliner Klinische Wochenschrift, 1899, Band xxxvi., p. 503.
34 C. E. Simon : American Journal of the Medical Sciences, 1890, vol. cxix., p. 39.

which it forms concretions. Its appearance in the urine is not associated with any primary morbid symptoms. All three conditions above referred to are extremely rare and all tend to advertise their presence in conspicuous manners. An albino cannot escape observation ; the staining of clothing and the colour of the urine of alkaptonurics seldom fail to attract attention, and the calculous troubles and the cystitis to which cystinurics are so liable usually bring them under observation sooner or later. May it not well be that there are other such chemical abnormalities which are attended by no obvious peculiarities and which could only be revealed by chemical analysis ? If such exist and are equally rare with the above they may well have wholly eluded notice up till now. A deliberate search for such, without some guiding indications, appears as hopeless an undertaking as the proverbial search for a needle in a haystack.

If it be, indeed, the case that in alkaptonuria and the other conditions mentioned we are dealing with individualities of metabolism and not with the results of morbid processes the thought naturally presents itself that these are merely extreme examples of variations of chemical behaviour which are probably everywhere present in minor degrees and that just as no two individuals of a species are absolutely identical in bodily structure neither are their chemical processes carried out on exactly the same lines. Such minor chemical differences will obviously be far more subtle than those of form, for whereas the latter are evident to any careful observer the former will only be revealed by elaborate chemical methods, including painstaking comparisons of the intake and output of the organism. This view that there is no rigid uniformity of chemical processes in the individual members of a species, probable as it is *a priori*, may also be arrived at by a wholly different line of argument. There can be no question that between the families, genera and species both of the animal and vegetable kingdoms, differences exist both of chemical composition and of metabolic processes. The evidences for this are admirably set forth in a most suggestive address delivered by Professor Huppert[33] in 1895. In it he points out that we find evidence of chemical specificity of important constituents of the body, such as the hæmoglobins of different animals, as well as in their secretory and excretory products such as the bile acids and the cynuric acid of the urine of dogs. Again, in their

behaviour to different drugs and infecting organisms the members of the various genera and species manifest peculiarities which presumably have a chemical basis, as the more recent researches of Ehrlich tend still further to show. To the above examples may be added the results of F. G. Hopkins's[36] well-known researches on the pigments of the pieridæ and the recent observations of the precipitation of the blood proteids of one kind of animal by the serum of another. From the vegetable kingdom examples of such generic and specific chemical differences might be multiplied to an almost indefinite extent. Nor are instances wanting of the influence of natural selection upon chemical processes, as for example, in the production of such protective materials as the sepia of the cuttle-fish and the odorous secretion of the skunk, not to mention the innumerable modifications of surface pigmentation. If, then, the several genera and species thus differ in their chemistry we can hardly imagine that within the species, when once it is established, a rigid chemical uniformity exists. Such a conception is at variance with all that is known of the origin of species. Nor are direct evidences wanting of such minor chemical diversities as we have supposed to exist within the species. Such slight peculiarities of metabolism will necessarily be hard to trace by methods of direct analysis and will readily be masked by the influences of diet and of disease, but the results of observations on metabolism reveal differences which are apparently independent of such causes, as for example, in the excretion of uric acid by different human individuals. The phenomena of obesity and the various tints of hair, skin, and eyes point in the same direction, and if we pass to differences presumably chemical in their basis idiosyncrasies as regards drugs and the various degrees of natural immunity against infections are only less marked in individual human beings and in the several races of mankind than in distinct genera and species of animals.

If it be a correct inference from the available facts that the individuals of a species do not conform to an absolutely rigid standard of metabolism, but differ slightly in their chemistry as they do in their structure, it is no more surprising that they should occasionally exhibit conspicuous deviations from the specific type of metabolism than that we should meet with such wide departures from the structural uniformity of the species as the presence of supernumerary digits or transposition of the viscera.

---

[33] Hupp. ' Ueber die Erhaltung der Arteigenschaften, Prague, 1896.

[36] F. G. Hopkins: Philosophical Transactions of the Royal Society, 1895, vol. clxxxvi., B., p. 661.

2

Reprinted from *Reports to the Evolution Committee, Royal Society*, Harrison and Sons, 1902, pp. 132–134

EXPERIMENTAL STUDIES IN THE PHYSIOLOGY OF HEREDITY

W. Bateson and E. R. Saunders

[*Editor's Note:* In the original, material precedes and follows this excerpt.]

Selection and the Phenomenon of Dominance. We have seen that the want of fixity in certain forms, though continually selected, may at once be explained by the hypothesis that they are heterozygous only, and have no gametes corresponding to them. Another illustration of the failure of selection is the constant recurrence of a particular "rogue" in the best strains. Seed is never taken from such rogues. Every year they may be pulled up as soon as detected, but they continually reappear.

The hypothesis that such a "rogue" is a recessive form *may* give a complete explanation of this phenomenon in many cases. Selection from *individuals* of known fertilisation would at once test the truth of this view, and might provide a means of producing a pure strain once and for all from the pure dominants.

It is well known that some of the best modern beardless wheats which have been raised of late years by crossing distinct varieties will give a small proportion of bearded plants. This is, of course, called "reversion" to a bearded ancestor used in the original cross.

From the experiments of Rimpau,* we find that when bearded and beardless varieties are crossed, beardlessness is dominant, and the bearded character is recessive. By subsequent breeding a form is produced with a desirable character, and after a few years of selection it is found to give this character with sufficient purity and it is put on the market. It may be a bearded or a beardless form, but if the latter, the chances are that it will always produce a certain proportion of bearded plants.† This may happen in every case where there has been a *promiscuous* selection of many dominant plants, for any one of these may be a heterozygote and bear in each year both dominant and recessive germs.

The fact that the hornless breeds of goats still give some horned offspring is probably referable to the same cause. The point is of course not certain, but from the analogy of cattle we may anticipate that the hornless form is dominant. In the polled breeds of cattle, which are never *promiscuously* selected, the polled character has naturally been easily fixed pure, but in goats selection among the *ewes* has been probably to a large extent promiscuous.

---

*'Landw.J.B.,'20.
†Such a variety is Garton's Red King.

The phenomenon is without doubt occurring very widely in nature. To it we may perhaps attribute the undiminished persistence of some weakly varieties, which are unceasingly exterminated by natural or artificial selection without ever leaving offspring. Cases have only to be looked for to be found in abundance. We may note the paradox that, for anything we know to the contrary, a recessive allelomorph may even persist as a gamete *without the corresponding homozygote having ever reached maturity in the history of the species*.* It would be premature to trace out the deductions to which this suggestive fact points, but we see at once that it may give the true account of the phenomenon that domesticated forms constantly give rise to varieties not met with in the wild state, a fact often ascribed on insufficient grounds to the action of changed conditions in producing greater *variability*.

It will be clear — a point which may have some economic importance — that in any such case the recessive "rogue" can be eliminated by selection from *individual* plants or animals, breeding only from those which give no recessives on being self-fertilised, if hermaphrodite. If the organism be dioecious the process will be more elaborate, for it will be first necessary to test for recessive allelomorphs by fertilising with a recessive, and afterwards to fertilise those that gave no recessive offspring, with a dominant similarly proved to be free from recessive influence. Nevertheless it is certain that by this process alone can a strain of pure dominants be readily made.

"Purity" then acquires a new and more precise meaning. An organism resulting from an original cross is not necessarily pure when it has been raised by selection from parents similar in appearance for an indefinite number of generations. *It is only pure when it is compounded of gametes bearing identical allelomorphs, and such purity may occur in any individual raised from cross-bred organisms.*

An organism can be strictly defined as genetically pure if all its gametes when mated with similar gametes reproduce the parent identically; and in practice the only way in which such purity can, by one breeding, be tested, is by crossing the organism in question with pure recessives.

---

*[In illustration of such a phenomenon we may perhaps venture to refer to the extraordinarily interesting evidence lately collected by Garrod regarding the rare condition known as "Alkaptonuria." In such persons the substance, alkapton, forms a regular constituent of the urine, giving it a deep brown colour which becomes black on exposure. The condition is exceedingly rare, and, though met with in several members of the same families, has only once been known to be directly transmitted from parent to offspring. Recently, however, Garrod has noticed that no fewer than five families containing alkaptonuric members, more than a quarter of the recorded cases, are the offspring of unions of *first cousins*. In only *two* other families is the parentage known, one of these being the case in which the father was alkaptonuric. In the other case the parents were *not* related. Now there may be other accounts possible, but we note that the mating of first cousins gives exactly the conditions most likely to enable a rare and usually recessive character to show itself. If the bearer of such a gamete mates with individuals not bearing it, the character would hardly ever be seen; but first cousins will frequently be bearers of *similar* gametes, which may in such unions meet each other, and thus lead to the manifestation of the peculiar recessive characters in the zygote. See A. E. Garrod, 'Trans. Med. Chir. Soc.,' 1899, p. 367, and 'Lancet,' November 30, 1901.]

3

Reprinted from *Papers, Peabody Museum, Harvard University* 3:69–77 (1905)

INHERITANCE OF
DIGITAL MALFORMATIONS IN MAN

William C. Farabee

A great deal has been written concerning the abnormities of the hands and feet, and numerous cases of polydactylism, syndactylism, ectrodactylism, and macrodactylism have been recorded; but, very few cases of hypophalangia or hyperphalangia have ever been noted. The following case of hypophalangia, or diminution in the number of phalanges, came under my notice a few years ago in Pennsylvania. After careful examination the anomaly was found to be interesting not only as a case of anatomical variation, but also as a study in heredity. All the digits of all extremities of thirty-seven persons are affected and the anomaly is inherited in conformity with Mendel's law for five generations. Measurements, tracings, photographs, radiographs, plaster casts, and complete genealogical tables, were made.*

As shown by the photograph (Plate XXIII), the people appear perfectly normal in every other respect and seem to suffer very little inconvenience on account of the malformation. The ladies complain of but one disadvantage in short fingers, and that is in playing the piano; they cannot reach a full octave and hence are not good players. Among the men are farmers, mechanics, business men, and school teachers. One man is at the head of a commercial school and a very excellent penman; another is catcher for the city baseball team. The hands and feet have the normal number of digits and the digits have the normal proportions, each to each. The thumbs and great toes have the normal number of phalanges, but the first phalanx

---

*I acknowledge my especial obligation to Prof. F. W. Putnam under whose direction this study was carried on.

in each case is greatly reduced in length,—so much so, especially in the thumbs, that they are said to have "double jointed thumbs." The radiograph (Plate XXVI) shows the first phalanx to be about 12 mm., and the distal 22 mm., respectively in length. Hence, the thumbs thus shortened have the same relation to the other two phalanged digits that exists in the normal hand. Each of the fingers has but two phalanges instead of three. The metacarpal bones are normal except in length, being reduced in relative proportion to the length of the digits. The following table gives the length in millimeters of metacarpals and phalanges.

TABLE I.

Length of Metacarpals and Phalanges.

	METACARPAL.	PROXIMAL.	MIDDLE.	DISTAL.
Thumb	34 mm.	12 mm.	– –	22 mm.
Index	55 "	30 "	– –	15 "
Middle	55 "	40 "	– –	15 "
Ring	46 "	32 "	– –	15 "
Little	42 "	22 "	– –	18 "

The length of the hand is 162 mm., and the width 94 mm. In a few cases the distal phalanx of the ring finger is not in line with the proximal, but inclines toward the middle finger.

The hands of all are broad, thick, and pulpy, as is seen in the photographs of the hands (Plates XXIII, XXIV, XXV). The joints of fingers and toes, as appeared on examination and as shown by the radiographs (Plates XXVI, XXVII), are loosely articulated. This may account for the lack of strength in the hands which was spoken of by many. One man, who has been a noted wrestler, said that his defeats were due to his weak hands. The table II also shows a weak grip. This, however was not a fair test as the instrument used was too wide for their short hands.

The feet, as shown in the cast and the outlines of the radiograph (Plate XXVII), do not outwardly appear abnormal. The toes are slightly shorter and the foot a little thicker than normal but not enough so to attract attention. The bones, however

17

as seen in the radiograph, present precisely the same condition as already noted in the hands and all that has been said in regard to the relative length of digits and phalanges in the hands applies equally to the feet.   No dissections have been made of the hands or feet, hence we are deprived of the light that the musculation might throw upon the question of which is the missing phalanx; but, judging from the length, size and form of the proximal bones, especially the size and form at the base, it appears that the bases of the distal phalanges articulate with the heads of the first row.   Yet one is hardly justified in saying that either the one or the other segment is missing.   It is safer to say, simply, that there is a reduction in the number of phalanges.

On account of the reluctance to submit to examination, I was able to take measurements of only three adult males, one adult female, and some children.   The numbers measured are too few for the results to be of any particular value, except to give some notion of their relation to the measurements of normal individuals.   Those measured are fairly representative of all families.   We give, in table II, the measurements of the female and the average of the three males.   There was very little variation in the males.   The height of the males, 159 cm. or 5 ft. 3 in., is much below the average height of normal men. In the table of measurements, I have placed normal measurements secured by calculating the proportions of a normal individual of the same height.   The average span of the arms, or reach of the males is but 146 cm. while normally it should be 165 cm.   The reach is 92% of the height against 104% in normal man.   The reach of the female is but 86.6% of the stature. As will be seen, the reduction in the number of phalanges does not account for all this difference in reach.   It is distributed almost equally between the arm, forearm, and hand.   The difference in reach is 19 cm.; in whole arm, 9.5 cm.; in upper arm, 3.1 cm.; in forearm, 3.2 cm.; and in the hand 3.2 cm. The reach diminished by the sum of the lengths of the arms would leave the width of the body about normal.   On account of the short arms the body has the appearance of being very long, but by consulting the table it will be seen that the height sitting is very nearly normal. The length of the foot is only 2 cm. short. The weight, however, is 16 lbs. heavier than normal. There does not appear to be complete correspondence between the upper and lower extremities. The upper are shorter than normal in every part whereas the lower are about normal in every way except the number of phalanges. There has never been a single instance of partial inheritance, but in all cases all extremities have been affected in precisely the same way. This is a most excellent example of similar and simultaneous variation in both extremities.

TABLE II.

MEASUREMENTS.

	Males.		Females.	
	NORMAL.	ABNORMAL.	NORMAL.	ABNORMAL.
Length of head		18.8 cm.		18 cm.
Width " "		15.0 "		15.2 "
" " face		13.5 "		13.6 "
Cephalic index		80		84
Height	159 cm.	159 cm.	150 cm.	150 cm.
" sitting	85 "	83 "	82 "	78 "
Reach	165 "	146 "	156 "	130 "
Length of arm	71.5 "	62 "	67 "	58 "
" forearm & hand	43.4 "	37 "	41 "	34 "
" 2nd finger	7.9 "	6.4 "	7.4 "	5.7 "
" Hand	18.8 "	15.6 "	17.1 "	14.2 "
Width of hand	8 "	9.3 "	7 "	7.6 "
Length of foot	25 "	23 "	23.6 "	20 "
Width " "	9.2 "	9.9 "	8.2 "	8.6 "
Grip	48. kgm.	28 kgm.	25 kgm.	12 kgm.
Weight	139 lbs.	155 lbs.	129 lbs.	144 lbs.

The family tradition is, that the first person having short digits came from Normandy in the army of William the Conqueror, and remained in England; that persons with short fingers have never intermarried; that every other child born

19

of a short fingered parent has short fingers; and that no long fingered descendant of a short fingered parent ever had short fingered children.   There is no historic evidence, so far as I can learn, to support the first part of the tradition; it may, or may not, be true.   The fact that there is a tradition concerning the anomaly, without accounting for its origin, may be taken as partial proof that the origin is so remote that it has been forgotten.   The second part that exogamy has been the custom, is true for at least five generations, as will be seen in table V. It would be very interesting indeed if this part of the tradition should be violated.

HEREDITY.

Probably the most important part of this study is that relating to the remaining portions of the tradition concerning heredity.   At present the question of heredity is one of live interest on account of the testing of Mendel's discovery, — the law of heredity.   The present case demonstrates the fact that the law operates in man as well as in plants and the lower animals.   The abnormality here is shown to be the dominant character.   The tradition that "*every other* child has short fingers," is not quite true; yet, as nearly as possible, half the offspring have the anomaly.   This is in perfect conformity with the law, the underlying principle of which is the purity of germ-cells and their production in equal members.   When there is a union of normal and abnormal individuals, the abnormals producing germ-cells N and A in equal numbers, the chances are equal that germ-cell N of one sex may unite with germ-cell N of the opposite sex, or that germ-cell A of one sex may unite with germ cell N of the opposite sex.   Since the abnormal character is shown to be dominant, the chances are even that the offspring may be normal or abnormal.   According to the laws of chance we should not expect that every other child would be abnormal, as in the tradition, but we should expect the total number of normals and abnormals in a large series to be very nearly equal, and that is what we find to be true here.

20

TABLE III.

ORDER OF BIRTH AND SEX.

GENERATION.	FAMILY.	ABNORMAL PARENT.	OFFSPRING.
I		Unknown	♀
II	1	Female.	♂ ♂ ♂ ♂ ♂ ♀ ♀ ♀
III	2	Male.	♀ ♀ ♀ ♂ ♀ ♀ ♂ ♀ ♀ ♀ ♂ ♀ ♂
IV	3	Female.	♀ ♂ ♂ ♀
	4	Male.	♀ ♀
	5	Female.	♀ ♀ ♀ ♂ ♀
	6	Female.	♀ ♂ ♀
	7	Male.	♀ ♀
V	8	Male.	♂ ♀ ♀ ♂ ♂ ♀ ♀ ♀ ♂ ♀
	9	Male.	♀ ♂ ♂ ♀ ♂
	10	Female.	♂ ♀ ♂ ♂ ♂ ♀ ♀
	11	Female.	♀ ♂ ♂ ♀ ♂ ♂
	12	Female.	♂
	13	Male.	♀ ♂
	14	Female.	♂ ♂

Explanation of characters:—

    ♂ normal male.   ♀ normal female.

    ♂ abnormal male.   ♀ abnormal female.

By referring to table III above it will be seen that normals and abnormals do alternate in a few instances; as in family 6 where there are three children; in families 7, 13. and 14, where there are but two each; and in 8 where the alternation is continued until the eighth child. But this rule does not hold in the other families. In families 9 and 10 the first three children in each are normal, while in 11 the first three are abnormal and the last three are normal. It will be noticed that the *first* child is abnormal in but three of the fourteen families, and the *second* abnormal in nine of the thirteen families. The others are about evenly divided. The total number of offspring descended from abnormals is sixty-nine of whom thirty-three are normal and

21

thirty-six abnormal, distributed as follows: in the second generation, four normals and four abnormals; in the third, five normals and seven abnormals; in the fourth, seven normals and nine abnormals; and in the fifth, seventeen normals and sixteen abnormals.

TABLE IV.

Sex Relations.

GENE-RATION.	SEX OF ABNORMAL PARENTS.		NUMBER OF OFFSPRING.	NORMAL		ABNORMAL	
				MALE.	FEMALE.	MALE.	FEMALE.
I	?		?				1
II	1	female	11*	4	0	1	3
III	1	male	12	2	3	3	4
IV	2	males	4	0	1	0	3
	3	females	12	1	5	3	3
V	3	males	17	4	4	4	5
	4	females	16	6	3	4	3

Table IV shows that the abnormality is inherited through both sexes. Six abnormal male parents have twelve males, six normal and six abnormal, and twenty-one females, eight normal and thirteen abnormal; while eight abnormal female parents have nineteen males, eleven normal and eight abnormal, and seventeen females, eight normal and nine abnormal. Of the descendants of the six males, fifty percent of the males and sixty-two percent of the females are abnormal; while, among the descendants of the eight abnormal females, only forty-two percent of the males and fifty-three percent of the females are abnormal. Fifty-eight percent of all the descendants of males are abnormal, whereas only forty-seven percent of the descendants of females are abnormal. There are five and a half times as many offspring as abnormal male parents and only four and a half times as many offspring as abnormal female parents. Forty-five percent of all descendants are males and fifty-five percent, females. Yet the whole number of abnormal males

*Three are unknown.

is less than two-thirds the number of abnormal females. Of the thirty-six descendants affected, twenty-two are females and but fourteen are males, or sixty-one and thirty-nine percents respectively. Forty-five percent of all males and fifty-eight percent of all females are abnormal. It thus appears that the males are more prolific than the females; that a higher percent of the offspring of males than of females are abnormal; and the female offspring of both male and female parents are more often abnormal. The numbers here compared are too small to base conclusions upon, yet the sexual differences are so marked that they must be of some significance.

Table v gives the genealogy of the abnormals only — the number and sex of normal and abnormal offspring in each generation. It shows that exogamy has been the custom in all these generations. Table vi gives all the known normals and abnormals and proves the last part of the tradition, — that no normal descendant of an abnormal parent has had abnormal offspring. We have here three complete lines of descent from the second generation to the fifth and all descendants are known in two of these lines. In all, twenty-one normals have married other normals outside the family and have had born to them seventy children, not one of whom is abnormal. According to former theories it should be expected that the character would reappear somewhere in these lines; but according to Mendel's law, even if the character were recessive, we should not expect it to reappear at all, since these families practised exogamy. Luckily, for the testing of recession, two cousins in the third generation married and had only normal offspring. This is shown in the table by uniting two of the lines of descent to form a new one. If the character were recessive it should certainly appear here, but it does not.

It will be noticed that fourteen normal parents in the fourth generation have but twenty-eight offspring, whereas seven abnormal parents have thirty-three. This does not signify that the abnormals are prepotent, as might be inferred. The cause was explained by one of the abnormal ladies, who said: "They always pick us up first." The abnormals all along the line have married earlier in life than the normals, so that when

TABLE V.

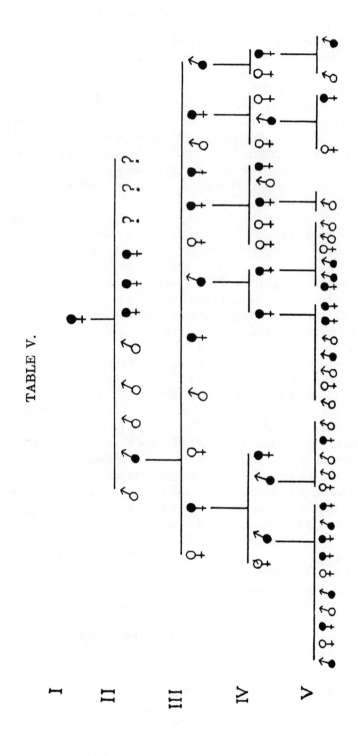

24

TABLE VI.

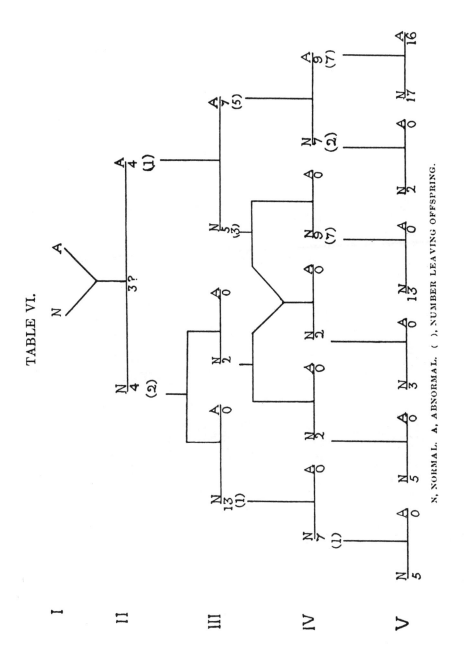

N, NORMAL. A, ABNORMAL. ( ), NUMBER LEAVING OFFSPRING.

25

the fifth generation is reached their families number ten, seven, five, etc., while the normals in many cases have but one child. In a short time, at the present rate, the abnormals will have gained a generation.

A very careful study was made to ascertain whether or not any other characteristics of an abnormal parent were inherited by the abnormal offspring. Besides the measurements taken many other things were noted; as, color of eyes and hair, form of head, facial expression, and other individual characters. The only constant accompanying characters found were the short arms and short stature. In every case the abnormal man is shorter and stouter than his normal brother, and the abnormal woman, than her normal sister. I regret that it was impossible to get a photograph of a group of normals and abnormals to show this difference in stature.

MR. A., SHOWING STATURE AND SHORT HANDS.

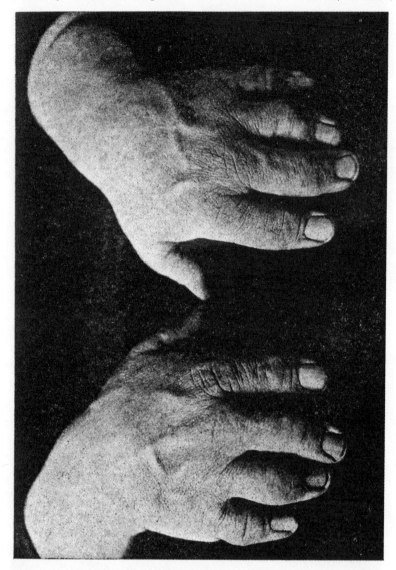

HANDS OF MR. A., SHOWING THEIR BREADTH AND THE LENGTH OF THE
FINGERS.

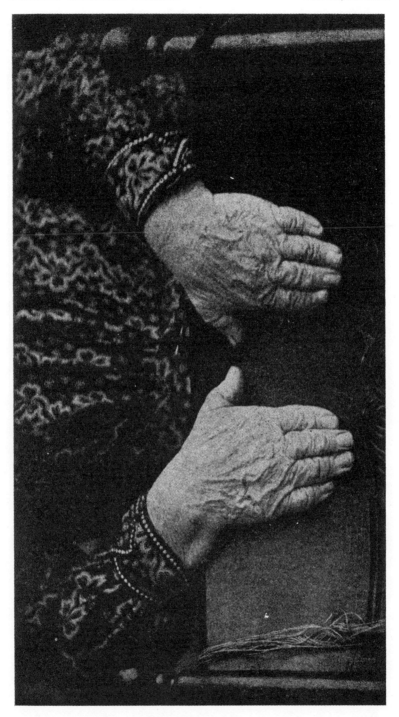

HANDS OF MR. A'S MOTHER.

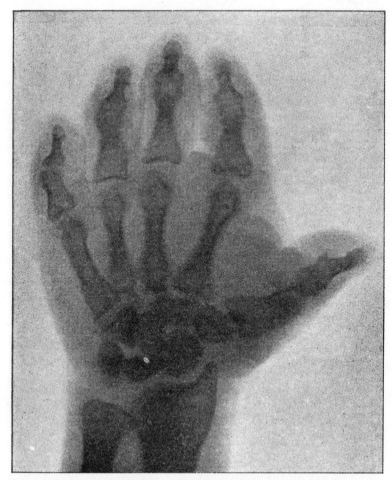

RADIOGRAPH OF THE LEFT HAND OF A'S BROTHER, SHOWING THE TWO-
PHALANGED DIGITS AND THEIR LOOSE ARTICULATION. 3/5 SIZE.

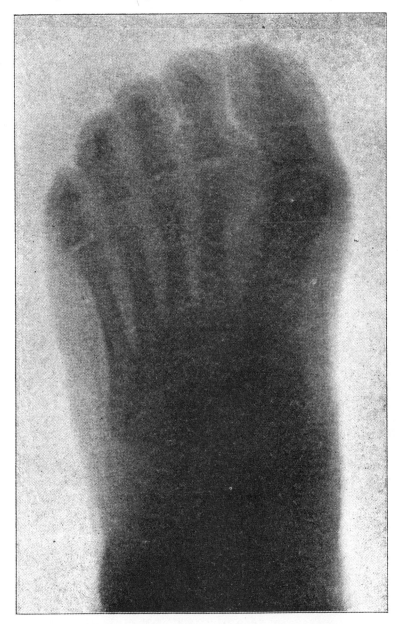

RADIOGRAPH OF LEFT FOOT OF MR. A., SHOWING THE TWO-PHALANGED
DIGITS. ¾ SIZE.

4

Reprinted from *Proc. N.Y. Path. Soc.* **9**:117–123 (1908)

A SIMPLE METHOD OF PERFORMING SERUM REACTIONS.

ALBERT A. EPSTEIN, M. D., AND REUBEN OTTENBERG, M. D.

Dr. Reuben Ottenberg presented for Dr. Epstein and himself a simple method for performing serum reactions, and added a few remarks on some interesting results which had been obtained.

The method arose in the first place from the necessity of testing for hemolysis in cases in which transfusion of blood was proposed. Since the work of Drs. Beebe and Weil on the hemolytic effects of malignant tumors, since Crile's report of very frequent hemolysis by sera of cancer patients, and since Pepper's report of a fatal case of hemolysis following transfusion, it has been well recognized that hemolysis, though not common, does occur on mixing human bloods. Pathologists are now called upon to examine the blood of persons on whom a transfusion is to be performed. To do this test it has usually been necessary to secure several cubic centimeters of blood. In practice, however, it has often been impossible to do this, either on account of the urgency of the case, or of the fact that the patient was an infant. For this reason, the present method, which is practically an application of Wright's opsonic technique, has been evolved. The method requires but a few drops of blood, and can be carried out very quickly.

The things required are:

1. Glass capsules made of small caliber tubing (4 to 5 mm. in diameter) drawn out at both ends into fine capillaries.

2. Pipettes made of glass tubing (6 to 8 mm. in diameter). These are fitted with nipples.

3. Pipettes like the above but made of tubing of 4 to 5 mm. in diameter.

4. Normal salt solution (0.85 per cent.) to which 1 per cent. of sodium citrate has been added.

In obtaining the blood, the lobule of the ear is pricked with a needle. To prepare the suspension of red blood cells, the blood which oozes out is drawn up to an arbitrary mark on the first stem of one of the larger pipettes. This blood is then diluted by drawing into the same pipette successive portions of the salt-citrate solution up to the mark, allowing a small bubble of air to intervene between the portions. The blood can thus be diluted to any desired extent; usually a 1-10 dilution is used. After the blood and the citrate solution have been thoroughly mixed by running the two up and down in the pipette, the tip of the pipette is sealed in the flame. If desired, this pipette can then be used as a centrifuge tube and the blood cells washed in salt solution until free of serum. In practical tests before transfusion, however, this is not necessary, as after transfusion itself both sera are present in the vessels of the recipient.

The serum for the test is obtained by immersing the tip of the glass capsule in the issuing drops of blood. The capsule fills itself by capillary action. When the capsule is about three-quarters full, its free end is sealed in the flame. The capsule is then centrifuged, and clear serum is obtained in a few minutes.

(To make the test even more practical, so that it can be carried out at the bedside, Dr. Epstein has devised a small portable hand centrifuge.)

To carry out the test, equal volumes of serum and red blood cell suspension are mixed in the smaller pipettes described above; the capillary ends of the pipettes are sealed; and the mixture is then incubated at 37.5° C. for one to two hours. Observations are then made.

Of the ninety cases tested up to the present the results so far as hemolysis is concerned will be presented in a future communication.

One thing, however, seems to deserve discussion here; namely, the possible role of agglutination in transfusion. Up to the present, this point, though mentioned by Hektoen, has not received sufficient consideration. The reaction of agglutination is so striking as to lead to the thought that agglutination might be a danger in transfusion. Though there is no conclusive evidence on the subject, this might account for some of the unfavorable results in transfusion. Dr. Pearce has shown in experiments on dogs, that the injection of agglutinative sera causes severe symptoms and frequently death, due to thrombi of agglutinated red blood cells in the liver. Whether or not this is a real danger in transfusion remains to be settled by further investigation.

Tabulation of the agglutinations was begun simply to find out whether the results of the tests were reliable and would coincide with those of other observers. The beautiful way in which the bloods of various persons arranged themselves in sharply defined groups was very striking.

It was first pointed out by Landsteiner, and later by Hektoen, Gay, and others, that according to the mutual agglutinations of their sera, bloods could be divided into three classes, defined as follows:

1. In the first group the red blood cells are not agglutinable by any other blood, although the serum is found to agglutinate the red blood cells of all persons not of this group.

2.   The second group is agglutinated by the first and third groups, but it agglutinates only the third group.

3.   The third group is agglutinated by the first and second; but it agglutinates only the second group; occasionally also members of the third group.

This grouping was remarkably regular. Out of all the tests there was really not one blood which did not fit into it. There was one exception in which with repeated trials the red blood cells of group 2 were not agglutinated by the serum of a blood which clearly belonged to group 1, since it was not agglutinated by any other blood.

The explanation of this grouping is very much in dispute, and several theories have been put forward. Gay has tried to prove that the grouping is due to differences in tonicity of the serum. Hektoen, on the other hand, believes that there are different specific agglutinating substances, and that a given blood may possess one or two or none of them. His view, which seems the more reasonable, is supported by his absorption experiments, in which after the agglutinating power of a serum for the red blood cells of one class had been absorbed by an excess of red blood of that class, the serum still retained its agglutinating power for red blood cells of another class.

The coincidence of a brother and sister, whose bloods were examined, belonging to the same agglutination group, led the authors to inquire whether this blood characteristic, which from the work of Hektoen and Gay seems to be a permanent characteristic of the indivdual, is hereditary. Hektoen tested a family, and found that the mother and three of the children belonged to group 1, and the remaining child to group 2. The authors tested two families. In the one the mother and seven children were all found to belong to group 2; the father could not be examined. In another family, mother, father, and four children all belonged to group 3. It seemed probably a coincidence that the father and mother were of the same group, but possibly a matter of heredity that the children were.

Before any definite conclusions can be reached on this point

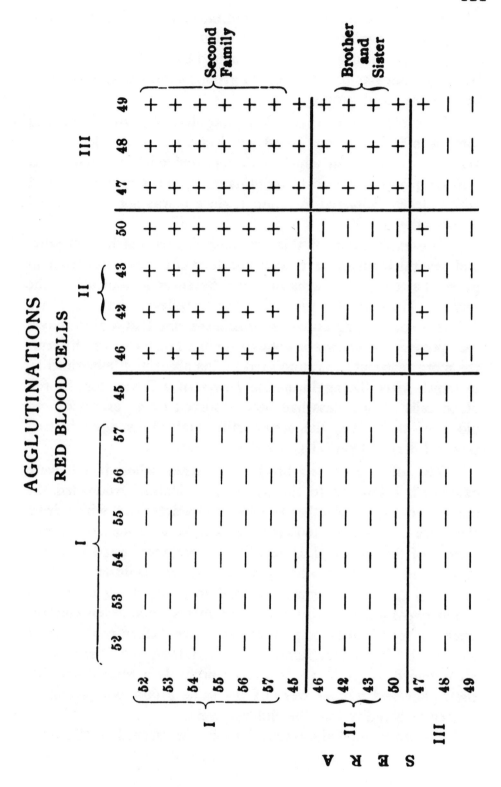

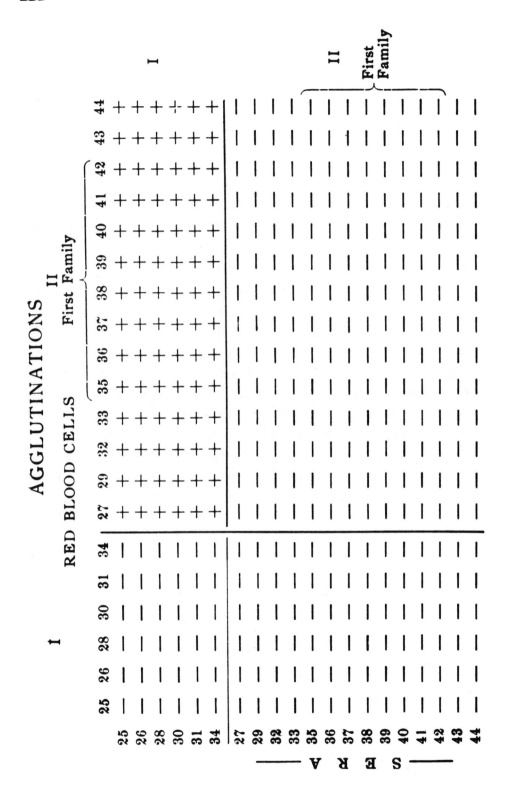

a great deal of careful work must be done, and the authors hope to present further studies later on. It seems, however, from the sharply opposed nature of these blood characteristics that if they are inherited at all they will form a very good example of the Mendelian law of heredity.

Discussion.

DR. S. P. BEEBE asked whether or not it had been possible by the use of this method to detect any dangerous elements in transfusion and prevent dangerous transfusions; or whether any transfusion had been made in a case where this test showed hemolysis to exist, and if so, what results had followed.

DR. OTTENBERG said that they had not yet been able to use this method in any cases which had been followed by transfusion. They had been working on it for only three weeks. Before that the tests were done in the ordinary way. In one case in which hemolysis was detected the proposed transfusion was not done. Dr. Ottenberg said that he knew of one case in which slight hemolysis was detected, and yet the transfusion was done. It was followed by some hemoglobinuria, etc., but nothing serious resulted. Evidently hemolysis was not always a contraindication. Nevertheless tests should always be made and a marked hemolysis should be regarded as a distinct contraindication to transfusion.

5

Reprinted from *J. Exp. Med.* **48**:731–749 (1928)

ON THE INHERITANCE OF AGGLUTINOGENS OF HUMAN BLOOD DEMONSTRABLE BY IMMUNE AGGLUTININS.[1]

By K. LANDSTEINER, M.D., and PHILIP LEVINE, M.D.

(*From the Laboratories of The Rockefeller Institute for Medical Research.*)

Studies on the inheritance of serological properties were first undertaken systematically by von Dungern and Hirschfeld with the agglutinable substances in the blood of dogs (2) and with the human isoagglutinogens (3).[2] The authors named discovered the fact that the isoagglutinogens A and B are inherited as Mendelian dominants and this result has been amply confirmed by numerous workers.

According to their hypothesis there are two pairs of allelomorphic genes, Aa, and Bb, where A and B, the dominant genes, determine the presence of the corresponding agglutinogens, and a and b, the recessive genes, their absence. The genes for the blood groups are the following; group O:[3] aabb; group A: AAbb or Aabb; group B: aaBB or aaBb; group AB: AABB or AABb or AaBB or AaBb.

Another hypothesis has been advanced by Bernstein (6). He assumes multiple (three) allelomorphs, R, A, and B. The genetic formulæ accordingly are; group O: RR; group A: AA, AR; group B: BB, BR; group AB: AB. The theory of Bernstein does not involve a deviation from the older theory in the types of offspring except in the cases of parents belonging to group AB. According to the older view there may be children of any group in unions where one or both of the parents are in group AB; Bernstein's hypothesis, on the other hand, excludes children of groups O and AB in unions O × AB, and children O in unions A × AB, B × AB, or AB × AB. The recent work especially of Schiff (7), Thomsen (8), Preger (9), and Sievers (10) supports the opinion of Bernstein.[4,5]

---

[1] See the preliminary report (1).

[2] The problem of the inheritance of the human blood groups and a few results had been mentioned by Ottenberg and Epstein (4).

[3] The nomenclature of the blood groups by letters instead of numerals has been recommended both by the American Association of Immunologists and by the National Research Council (5) and is used in the present publication.

[4] The objection of Mendes-Correa (11) to the theory of Bernstein would imply

TABLE I.

Tests for M and N in Several Families. F = Father; M = Mother.

Family No.............................	143						144					
Blood No.............................	F 298	M 299	300	301	302	303	F 304	M 305	306	307	308	309
Group........................	A	A	A	A	A	A	O	O	O	O	O	O
Reaction for M................	+++	+++	++±	++±	+++	+++	++±	++±	++±	++±	+++	++±
Reaction for N................	−	−	−	−	−	−	+±	+±	+±	+±	−	+±

The strength of the reactions is indicated by the signs +, +±, ++, ++±, +++.

TABLE II.

Heredity of the Agglutinogen M.

Type of parents	No. of families	No. of children of type		Per cent of children of type	
		M+	M−	M+	M−
M+ × M+	101	403	33	92.4	7.6
M+ × M−	59	165	85	66.0	34.0
M− × M−	6	0	29	0	100

145						146						147					
F 310	M 311	312	313	314	315	F 316	M 317	318	319	320	321	F 322	M 323	324	325	326	327
O	A	A	A	A	A	O	A	O	A	A	O	A	O	A	A	O	A
−	++	++	++±	−	++±	++	++±	++±	++±	++±	++±	−	++±	−	++±	−	−
++	+±	+	+	++	+±	−	++	+±	+±	−	−	++±	++	++±	+±	++±	++

TABLE III.

Heredity of the Agglutinogen N.

Type of parents	No. of families	No. of children of type		Per cent of children of type	
		N+	N−	N+	N−
N+ × N+	31	130	18	88.5	11.5
N+ × N−	29	81	40	66.9	33.1
N− × N−	4	0	17	0	100

41

The investigations outlined on the heredity of human blood groups are not only of theoretical interest[6] but they have attracted much attention because of their practical application in forensic medicine A certain limitation lies in the fact that only two properties could be utilized. It is true that some experiments pointed to the existence of differences in human blood aside from the blood groups (13–16), but as a result of these studies no genetic investigations worthy of notice have been reported although obviously such would have been desirable. The reason for this is to be seen in the lack of workable methods.

The observations reported in previous publications (1, 17, 18) enabled us to undertake a study of the heredity of serological properties of human blood other than those determining the four blood groups.

With regard to the property designated as M there was no difficulty in selecting immune sera and absorbing bloods in such a manner that the reactions were either entirely negative on microscopic examination or so strong that clumps were visible to the naked eye.

The results with a second property, N, whose heredity was studied, varied depending upon the particular immune serum used. Although the strongest agglutinations occurred with the same bloods, there were differences in the reactions of minor strength so that the number of positive tests was greater with some sera than with others. In the following experiments two sera were selected which behaved identically and gave the smallest number of positive reactions; *i. e.*, the bloods acted upon by these sera reacted positively with all sera. Moreover with the sera chosen there was a distinct break between positive and negative tests, a point of significance for the present issue.

The frequency of the types M+ and M−, and N+ and N−, as already stated, is approximately the same in the four blood groups. According to our present results, there were in 1708 white individuals 326 (19.1 per cent) with negative reactions for M, and in 532 white individuals there were 139 (26.1 per cent) negative for N.

---

that the formula $p + q + r = 1$, holds for arbitrarily chosen values, which is obviously not the case.

[5] While this paper was in press, another explanation based upon the assumption of incomplete linkage was proposed by Bauer (*Klin. Woch.*, 1928, vii, 1588).

[6] *Cf.* Morgan (12).

The technique of performing the tests has been described (18). The absorptions and tests for N were made at about 40°C.

It should be stated that the technique offers some difficulties as compared to the common isoagglutination tests. It is necessary to become well acquainted with the method and to know the properties of each serum in order to absorb completely all agglutinins but those in question.

The material for this study was obtained from two maternity clinics in the City of New York. Altogether 166 families were studied; in most of these (122) there were four or more children. Several families were always included in one experiment and also a number of control bloods with known properties.

TABLE IV.

Unions No.	Type of parents	No. of families	No. of children of types		
			M+N+	M+N−	M−N+
1	M+N+ × M+N+	11	31	17	7
2	M+N+ × M−N+	17	40	1	34
3	M+N+ × M+N−	24	60	40	3
4	M+N− × M−N+	5	17	0	1
5	M+N− × M+N−	4	0	17	0
6	M−N+ × M−N+	3 (6)	0	0	18 (29)

The figures in parentheses in unions of type 6, include the three families tested only for M but which, according to our experience, would be of the type M−N+.

In 166 families only the property M was investigated; 64 families were examined for M and N. A representative experiment is given in Table I.

The results for M are summarized in Table II and are arranged in three classes corresponding to the three types of unions and those for N in Table III are similarly arranged.

64 families were examined both for M and N. The results (Table IV) are arranged according to the six sorts of matings and the three types of offspring that have been observed.

A list of the tests is given in Tables Va and Vb for the individual families, the former showing the tests for M (102 families) and the latter tests for both M and N (64 families). In each case the children are

TABLE V*a*.**

Reactions for M.

Family No.	Father	Mother	Children				
1	O+	B+	O+ ♂	O+ ♂	O+ ♀	O+ ♀	
2	O+	O+	O− ♀	O− ♂	O+ ♂		
3	AB−	AB+	AB− ♀	B+ ♂	B− ♂	A− ♀	
4	A+	A+	A+ ♂ A+ ♂	A+ ♀	A+ ♀	A− ♂	A− ♂
5	A+	A+	A+ ♀	A+ ♂	O+ ♀	A+ ♂	
6	B−	A+	B+ ♀	B+ ♀	B+ ♂	B+ ♂	
7	O+	O+	O+ ♂	O+ ♀	O+ ♂	O+ ♂	
8	AB+	A+	A+ ♂	B+ ♀	A+ ♂	A+ ♂	
9	A+	O−	A+ ♂ O+ ♀	O+ ♀	A+ ♀	O+ ♂	A+ ♀
10	B+	B+	O− ♂	B− ♀	B+ ♂		
11	O+	AB+	A− ♀	A+ ♂			
12	O+	A+	A+ ♀	O+ ♀	A+ ♀	A+ ♂	
13	O−	O−	O− ♀	O− ♀	O− ♀	O− ♀	
14	AB+	A+	A+ ♂	A+ ♂	A+ ♀	AB+ ♂	A+ ♂
15	AB+	A+	A+ ♀	AB+ ♀	AB+ ♂		
16	B+	O−	B− ♀	O− ♂	O+ ♀	O+ ♂	O− ♂
17	A+	O+	A+ ♂	O+ ♀	A+ ♂	O+ ♂	
18	A+	O+	O+ ♀	A+ ♀	A+ ♂*	A+ ♀*	O+ ♀
19	A+	O−	O+ ♀	O+ ♀	A+ ♀	O+ ♀	
20	A−	B+	O+ ♀	AB+ ♀	AB+ ♀	A− ♀	

* Twins.
** 20 of the 166 families examined were negro families.

TABLE V*a*—*Continued.*

Family No.	Father	Mother	Children				
21	A+	O+	A+ ♀	A+ ♀	O+ ♀	O+ ♂	
22	O+	O+	O+ ♀	O+ ♂			
23	O+	O+	O+ ♂	O+ ♂	O+ ♂	O+ ♂	O+ ♂
24	O+	O−	O+ ♂ O+ ♀	O− ♂ O+ ♀	O− ♂	O+ ♀	O+ ♂
25	B+	A−	AB− ♀ B+ ♀	B+ ♀	B+ ♂	AB+ ♀	AB+ ♂
26	AB+	A+	AB− ♂	A+ ♀	AB+ ♂	B+ ♀	B+ ♂
27	O−	A+	O− ♂	A+ ♀	A− ♂	A− ♂	
28	B+	O+	B+ ♂	O+ ♂	B+ ♀	O+ ♂	
29	O+	A−	A+ ♂	O− ♀	O− ♀	A+ ♀	O− ♀
30	O−	O+	O+ ♀	O− ♀			
31	A+	O−	A− ♀	O− ♀	A+ ♀	A+ ♀	
32	A+	O−	A− ♀	A+ ♀	A+ ♂		
33	O−	O+	O− ♂	O+ ♀	O− ♀	O− ♀	
34	B+	A+	A+ ♂	AB+ ♂	AB+ ♀	AB+ ♀	
35	A−	B+	A− ♀	AB− ♂	B− ♂	AB+ ♂	
36	O−	O+	O+ ♂	O+ ♂	O+ ♂	O+ ♂	O+ ♂
37	O+	A+	A+ ♀	A+ ♂	O+ ♀		
38	A+	A+	A+ ♂	O+ ♂	A+ ♂	O+ ♂	
39	O+	A−	A+ ♀	O+ ♂	O+ ♂		
40	O+	A+	A+ ♀	A+ ♀	A+ ♂	A+ ♂	A+ ♀
41	A+	B+	AB+ ♀ AB+ ♂	AB+ ♂	A+ ♀	AB+ ♀	AB+ ♂

45

TABLE V*a*—*Continued.*

Family No.	Father	Mother	Children				
42	A+	B−	A+ ♂	A− ♀	AB+ ♀	A− ♂	
43	A+	B+	A+ ♀	A+ ♂	AB+ ♀	A+ ♀	
44	B+	A+	A+ ♀	A+ ♂	A+ ♀		
45	B+	O−	O− ♂	B+ ♂			
46	AB−	O+	B+ ♀	B+ ♀*	B+ ♀*	B+ ♀	
47	A+	O+	O+ ♂	A+ ♀	O+ ♂	A+ ♂	O+ ♂
48	A+	A+	O+ ♂	O− ♂	O− ♂		
49	A+	A+	A+ ♀	A+ ♀	O− ♀	A− ♀	
50	B+	A+	B− ♀	O+ ♂	AB+ ♀		
51	O+	O−	O+ ♀	O+ ♀*	O+ ♀*	O+ ♀	
52	O+	O−	O+ ♂	O− ♀	O+ ♂	O− ♂	O− ♂
53	A+	B−	A+ ♀	AB+ ♂	AB− ♂	A− ♂	
54	A+	A+	A+ ♀	A+ ♂	A+ ♀		
55	A+	A+	A+ ♂	A+ ♀			
56	O+	A+	O+ ♂	A+ ♀	O+ ♀		
57	O−	O+	O+ ♀	O+ ♂	O+ ♀	O+ ♂	
58	O+	A+	A+ ♀	A+ ♂	A+ ♀		
59	O+	A+	O+ ♂	O− ♂	O+ ♂	O+ ♀	A+ ♂
60	B−	O−	O− ♂	O− ♂	O− ♀		
61	O+	A−	O+ ♀	O+ ♂	A+ ♀		
62	A+	A+	A+ ♀	O+ ♂	A+ ♂	O+ ♀	
63	O+	A+	O+ ♀	O+ ♂	A+ ♂	A+ ♂	A− ♂

* Twins.

46

TABLE V*a—Continued.*

Family No.	Father	Mother	Children				
64	O+	O+	O+ ♂ O+ ♂	O+ ♂	O+ ♀	O+ ♀	O+ ♂
65	O+	O−	O+ ♂	O+ ♀			
66	O+	B+	B+ ♂	B+ ♂			
67	A+	O+	A+ ♂	O+ ♂			
68	O+	O+	O+ ♀	O− ♂	O+ ♀	O− ♀	
69	A+	O+	O+ ♀	O+ ♀	A+ ♀	O− ♀	A+ ♂
70	A+	O+	O+ ♀ A+ ♂	A+ ♂ O+ ♂	O+ ♀ O+ ♂	O+ ♀ A+ ♀	O+ ♀
71	AB−	O+	O+ ♀ O+ ♀	A+ ♀	B+ ♂	A+ ♂	A+ ♀
72	B+	A−	O+ ♀	O+ ♀	B− ♀	AB+ ♀	
73	A+	O+	O+ ♀	O+ ♀	A− ♂	O+ ♂	
74	O+	A+	A+ ♀	A+ ♂	O+ ♂	O+ ♂	
75	B+	O+	O+ ♂	O+ ♂	B+ ♂	O+ ♀	
76	O+	A+	O+ ♂ O+ ♂	O+ ♂	A+ ♀	A+ ♂	A+ ♀
77	O−	B+	O+ ♂ O+ ♀	O+ ♀	O+ ♂	B+ ♀	O+ ♂
78	O+	O+	O+ ♀ O+ ♀	O+ ♂ O+ ♀	O+ ♂ O+ ♂	O+ ♀	O− ♀
79	O+	B+	B+ ♂	B+ ♂	B+ ♂		
80	O+	B+	O+ ♂	B+ ♀	O+ ♂	B+ ♂	O+ ♂
81	O−	B+	O+ ♀	O+ ♂	B+ ♂		
82	O+	O+	O+ ♂	O+ ♂	O+ ♂		

TABLE V*a*—*Concluded.*

Family No.	Father	Mother	Children				
83	O+	O+	O+ ♂	O− ♀	O+ ♀		
84	O+	O+	O− ♀	O− ♂	O+ ♂	O+ ♂	
85	O+	B−	B− ♂	B+ ♂	B− ♂		
86	A+	B+	B+ ♀	O+ ♀	AB+ ♀		
87	O+	A+	O+ ♀ O+ ♀	A+ ♀	O+ ♂	O− ♀	A+ ♂
88	O+	O−	O+ ♂	O+ ♀	O+ ♂		
89	A+	O−	A+ ♂ O+ ♀	O+ ♂	O+ ♂	A+ ♂	O+ ♂
90	AB−	O+	A− ♂	B− ♀	A+ ♂		
91	O−	AB+	A− ♂ B− ♀	B− ♀	A− ♀	B− ♀	A+ ♀
92	O+	O+	O+ ♀	O+ ♂	O+ ♂	O+ ♂	
93	B+	B+	B+ ♂	B+ ♂	B+ ♂	B+ ♂	
94	O+	A−	A+ ♀	A− ♂	A+ ♂	A− ♂	A+ ♀
95	B+	A+	B+ ♀	B+ ♂	O+ ♂	A+ ♀	
96	A+	A−	O− ♂ O− ♂	A− ♀	A+ ♂	A+ ♀	O− ♂
97	A+	O+	A+ ♂ O+ ♂	A+ ♀	O+ ♀	O+ ♀	O+ ♂
98	O+	A+	O+ ♀	O+ ♀	O+ ♀	O+ ♂	O+ ♀
99	B+	O+	B+ ♂ O+ ♂	B+ ♀	O+ ♀	O+ ♀	O+ ♂
100	A+	A+	A+ ♂	A+ ♂	A+ ♂	A+ ♀	
101	B+	O+	O+ ♂ B+ ♀	B+ ♂ O+ ♀	B+ ♀	O+ ♂	O+ ♂
102	B−	A−	A− ♀	A− ♀	A− ♂	AB− ♀	

TABLE V*b*.

Reactions for M and N.

Family No.	Father	Mother	Children			
103	O++	A++	A++ ♂ O+− ♀ A−+ ♂ O++ ♂ A+− ♂			
104	O−+	A++	O−+ ♀ O++ ♂ O++ ♂ A−+ ♂			
105	O−+	O−+	O−+ ♀ O−+ ♂ O−+ ♂ O−+ ♂			
106	O++	AB+−	B++ ♂ B++ ♀			
107	O+−	A++	O+− ♀ A++ ♂ A+− ♀ A+− ♀			
108	O+−	A+−	O+− ♀ A+− ♀ O+− ♂ O+− ♂			
109	O+−	A++	A++ ♂ O++ ♂ A+− ♂ A+− ♂ O−+ ♂			
110	A++	O−+	A++ ♀ O++ ♀ O++ ♂ O++ ♀			
111	A+−	A−+	A++ ♀ A++ ♂ A++ ♀ A++ ♂ A++ ♂			
112	O++	O+−	O++ ♂ O++ ♀ O++ ♀ O+− ♂ O++ ♂			
113	A−+	A++	A++ ♂ A++ ♀ A−+ ♀			
114	A++	O++	A−+ ♀ O+− ♂ O−+ ♀ A++ ♀ O+− ♂ A++ ♂			
115	A+−	A++	A++ ♂ A++ ♀ A++ ♀ A++ ♂			
116	O++	B−+	B−+ ♀ O−+ ♀ B−+ ♂ O−+ ♂ O++ ♂ B++ ♀			
117	O−+	O++	O++ ♀ O−+ ♀ O−+ ♀ O++ ♂ O++ ♂ O++ ♂			
118	O+−	A++	O++ ♀ O++ ♂ O++ ♂ O+− ♀ A+− ♀ A++ ♀ A+− ♀ O++ ♀ A++ ♀			
119	O−+	A++	A−+ ♂ A−+ ♀ A++ ♀ O−+ ♀ A++ ♂ A++ ♀ O++ ♀			

TABLE V*b*—*Continued.*

Family No.	Father	Mother	Children			
120	A++	O−+	A++ ♂	A++ ♀	A++ ♂	A++ ♀
121	B++	A++	AB−+ ♂ A+− ♂	A++ ♂ A++ ♀	A++ ♀	A++ ♂
122	O+−	A++	O++ ♂ A++ ♂	A+− ♂	A+− ♂	A++ ♀
123	O−+	A+−	A++ ♂	A++ ♀		
124	A+−	A++	A++ ♂ A+− ♀	O+− ♀	A++ ♀	O+− ♂
125	O++	A++	A+− ♂ O++ ♀	A++ ♀	O++ ♀	A++ ♂
126	B++	B++	B++ ♀	B+− ♂	O++ ♂	
127	A+−	B++	B−+ ♂	A+− ♀	O+− ♂	
128	O+−	O++	O+− ♀	O++ ♀	O+− ♀	O++ ♂
129	O++	O+−	O+− ♂ O+− ♀	O+− ♂ O++ ♂	O++ ♀	O++ ♀
130	A−+	A−+	A−+ ♂ A−+ ♂	O−+ ♂ O−+ ♂	A−+ ♀	A−+ ♀
131	A+−	O−+	O++ ♀	O−+ ♀	O++ ♂	O+− ♂
132	A−+	O++	A++ ♀ O++ ♂	O−+ ♀	O−+ ♂	O++ ♂
133	O++	O+−	O+− ♂ O+− ♀	O++ ♀	O+− ♂	O+− ♀
134	O+−	A+−	A+− ♀	A+− ♀	A+− ♂	
135	O+−	O++	O++ ♀	O+− ♂	O++ ♀	
136	A++	A−+	A−+ ♂ O−+ ♀	A−+ ♂	A−+ ♀	A−+ ♀
137	B+−	O++	B++ ♂	B+− ♀		

TABLE V*b*—*Continued.*

Family No.	Father	Mother	Children			
138	A+−	O++	O++ ♀	A+− ♀		
139	A+−	B−+	B++ ♀	B++ ♀	B++ ♀	AB++ ♀
140	A++	B++	O++ ♀	O++ ♂	A++ ♂	AB+− ♂
			O++ ♀	O++ ♂		
141	O++	B++	B++ ♀	B++ ♂	O++ ♀	O−+ ♂
			O++ ♂			
142	O+−	O++	O++ ♂	O++ ♂	O++ ♀	O++ ♂
			O−+ ♂			
143	A+−	A+−	A+− ♂	A+− ♀	A+− ♂	A+− ♂
144	O++	O++	O++ ♀	O++ ♀	O+− ♀	O++ ♂
145	O−+	A++	A++ ♂	A++ ♂	A−+ ♀	A++ ♀
146	O+−	A++	O++ ♀	A++ ♂	A+− ♂	O+− ♂
147	A−+	O++	A−+ ♀	A++ ♂	O−+ ♀	A−+ ♂
148	A++	B−+	B−+ ♀	A++ ♂		
149	O−+	B−+	B−+ ♂	O−+ ♀	O−+ ♀	B−+ ♂
			O−+ ♂	B−+ ♂	O−+ ♂	B−+ ♀
150	O−+	A++	A−+ ♀	A++ ♀	A+− ♂	A−+ ♀
151	B+−	O−+	O++ ♀	B++ ♀	B++ ♀	
152	O−+	O++	O−+ ♂	O++ ♂		
153	A−+	O++	A++ ♂	A−+ ♀	O−+ ♂	O−+ ♂
			O++ ♀	O−+ ♀		
154	A++	B+−	O++ ♀	B++ ♀	O+− ♂	
155	A+−	A++	A++ ♂	A+− ♂	A++ ♂	A+− ♂
			A+− ♂	A++ ♂		
156	A++	O++	O++ ♀	O+− ♀	O+− ♂	A+− ♀

TABLE V*b*—*Concluded.*

Family No.	Father	Mother	Children			
157	O−+	O++	O++ ♀ O++ ♂ O−+ ♀ O−+ ♀ O++ ♀ O−+ ♂			
158	O++	A−+	A++ ♀ A++ ♂ O++ ♂			
159	O++	B+−	B++ ♀ O++ ♀ B++ ♀ B++ ♂			
160	A++	O++	A+− ♂ A++ ♀ O+− ♂ A−+ ♂ O++ ♂ O+− ♀			
161	O++	A++	A++ ♀ A++ ♀ A+− ♀ A+− ♂ A−+ ♀			
162	A+−	O++	A++ ♂ A++ ♂ A+− ♀ A+− ♂ O++ ♀			
163	B+−	O+−	O+− ♂ B+− ♂ O+− ♂ B+− ♂ O+− ♂ B+− ♂			
164	B+−	A++	O++ ♀ A++ ♂ A++ ♂ O++ ♂ AB++ ♂			
165	B+−	O++	O++ ♀ O+− ♀ B+− ♀ O+− ♀			
166	B+−	O++	B++ ♀ O++ ♀ B+− ♂			

recorded in order of decreasing age beginning with the eldest. The letters designate the groups, and the signs + and − the reactions for M (Table V*a*). In Table V*b* the first + or − sign designates the test for M and the second sign that for N.

As to the heredity of the factors A and B our results agree with the established fact that they are inherited as Mendelian dominants, except for three families in which A or B appeared in children when they were absent in the blood of the parents. These cases were considered as instances of illegitimacy and were excluded from the tabulations. One of these families was examined only for M and two for both M and N. The results were:

Father	Mother	Children			
O+	O+	O+ ♂	B+ ♂	O+ ♀	A+ ♀
O++	O++	A++ ♀	A++ ♀	O++ ♀	O+− ♀
O+−	B−+	AB++ ♀	O++ ♂	O++ ♀	O++ ♀

In family 71, AB × O, there were two children in group O; the mother refused reexamination.

From the data reported it is evident that the agglutinogens studied are inherited properties. If we consider M and N separately they would seem to behave like Mendelian dominants. The characters cannot be recessive since in unions + × + there are children whose blood lacks the property. This result is to be expected if there are individuals among the parents heterozygous for a dominant character. If the absence of the agglutinogens is recessive there should occur no positive reactions in children from unions − × −. This is actually borne out by the observations. Thus in six such families with M − parents all the children (29 in number) gave negative reactions for M; likewise in the four families with N − parents all the children (17 in number) belonged to the N − type. In this respect our findings are analogous to the rule established by von Dungern and Hirschfeld for the isoagglutinogens A and B, i. e., that these do not appear in the offspring if they are absent in both parents.

In order to discuss the numerical results for M alone in the three sorts of matings it is necessary to know the incidence of homozygous (MM) and heterozygous (Mm) individuals among the M+ parents. From a formula quoted by Johannsen (19), the following values are obtained: MM = 29.6; Mm = 49.6; mm = 20.8 (approximately 30, 50, 20, respectively). According to this formula the percentage of homozygous individuals equals $100 - 20 \sqrt{R} + R$, that of heterozygous $20 \sqrt{R} - 2R$; where R is the percentage of recessive (M −) individuals observed in the population. The figures of M+ and M − reactions are taken from all individuals, including the parents, of the 166 families examined for M.

Calculating from these figures the number of M − children in the unions M+ × M+ one has to consider only those in which both parents are heterozygous, i. e., approximately 5/8 × 5/8; since 1/4 of the children of these matings should be M −, 9.8 per cent of M − children are to be expected; the observed value is 7.6 per cent.

In the unions M+ × M − 50 per cent of the offspring of heterozygous M+ parents may be expected to be M −; i. e., 5/8 × 1/2 = 31.3 per cent; actually 34 per cent M − children were found.

In the smaller series where both factors were examined (see Table IV) the agreement between the calculated and the observed values is not so satisfactory.

Applying the formula of Johannsen for the factor N we obtain NN = 23.1; Nn = 49.8; nn = 27.1. Calculated as above the figures are in matings N+ × N+ 11.7 per cent N— children (observed 11.5 per cent) and in matings N+ × N— 34.2 per cent N— children (observed 33.1 per cent).

So far the cursory analysis of the results does not contradict the idea of two independent factors. However, there is evidence which does not seem compatible with this view. In the first place, if the factors were independent one would expect a certain, although small, percentage of bloods to lack both M and N, that is, if the genotype M — N— is not lethal, or its phenotype indistinguishable from one of the other types. In fact, as has been stated formerly (18) no such case has been found in the examination of more than 1200 specimens[7] and in each M— blood the reaction for N was found to be very strong. Further evidence emerges from an analysis of results in families examined for both properties. One sees that the frequency of M'— children in the three sorts of matings, Nos. 1, 3, and 5 (Table IV) of parents M+, varies greatly according to the N reactions of the parents and that likewise the occurrence of children N— in matings 1, 2, and 6 is influenced by the presence or absence of M in the parents. A similar statement holds for the appearance of M— or N— children in matings 2 and 4, and 3 and 4, respectively.

Actually in most of the six sorts of unions the observed figures do not tally satisfactorily with those to be expected on the basis of two independent factors if one computes the expectancy from the figures given above for heterozygous and homozygous individuals. Thus in matings 2 and 3 there are too many children of the type M—N+ or M+N—, respectively, and in union 4 there appear with one exception only children of type M+N+. These numerical results could be interpreted in various ways. One hypothesis consists in assuming two genes which, when homozygous, would determine the phenotype M+N— and M—N+ respectively, while the phenotype M+N+

---

[7] This number includes the blood of negroes and Indians. About 900 of these were tested with the improved technique, namely, performing the tests for N at about 37–40°C.

would correspond to the heterozygous gene constitution. This view accounts for the non-existence of the type $M-N-$. On the basis of this hypothesis the expected values for the types of offspring are: mating 1: $M+N+$ 50 per cent, $M+N-$ 25 per cent, $M-N+$ 25 per cent; mating 2: $M+N+$ 50 per cent, $M-N+$ 50 per cent; mating 3: $M+N+$ 50 per cent, $M+N-$ 50 per cent; mating 4: $M+N+$ 100 per cent; mating 5: $M+N-$ 100 per cent; mating 6: $M-N+$ 100 per cent.[8] Allowing for the relatively small number of individuals examined these figures agree fairly with those observed and better than the figures calculated for independent factors. Especially striking is the fact that in matings 2 and 3 there are almost no children $M+N-$ or $M-N+$, respectively, and in mating 4 only one child not of the type $M+N+$. Still, there are five cases which contradict the hypothesis mentioned, namely, the individual $M+N-$ in union 2, the three children $M-N+$ in union 3, and one child $M-N+$ in union 4. To attribute all these five exceptions to illegitimacy seems hazardous since only a proportion of illegitimate children would be detected by the tests employed and because the number of the exceptional cases is high compared with that of children which do not conform to the rule of von Dungern and Hirschfeld.

It may be pointed out that in all of the five exceptional cases in unions 2, 3, and 4, the father and not the mother is of the type opposite to that of the child, e. g., father $+-$, child $-+$.

On the basis of the assumption just discussed and with the aid of the formula used above, the frequency of one type could be used to calculate the frequency of the other two types. Starting from the figure 20.8 for $M-$ in a certain population the computation gave the value of 29.6 $N-$; similarly in a population of 205 Indians examined by us, the observed value of 4.9 $M-$ leads to an expectancy of 60.7 for $N-$. Both these figures are in good agreement with those observed, namely, 26.1 and 60.0 respectively.[9]

An alternative hypothesis would suppose a close linkage between M and N. If, then, the gene combinations Mn and mN are numerically predominant this could explain the observed figures and also the occurrence of the exceptional cases aforementioned, but unless one assumes a lethal effect there arises a difficulty from the following con-

[8] The numbers of the matings refer to those given in Table IV.
[9] These results will be discussed in another publication (20).

sideration. If the factors M and N have been in the race for a long time the occurrence of cross-overs should by now have reduced the assumed numerical difference. However, on account of the existence of agglutinogens similar to M and N in anthropoids (chimpanzees) (18) it does not seem likely that they are due to recent mutations.

If in the unions listed as No. 1, Table IV, one of the parents be homozygous with respect to both M and N, then all children of such a union would be of the phenotype M+N+. The fact that this was not the case in any of the eleven families shows that none of the parents can be homozygous for both M and N, but such homozygous parents may have occurred in unions 2 and 3.

In view of the limited number of families studied it would seem premature to attempt a final interpretation and to discuss further possibilities such as the existence of more than two allelomorphs. Also it has to be considered that the state of affairs may be complicated, *e. g.*, by interacting or modifying effects of factors determining hitherto unknown agglutinable structures.

It may be added that there is no indication of a linkage between M and N and the isoagglutinogens A and B, and, as in the case of A and B, no evidence of a sex linkage.

SUMMARY.

The heredity of two agglutinable structures demonstrable by immune agglutinins was studied in 166 families. From the data collected it is evident that one deals with a case of Mendelian inheritance. The main result of the studies is the demonstration that it is feasible to investigate the heredity of serological structures of human blood other than the group agglutinogens. Irrespective of the ultimate theory it seems very probable that the properties M and N do not appear in the offspring when they are absent in both parents—a conclusion substantiated by the examination of ten families with 46 children. These findings offer the prospect of forensic application to cases of disputed paternity and, in our opinion, a correct decision could already be given, at least with great probability, provided the reagents are available and the method properly applied. Of course further work is needed before the test can be adopted as a routine procedure.

56

For the material used in this study we express our thanks to Dr. Harold Bailey and Dr. H. C. Williamson and also to Miss M. C. Skelly, Superintendent of the John E. Berwind Free Maternity Clinic, and to Miss Agnes Martin, Superintendent of the Manhattan Maternity Hospital.

The authors are greatly indebted to Dr. C. B. Bridges and Dr. A. Weinstein of the laboratory of Professor T. H. Morgan, Columbia University, for their kindness in discussing with us the results reported.

REFERENCES.

1. Landsteiner, K., and Levine, P., *Proc. Soc. Exp. Biol. and Med.*, 1927, xxiv, 941.
2. von Dungern, E., and Hirschfeld, L., *Z. Immunitätsforsch.*, *Orig.*, 1910, iv, 531.
3. von Dungern, E., and Hirschfeld, L., *Z. Immunitätsforsch.*, *Orig.*, 1910, viii, 284.
4. Epstein, A. A., and Ottenberg, R., *Proc. New York Path. Soc.*, 1908, viii, 117.
5. Editorial, *J. Am. Med. Assn.*, 1927, lxxxviii, 1421.
6. Bernstein, F., *Z. indukt. Abstammungs. u. Vererbungsh.*, 1925, xxxvii, 237.
7. Schiff, D., *Deutsch. Z. ges. gerichtl. Med.*, 1927, ix, 369.
8. Thomsen, O., *Deutsch. Z. ges. gericht. Med.*, 1927, x, 1; *Compt. rend. Soc. biol.*, 1928, xcviii, 1270.
9. Preger, A., *Z. Immunitätsforsch.*, *Orig.*, 1927, liii, 192.
10. Sievers, O., *Acta path. et microbiol.*, 1927, iv, No. 4, 285.
11. Mendes-Correa, A. A., *Le Sang*, 1927, i, 322.
12. Morgan, T. H., Evolution and genetics, Princeton, Princeton University Press, 1925.
13. von Dungern, E., and Hirschfeld, L., *Z. Immunitätsforsch.*, *Orig.*, 1910, viii, 526.
14. Guthrie, C. G., and Pessel, J. F., *Bull. Johns Hopkins Hosp.*, 1924, xxxv, 81.
15. Landsteiner, K., and Witt, D. H., *J. Immunol.*, 1926, xi, 221.
16. Landsteiner, K., and Levine, P., *J. Immunol.*, 1926, xii, 441.
17. Landsteiner, K., and Levine, P., *Proc. Soc. Exp. Biol. and Med.*, 1927, xxiv, 600.
18. Landsteiner, K., and Levine, P., *J. Exp. Med.*, 1928, xlvii, 757.
19. Johannsen, W. L., Chapter on Allgemeine Vererbungslehre, in Die Biologie der Person, by Th. Brugsch and F. H. Lewy, 1926, i, 290, Berlin, Urban and Schwarzenberg.
20. Landsteiner, K., and Levine, P., *J. Immunol.*, in press.

6

Reprinted from *J. Genet.* **33**:411–434 (1936)

THE PROBLEM OF DOMINANCE IN MAN

By S. G. LEVIT

Maxim Gorky Medico-Genetical Research Institute, Moscow

(With Eight Text-figures)

Some peculiarities of pathological genes in man

THE opinion is rather widespread that the majority of mutant genes in man, as well as in most of the plants and animals which have been studied, are recessive. R. A. Fisher, in particular, emphasizes this view in working out his theory of the evolution of dominance. He cites *Homo sapiens* side by side with *Drosophila* and sweet peas as examples of species in which mutant genes are, as a rule, recessive. He does not deny that many *dominant* mutant genes in man have been described, but points out that this may be due to the fact that dominants in man are easier to discover than recessives. To prove this he calls attention to the fact that a comparatively large number of sex-linked recessive genes in man (which are discovered with relative ease) have been described, whereas we do not know as yet of any dominant genes in the sex-chromosome. Thus if most of the genes in the *X*-chromosome are actually recessive, then an analogous phenomenon should also be observed in the autosomes and only our imperfect technique prevents us from determining this. The same idea was pointed out by the author in 1929, and with this premise in mind then began a systematic investigation of pathological genes in man.

However, the hereditary diseases in man which the author and his co-workers have studied for the past six years have all proved to be *dominant*, with more or less limited penetrance. This was true for Basedow's disease (Levit *et al.*[1]), ulcus ventriculi (Levin & Kuchur), leukaemia (Ardashnikov), eunuchoidism (Likhziemr *et al.*), bronchial asthma (Malkova), paroxysmal tachycordia (Ryvkin, Gassko) and apparently also for pernicious anaemia (Presnyakov), hypertonus (Ryvkin, Malkova and Kantonova, unpublished data) and other diseases.

---

[1] In this paper the authors advanced the preliminary hypothesis that the genes for goitre and for the classic Basedow's disease are dominant, and that the gene for toxic struma is recessive. Later extensive research of I. A. Ryvkin (unpublished data) confirms the conclusions regarding the first two diseases but throws some doubt as to toxic struma.

In testing the dominance of the mutant genes in the above-mentioned cases it is necessary to know which of the allelomorphs should be called mutant, and also which should be called dominant. The first question cannot always be answered, particularly when we deal with genes which do not produce any pathological effect. For example, we cannot say which of the allelomorphs that determine the blood groups in man is the normal gene and which is the mutant. The same is true of the gene for the taste reaction to phenyl-thio-carbamide and of other genes. However, it is another matter when we deal with pathological genes. There can hardly be any doubt that the mutant is the pathological gene in the great majority of cases.

Of no less importance is the second question. When may a gene be recognized as dominant? It should be recognized here that we hardly ever know how new mutant genes which produce a pathological effect in the heterozygote manifest themselves in the homozygote. Not knowing this, we cannot call them dominants in the strict sense of the term, since this designation presupposes that the homozygote (**AA**) does not differ from the heterozygote (**Aa**) in its phenotypic expression. On the other hand, the fact that the mutant gene produces a pathological effect when in heterozygous condition proves that it has some degree of dominance, at least, and that this degree of dominance is of practical consequence. The author proposes, therefore, the term "conditionally dominant" for those genes which give clear pathological expression in the heterozygote and whose homozygote is as yet unknown. This group of "conditionally dominant" genes should be considered as in part a provisional one. In time, as the phenotypic expression of the homozygote becomes known, this group will naturally be converted into complete dominants (probably the minority), intermediates, and probably some others.

There is some degree of probability that for some of the cases referred to here the homozygote is lethal. This inference may be drawn from the genetics of other species, where it is quite often observed that genes producing pathological effects in the heterozygote are lethal in the homozygote. This would provide a possible explanation of our cases in which the mating of two affected persons failed to produce in offspring a more serious form of the disease.

Still more important is the question, on what grounds have we established the "conditional dominance" of the genes in question? As criteria of dominance or recessiveness we have taken the following: (1) The percentage of inbreeding among parents who produce affected offspring (as is known, this percentage is increased in cases of recessiveness).

Although we studied comparatively rare diseases, in some cases extremely rare ones, in no case did we discover a percentage of inbreeding higher than that which exists in the normal population. (2) The ratio between affected and healthy persons among the various types of relatives. First we should compare the incidence of affected sibs with the incidence of affected children or parents. In the case of recessive characters the greater incidence of affected individuals among sibs as compared with children or parents is well known. But such a condition was never observed in our material.

Of particular significance is the fact that for the above-mentioned diseases we found pedigrees manifesting the disease during three, four and even more generations. The supposition of recessiveness naturally requires that in all cases of the latter type the matings were between the homozygote (**aa**) and the heterozygote (**Aa**); this is extremely un-

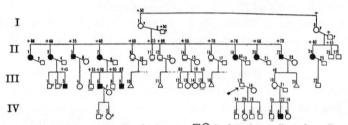

Fig. 1. Paroxysmal Tachycardia. Ryvkin's case. □○ Definitely unaffected. △ Both sexes. —⋯ Existence of more children unknown. + Dead. Figures above horizontal lines indicate age.

likely in so far as cases of this kind have taken place in the absence of inbreeding, and often even in matings between persons who are either of different nationalities or who were born in localities far removed from one another. The objection may be raised that in groups in which given genes attain a relatively high concentration, pedigrees of the above kind should quite often be observed. But this objection does not hold for cases in which we deal with extremely rare diseases. Thus, the transmission of a very rare disease for a number of generations in the absence of inbreeding (see for example Fig. 1, concerning paroxysmal tachycardia, observed in our Institute by I. A. Ryvkin) can hardly leave a doubt that we are dealing with a "conditionally dominant" gene.

Special consideration should also be given to the question of the polyfactorial nature of the character, i.e. as to whether the latter may be conditioned by two or more rare genes acting jointly. However, where information can be obtained not only of near relatives but also of distant relatives, this possibility may be excluded. As an example we present the data of Levit & Pessikova for diabetes mellitus (analysis of 222 pedigrees, see Table I):

TABLE I

Type of relationship	Unaffected	Affected	% affected
Parents	397	18	4·34
Sibs	445	16	3·59
Children	33	1	2·94
Grandparents	235	5	2·08
Uncles, aunts	309	12	3·74
Nephews, nieces	58	1	1·73
Cousins	220	2	0·91

As this table indicates there are approximately equal percentages of affected parents, sibs and children (the difference between them can easily be explained as a difference in age[1]); approximately half as many grandparents as parents (the somewhat higher percentage of affected aunts and uncles may be explained by the poor reproductivity of diabetics); approximately half as many nephews and nieces as children; and about one-fourth as many cousins as sibs. These proportions are characteristic of monofactorial dominant heredity when the pathological gene is sufficiently rare. If, on the contrary, we were dealing with a polyfactorial disease the decrease in the incidence of affected individuals among more distant relatives would proceed at a far more rapid rate (for example, if diabetes were due to the combined action of two rare dominant genes, the percentage of affected grandparents should be about one-fourth the percentage for the parents, etc.).

The question arises as to whether or not the fact that all the mutant genes investigated by us happened to be dominants was due to chance. This is hardly possible since we are dealing with a comparatively large number of diseases. Moreover, a critical review of the literature has convinced us that the phenomenon observed by us is characteristic for man in general. In this connexion it should be emphasized that it is exceedingly difficult to study the literature on the genetics of man from this standpoint. The difficulties are of two kinds. First, the literature itself is immense. Second, the greater part of it concerns investigations carried out by inexact methods. Hence the available literature, and in particular the conclusions of individual authors, must be used with great caution. For this reason we chose four collected works: Davidenkov (nervous diseases), Siemens (skin diseases), and Franceschetti and Waardenburg (both eye diseases). We approached the facts presented in these summaries critically, and did not recognize as recessive and

[1] Penetrance of diabetes increases with age.

dominant genes nearly all that the authors designated as such. We excluded particularly those diseases which were described as casual observations and especially such of these as gave contradictory results leading the authors to infer dominance in certain cases and recessiveness in other cases of apparently the same disease. We, therefore, fixed our attention primarily on those diseases concerning which there are systematic investigations. In many cases we became familiar not only with the detailed data presented in the summaries enumerated above but also with the original investigations before we finally decided to interpret the hereditary disease in question as "conditionally dominant" or recessive. This critical examination of the data in the literature resulted in the following classification for autosomal genes:

Conditional dominants	Recessives	Intermediates
38	14	3

These facts agree with our own findings and confirm our hypothesis that *the majority of mutant pathological genes in man are "conditionally dominant"*.

It should be emphasized once again that neither we nor the authors whose data we used made any special selection of dominant diseases. As a matter of fact a certain unconscious selection probably exists in the literature, but not because the authors concentrate their attention on "conditionally dominant" diseases. At first glance this sounds paradoxical, for it is usually supposed that as a consequence of man's small reproductivity it is more difficult to study recessive than dominant diseases. It seems to me that the contrary is true, namely, that recessive genes are easier to investigate than dominants, since increased incidence of inbreeding is usually already to be observed in the first few families which the investigator encounters. Moreover, there is a marked prevalence of affected sibs with an almost complete absence of other affected relatives. Then again, the penetrance of the recessive in the homozygotes is usually far better than the penetrance of "conditional dominants" in the heterozygotes. This is very important in view of the difficulties that poor penetrance raises in genetic analysis. Possibly all these conditions have influenced the selection of material previously studied (how otherwise can one explain the fact that such comparatively insignificant diseases as albinism, amaurotic idiocy, alkaptonuria and others have been so well investigated, while till recent times there has been a lack of studies of the genetics of such significant diseases as stomach ulcer, Basedow's disease, etc.?). At any rate, there is no doubt that there was no selection in favour of dominant diseases for the investigations.

From the data presented, then, we are led to conclude that there is a definite preponderance of "conditionally dominant" pathological traits in man.

A second trait, no less characteristic than dominance, is the *extreme variability (poor penetrance) of most pathological mutant genes in man.* We may say with more or less certainty that penetrance varies, in the diseases investigated by us, from a fraction of 1 per cent (leukaemia, pernicious anaemia) to 10–20 per cent and occasionally higher. The same poor penetrance is also to be seen when we make a careful study of the literature. Conditionally dominant genes give especially convincing evidence on this point when they are very rare. The skips found in the latter cases, especially where the disease is transmitted for several generations without inbreeding, leave no doubt that we are dealing here with a failure of penetrance (see for example Fig. 1). Recessive genes, on the contrary, probably show, as a rule, good penetrance in the homozygotes.

The following case is of interest in connexion with what has been said. Hogben (1932), using material gathered by Bulloch, came to the conclusion that diabetes insipidus is due to a good dominant gene. But from the point of view stated above we were in doubt about this, and so the author, together with L. N. Pessikova, made a study of this disease in their own material. The study clearly demonstrated that the gene for diabetes insipidus is not a good dominant gene. Among the twelve systematically gathered cases of diabetes insipidus included in this study only two were familial, and even these did not show perfect penetrance. Moreover, there are skips recorded in the material on which Hogben himself bases his conclusions (the cases of Wile, Case and Clay).

In this connexion the danger must be emphasized of formulating genetic generalizations on the basis of data taken from the literature, for such data are often "selected data". Thus all the cases gathered by Bulloch are selected ones, in that they deal with "interesting cases", that is, with cases in which the phenomenon of inheritance is expressed most distinctly. Our material, on the other hand, differs from that used by Hogben in that our investigation was systematic: that is, familial as well as non-familial cases were taken without selection and subjected to investigation.

A third characteristic is suggested by the study of pathological genes in man. This is that recessive diseases have a tendency to manifest themselves at an earlier age than "conditionally dominant" diseases. We frequently encounter the latter in persons of fairly advanced age, and sometimes the manifestation is seldom to be found before the end of the reproductive period—a circumstance often observed in diabetes mellitus, for example, as well as in other diseases.

Such are the rules which may be formulated on the basis of the data obtained in our investigations. But here another question arises. If they are correct, and if, in particular, the basic thesis is correct, that most pathological mutant genes in man are "conditionally dominant", then how is all this to be reconciled with the statement cited above that almost all genes localized in the X-chromosome are recessive? With this question in view we have conducted a detailed study of the latter problem. Our results show that just the reverse conclusion is the correct one; namely, that only a small minority of genes which can be proved to be sex-linked have given definite evidence of being recessive. We may now proceed to examine the evidence on this point.

Sex-linked genes in man (in relation to the problem of dominance)

About thirty genes located in the X-chromosome of man have been described. The large majority of these genes have been considered to be recessive. When the material is reviewed in detail, however, one is surprised at how uncritically the authors consider their cases, both in regard to the evidence for sex-linkage, and in regard to that for recessiveness. The criterion usually employed for sex-linkage consists in the manifestation of the disease only or chiefly by males and its transmission by females. Detailed proof should hardly be necessary to demonstrate that this criterion alone is inadequate, for autosomal dominant genes *limited* (entirely or in part) to the male sex meet these conditions and behave in very much the same way. To be sure, the authors frequently point out that the *non-transmission of the character from father to son* is another and necessary criterion for recognizing a sex-linked gene. This criterion can be correctly applied, however, only when there is a large enough number of sons of a male carrier of the character. But strangely enough no attention has ever been paid to this condition.

The following case is presented as an illustration. Lenz (1932), in discussing the genetics of nystagmus, cites a pedigree of Nodop, which is supposed to illustrate the sex-linked type of inheritance. This pedigree is given by Lenz in an abbreviated form. But it is characteristic that Lenz abbreviated just that part of the pedigree in which a healthy son was born of an affected father, although this was really important contributory evidence for his own hypothesis.

The treatment of the genetics of Pelizaeus-Merzbacher disease offers another example of exceptionally careless handling of this problem. This is an extremely rare disease of the nervous system and, so far as I know, its genetics is based on observations of only one family (see Fig. 2). But this family record contains no mating of an affected male, and consequently no evidence of non-transmission from an affected father to his sons. Hence, as a matter of fact, it offers no proof of the presence of sex-linkage in this case. We might just as well assert that it is a case of autosomal inheritance. But let us assume that the gene in this case is actually located in the X-chromosome. Then why should it be regarded as recessive? As is indicated in Fig. 2, there are two affected females in this case, who obviously cannot be regarded as homozygotes. Therefore, even if we regard this as a sex-linked gene, it should be treated as "conditionally dominant", and perhaps partially limited to the male sex.

A large number of similar examples may be cited.

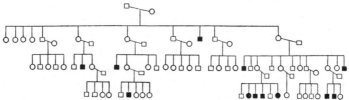

Fig. 2. Pelizaeus-Merzbacher disease.

All of the foregoing indicates the need for a reconsideration of the material presented in the literature as recessive sex-linked inheritance in man.

The criterion we have used consists of the absence of the trait in question among the sons of fathers carrying it. However, we have treated as sex-linked only those cases in which the sons of affected fathers numbered not less than nine, for only in these cases is the probability of such a result on the basis of autosomal inheritance (which we will call PA) sufficiently small $(1/2^9)$[1]. Now in order to affirm the recessiveness of a gene already proved to be sex-linked we should have more than this criterion as well as more than the absence of this trait in female heterozygotes in one or several observed families. The latter condition is inadequate, because the absence of affected females may result from occasional non-expression of the dominant gene in a limited number of female heterozygotes, which may be either accidental or due

[1] The probability $1/2^9 = 0.00195$ corresponds approximately to that accepted in genetics as statistically significant (3σ).

to a partial limitation of the trait to the male sex.[1] To prove that a sex-linked gene is not "conditionally dominant" but strictly recessive, the thorough investigation of a large population is necessary. In that case, the ratio of the male to the female bearers of a recessive trait should be as $q : q^2$. This naturally follows from the distribution of sex-linked genes in the population, which, as is known, may be expressed by the formula $p\mathbf{A} : q\mathbf{a}$ (in males) and $p^2\mathbf{AA} : 2pq\mathbf{Aa} : q^2\mathbf{aa}$ (in females). If the proportion of females is greater than q^2 we may conclude that the sex-linked gene is partially dominant, manifesting itself not only in female homozygotes, but partly also in female heterozygotes. When the proportion of females is less than q^2 we may assume partial limitation to the male sex, i.e. partial non-expression in females. In these cases, however, we assume that there is no differential viability—a condition not to be taken for granted.

However, investigations of this kind have been made only in relation to colour-blindness. Other sex-linked genes have not been studied from this point of view. For this reason alone they cannot be strictly classified as recessive. If this is too rigorous a requirement, then the minimum requirement which a sex-linked gene must fulfil to be classified as recessive is that at least one affected homozygous female must be known (with the absence of the latter trait in female heterozygotes).

The necessity of the latter requirement follows from the possible complete limitation of given traits to the male sex. In cases in which sex-linked genes are wholly limited to the male sex, it is impossible to differentiate dominants from recessives. I propose calling such genes "indeterminates". It should be emphasized that these genes are "indeterminates", not because we know so little about them, but because the question of whether they are dominants or recessives is meaningless. Along with them we must also provisionally class as "indeterminate" all sex-linked genes in respect to which the data are still insufficient to show that they can express themselves at all in females.

From the above point of view the author has found that in the case of only twelve of the genes described as sex-linked can the evidence for sex-linkage be considered as fairly conclusive. In respect to dominance these genes may be divided into the following four categories, with the behaviour of the genes in females as the criterion for this division.[2]

---

[1] We have some evidence that apparently the "conditionally dominant" pathological genes are limited more frequently to the male sex than to the female. This still awaits an explanation.

[2] The symbols designate: – – absence of trait, – + weak expression of trait, + + definite presence of trait.

Classification of genes	Heterozygote	Homozygote
Indeterminate	– –	– –
Recessive	– –	+ +
Intermediate	– +	+ +
Conditionally dominant	+ +	?

We may now proceed to a separate consideration of each of these categories.

A. *Indeterminate genes*

Here are included those genes whose manifestation has not yet been observed in either heterozygous or homozygous females. These comprise:

(1) Haemophilia. Strangely enough this gene is always treated as recessive. However, we do not know, as yet, of the existence of this trait in females, though matings from which we would expect homozygous females have apparently taken place. Consequently, the complete limitation of this gene to the male sex is to be presumed. The data of Schlössmann (1930) concerning the intermediate expression of haemophilia in female heterozygotes (mild bleeding, slight retardation of blood coagulation and other symptoms) could not be confirmed by a test investigation made in our institute by Freidberg.

(2) Hemeralopia, accompanied by myopia (Varelmann, Kleiner and others). A very rare anomaly.

(3) A peculiar form of juvenile glaucoma, accompanied by coloboma and ending with blindness. This character was described by Frank-Kamentzky in the former Irkutsk district. He observed three families.

The comparatively large number of indeterminate genes (one-fourth of the total number of sex-linked genes) may probably be explained by the small amount of material investigated in the given cases. That is, the inclusion of these genes in this category is still for the most part provisional. The possibility is not excluded that when more cases have been observed the characters here mentioned will be found in heterozygous females, so that some of the characters described here will then be of the "conditionally dominant" type. But if any of the characters are observed only in the female homozygotes, then the corresponding gene will be classified as recessive.

B. *Recessive genes*

These affect homozygous females but, so far as is known, are unexpressed in the heterozygote. The known cases are:

(1) Megalocornea. The most convincing cases are Kayser (1914) and Grönholm (1921). In the latter case, this trait was observed in two

female homozygotes. We cannot be certain that megalocornea is not expressed in the heterozygote, since it is a very rare disease.

(2) Ichthyosis vulgaris. A total of ten cases, in which inheritance can be explained as sex-linked, have been described (see Orel, 1929). The affected males in five of these cases had no sons at all. The PA of the others equalled respectively $1/2^2$, $1/2^5$, $1/2^6$, $1/2^7$. In the case of Csörsz[1] there was a mating between an affected male and a heterozygous female. Among the offspring there were two affected females (apparently homozygotes). However, we cannot be certain that in the case of ichthyosis vulgaris there is complete recessiveness because the number of investigated cases is small. Moreover, it should be pointed out that this disease is chiefly inherited as a dominant autosomal trait. Hence, we cannot be sure that these ten cases are due to a sex-linked gene.

C. *Intermediate genes*

The genes referred to here are expressed in the homozygote as in group B, but differ by being also expressed to a degree in heterozygous females.

These comprise:

(1) Colour-blindness, the intermediate expression of which in female heterozygotes was recently shown by Wieland (1933). The question still remains as to whether the various forms of colour-blindness are due to different allelomorphs of one gene or to different genes in the X-chromosome. Besides, there are relevant cases of high expression of colour-blindness in individual female heterozygotes (see Wieland). This gives ground for considering the question of transferring colour-blindness from this group to the group of "conditionally dominant" genes.

(2) Keratosis follicularis spinulosa. A case of this kind (Fig. 3) was described by Siemens (1925). It should be pointed out that since the PA in this case is equal to only $1/2^7$, this case is not very convincing from the viewpoint of sex-linkage. However, Siemens treats also the case of Laméris, in which the PA is equal to $1/2^3$, in the same manner. These two cases taken together give adequate ground for regarding this character as due to a sex-linked gene. A milder form of the disease in female heterozygotes than in males was observed in Siemens's case. Hence we include this character in group C. However, as we do not yet know the phenotype of the homozygote for this trait, we cannot exclude the possibility that in the future (if the homozygote proves indistinguishable from the heterozygote) this character will prove to belong to

---

[1] Cited from Orel.

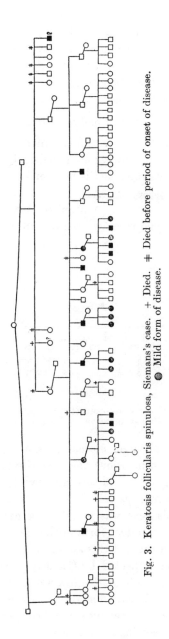

Fig. 3. Keratosis follicularis spinulosa, Siemans's case. + Died. ‡ Died before period of onset of disease. ● Mild form of disease.

the group of dominant sex-linked genes, perhaps with an expression limited, in degree, in females. Hence, this gene may be more accurately called "conditionally intermediate".

Summing up the above we may observe that the two genes which we classified in this group differ from one another, colour-blindness being closer on the whole to the recessive type, and keratosis follicularis to the dominant.

D. *Conditionally dominant genes*

Here we include those genes which have a definite effect, at least in certain cases in heterozygous females. The behaviour of these genes in the homozygous condition is not yet known.

These comprise:

(1) Leber's disease. A number of cases in which this is present in female heterozygotes have been described.

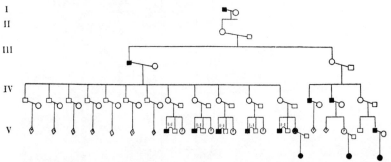

Fig. 4. Thadani s case (reconstructed by author). Anidrosis, adentia and hair defect. ? Number of siblings unknown. |:| Ratio of affected to unaffected. ⊕ Mild form.

(2) A complicated anomaly, including anidrosis, adentia, and hair defects. Darwin mentions one such Hindu family. Thadani (1921, 1934) has apparently described this same family twice (Fig. 4). Roberts (1929) described a similar case in U.S.A. Inbreeding was absent in both these cases, and this gene was found to be expressed in some of the female heterozygotes (a milder form in the case of Roberts).

Some other cases of this anomaly have been reported in the literature and have been collected by Cockayne. Some of these cases (Guilford-Atkinson, Cunningham, Kerley, Cockayne, Goeckermann, and others) also show some affected females. Cockayne distinguishes these from the other cases of the "recessive sex-linked type", as if they formed a separate class of cases showing dominance. We believe it probable that these all represent the same class of cases of conditionally-dominant with sex-linkage, and in support of this note that there is not one case even among the pedigrees showing dominance, of an affected male having affected sons. The fact that when females are affected they usually have a more "attenuated form" of the anomaly agrees with this interpretation (it will be explained below why this should usually occur in sex-linkage).

(3) The absence of the central incisors. Only one case has been described (Huskins, 1930). The *PA* was rather small[1] in this case (see Fig. 5), and the character was expressed in females.

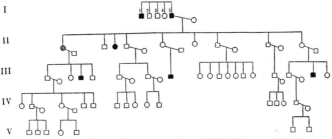

Fig. 5. Absence of central incisors. Case of Huskins. ◉ Mild form.

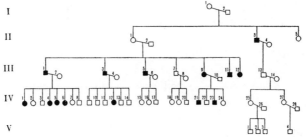

Fig. 6. Nystagmus. Waardenburg's case.

(4) Nystagmus. The genetics of this trait is very complex and as yet not thoroughly studied. In the literature nystagmus conditioned by a recessive sex-linked gene and that conditioned by an autosomal dominant are differentiated from one another. Lenz (1932) considers them as two allelomorphs (recessive and dominant) of one sex-linked gene. Moreover, without advancing any reasons he assumes the existence of a third allelomorph of recessive type, which simultaneously causes isolated albinism of the eye. I do not consider Lenz's arguments convincing, because he denies the autosomal inheritance of nystagmus while we have many cases of transmission from father to son (we have observed

[1] In 1, 1 there were six sons, who had many healthy offspring (footnote of the author in describing his case).

several cases of this type). Nor do I find convincing the differentiation of recessive sex-linked from dominant autosomal nystagmus, since among the pedigrees illustrating the second mode of inheritance we often find a very low PA (see, for example, of Waardenburg's case, Fig. 6; Hemmes treated this case as autosomal, whereas the PA here is equal to $1/2^9$). Moreover, in pedigrees usually considered as recessive sex-linked, there are cases of penetrance of the trait in female heterozygotes (see, for example, Engelhard's case). The case of Hemmes (Fig. 7) is especially odd in this sense. The author interprets it as a case including two types of inheritance: autosomal dominant (progeny 1, 1) and recessive sex-linked (progeny 1, 3). But it is rather clear that there are no grounds for such an interpretation, and that we have here a typical case of conditionally dominant sex-linked inheritance (PA is equal to $1/2^{11}$). Even if one agreed with Hemmes's interpretation, one would nevertheless have to regard the inheritance in progeny 1, 3 as being not recessive sex-linked but conditionally dominant sex-linked (due to the presence of an affected heterozygous female in 1, 3).

Considering the above, I am inclined to believe that the data prove the existence of at least two forms of primary nystagmus, one due to a conditionally dominant sex-linked gene of a phenotypically varying type (especially in females), and the second due to an autosomal dominant gene. The first form of nystagmus would include those cases previously described as sex-linked recessive, and in addition some of those cases which were previously described as autosomal dominants, and which have a sufficiently low PA (as, for example, the case of Waardenburg, presented in Fig. 6). In general this character demands further investigation. But at any rate the existence of a conditionally dominant sex-linked form of nystagmus can hardly be disputed.

(5) Retinitis pigmentosa. This character is usually inherited autosomally (apparently sometimes as a dominant and sometimes as a recessive). At the same time, McQuarrie recently described a case (Fig. 8) which makes sex-linked inheritance extremely probable for individual pedigrees (PA equal to $1/2^{14}$). The old cases of Usher and others, which were also interpreted as sex-linked but which were not convincing because the PA was too high, become more probable in the light of this observation.[1] In addition to the case presented in Fig. 8 there is at least one affected female, which gives ground for considering the character

---

[1] It should be emphasized that this case of McQuarrie's, the first case of retinitis pigmentosa which could be interpreted with certainty as sex-linked, proved to be incompletely recessive.

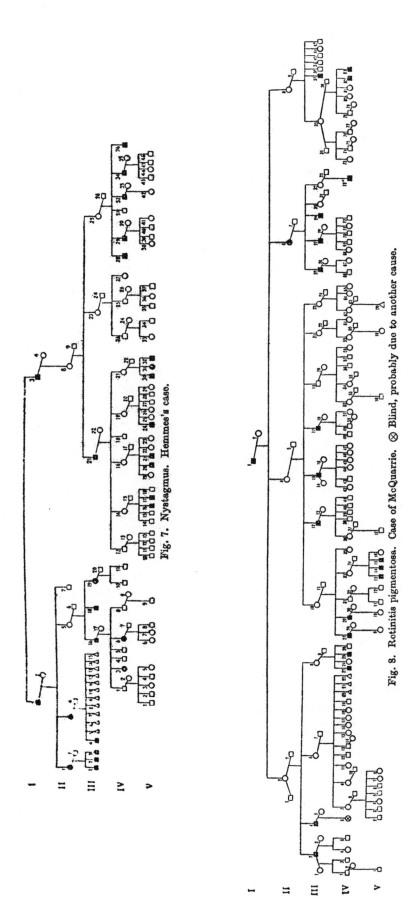

Fig. 7. Nystagmus. Hemmes's case.

Fig. 8. Retinitis pigmentosa. Case of McQuarrie. ⊗ Blind, probably due to another cause.

73

in question as conditionally dominant. Supplementary investigation of the existence of conditionally dominant sex-linked forms of retinitis pigmentosa is required before a final conclusion is reached.

None of the other characters described as sex-linked satisfy our criterion of having a PA as low as $1/2^9$.

Thus, it is seen that the supposition regarding recessiveness of the great majority or even all (Fisher) of the sex-linked genes in man is incorrect. The contrary is rather the case. Recessive genes are actually in the minority among sex-linked genes (2 out of 12, and if colour-blindness is provisionally added to group B, then 3 out of 12).

It may be repeated that it is much easier to recognize recessive than dominant genes located in the X-chromosome, since the latter may easily be confused with autosomal genes.

The problem of the more marked expression of the genes in group D (and possibly also A) among males, and of reduced expression among heterozygous females, requires special consideration. This may be explained simply by the fact that the effect of a sex-linked gene in the male is about equal to the effect of this gene in the female homozygote (as found in *Drosophila*). On the other hand, the expression of a conditionally dominant sex-linked gene in a female heterozygote must naturally be comparable to the expression of a conditionally dominant autosomal gene in a heterozygote of either sex. But in this respect we see in both these groups (that is, in the autosomal as well as in the sex-linked genes) a similarly low frequency of expression of conditionally dominant genes in the heterozygote.

Summing up, we come to the conclusion that the data in regard to sex-linked genes not only do not contradict the hypotheses suggested here for autosomal genes in man, but, on the contrary, *confirm* these hypotheses.

DISCUSSION

It has become generally recognized in recent years that wild genes (normal allelomorphs) did not always so generally dominate over their mutant allelomorphs as they do now, but that at least some of them became dominant through natural selection. There is a difference of opinion, however, as regards the question of the mechanism of this evolutionary process. Three independent theories of the mechanism (R. A. Fisher, 1928; J. B. S. Haldane, 1930; Plunkett, 1933, and Muller, 1932) have been proposed. Wright (1934), although having a different viewpoint on dominance, nevertheless admits the possibility of the hypothesis of Plunkett and Muller. This is not the place to discuss these theories. Moreover, it really does not matter, for the considerations which follow, which theory of evolution of dominance we accept for our hypothesis.

It need only be emphasized here that the fact that wild genes generally dominate over their mutant allelomorphs is the basis of *all* these theories. It was just this fact which led Fisher to the conclusion that wild genes *become* dominant by selection. But, *mutatis mutandis*, the same reasoning enables us to understand the peculiarities of the pathological genes described above. If the dominance of wild, normal allelomorphs brings us to conceive of the evolution of dominance, then naturally incomplete dominance may be presumed to be the result of an incomplete evolutionary process. In other words the pathological mutations of man have not yet completed the development which leads to absolute recessiveness. Therefore, a considerable majority of genes in man are *as yet not recessive.* However, *they are no longer strictly dominant,* because, as was demonstrated above, they are extremely variable phenotypically (in respect to penetrance). But genes which are sometimes expressed and sometimes not expressed in the heterozygote (**Aa**) should, as a matter of fact, be regarded as recessive in the cases of non-expression and as dominant in the cases of expression, or, better still, as "conditionally dominant". In other words, a considerable majority of genes in man have not as yet become fully recessive, but have already lost a large part of their pathological effect in the heterozygous condition. And from the point of view of evolution of dominance it is natural to suppose that this evolution in man should not yet have completed its course.

Can we propose a hypothesis which would explain this retardation of the evolution of dominance in man? I think that we can. It should be recalled, in this connexion, that natural selection, as a factor influencing the evolution of man, has become a great deal less significant than it was in his savage ancestry, being slowly replaced by new factors—social forces. Natural selection has played a progressively decreasing rôle since man's ancestors first began to use tools even of the most rudimentary kind in work activity. The rôle of natural selection decreases in proportion as man is farther removed from his ape-like ancestors. We may assume, therefore, that the process of evolution of dominance has been retarded along with this general retardation. For while the evolution of dominance is based upon the fact that the mutant gene reduces the vitality of its carrier, it is also true that the use of implements of labour (and later the development of medicine) fully or partly compensates for this deficiency, and the individual who was formerly mercilessly destroyed by natural selection is by this means enabled to live without impediment and to reproduce. With increased development in technique and industry, with further advancement in medicine, the greater may be the defects (results of pathological mutations) which *Homo sapiens* is able to bear. Since the basic stimulus for the evolution of dominance or for the conversion of mutant genes into recessives has thus disappeared, we come to the conclusion that the recessive pathological genes under present observation arose very long ago, and have in consequence already had time to lose their expression in the heterozygote. Those genes which, on the contrary, have arisen comparatively recently,[1] have not as yet passed through a corresponding period of development and consequently have preserved some degree of dominance.

H. J. Muller, after acquainting himself with the facts presented above and with their interpretation, has proposed to me a further possibility, with which I fully agree. His idea is that the retardation of natural selection in human society could have another result, namely, loss of recessiveness (because of mutations acting oppositely to those which cause dominance) in some of those genes which previously had time to acquire it. Should a loss like this accidentally affect a wild species, the process of redevelopment of the recessiveness of the mutant gene would immediately begin again. But as a consequence of the decline of natural selection in man, the latter can fully stand such a loss, and the loss of dominance would eventually become established. Thus, not only the development of dominance but also its maintenance depends upon natural selection not being relaxed.

The hypothesis presented by the author throws some light on the later ontogenetic expression of "conditionally dominant" genes, mentioned above, as compared to recessive genes. The possibility is not excluded that dominance, at least in individual cases, is conditioned by age, in such wise that the expression of the mutant gene is postponed till the end of the reproductive period. As was stated above, a number of "conditionally dominant" genes are expressed at 30–40 years of age and later. These genes may thus be interpreted as having already become recessive for the reproductive period. And yet they preserve their patho-

---

[1] It is understood that the expressions "long ago" and "recently" refer chiefly to the period of the origin of man and to the very early period of the history of the species *Homo sapiens*.

logical effect (in the heterozygote) for later life. Hence, it may be supposed that many diseases characteristic of old age are the expression of dominant or conditionally dominant genes which have already become recessive in early life.

The genetics of diseases of old age, which has hitherto been very superficially investigated, acquires special interest in the light of what has been said.

It is worth while to discuss another point in connexion with the question of the age when the mutant gene attains expression. As a matter of fact, one can really regard the tendency of conditionally dominant genes to be expressed at a later age than recessive genes as a particular case of a more general rule which is frequently mentioned in human genetics, namely, that recessive diseases in general are more serious than dominant diseases. However, on the basis of the above a more careful analysis may lead to the very opposite conclusion. This follows, not from general comparisons of recessive with dominant diseases, but from separate comparisons of heterozygous dominants with heterozygous recessives, and of the homozygous dominants with homozygous recessives. Thus, genes express themselves in immeasurably milder form in the heterozygotes in the case of recessive diseases (where they practically do not express themselves at all) than in the case of conditionally dominant diseases. This may be explained by the fact that the recessive gene has completed its evolution, and has lost its pathological effect, while the dominant gene has still to pass through this stage of its evolution. There are grounds for supposing that the homozygote also shows a milder effect in the case of recessives as compared to dominants. For, as was pointed out above, one can presume that at least part of the conditionally dominant genes are lethal in the homozygote (that is, they cause a more serious form than the most serious known recessive diseases). Secondly, as will be noted later, it is possible that some of the recessives also are no longer expressed pathologically even in the homozygote, since their effect also has probably been lost, due to the same evolutionary process. Such characteristics naturally cannot be ascribed to the dominants, since even their heterozygotes have as yet far from completed their evolution.

Thus when we come to the general conclusion that known recessive genes in man tend to a milder (less harmful) phenotypic expression than dominant genes, this in turn requires an explanation. This explanation may easily be formulated as follows: *a gene has a milder effect not because it is recessive but it is recessive because, either from the moment of its origin, or by reason of more selection, it gave a milder effect.* This is in accordance with the view expressed by Fisher, that the genes with less pathological effect naturally reach the limits of their evolutionary development sooner, lose their harmful effect in the phenotype more quickly, give way to the influence of gene modifiers more easily, than do genes with greater pathological effect.

The latter conclusion is important for two reasons. First, it affords another interpretation for the fact that the penetrance of conditionally dominant genes occurs later than that of recessive genes. Secondly, while the view that recessive genes tend to express more serious cases of disease than dominant genes is purely empirical and has no explanation, the hypothesis formulated by the author has the advantage of yielding to theoretical interpretation.

Classification of genes in man

On the basis of the data presented above, the author proposes the following classification of genes in man, subdivided with respect to dominance.

(1) Complete dominants. Evidently there are very few of these.

(2) Conditional dominants, the majority of which show rather poor penetrance. Most mutant pathological genes in man belong to this group.

(3) Intermediates. This designation should not be taken in the strict sense of the word, that is, in the sense that the phenotypic expression of the heterozygote is exactly intermediate between the expressions of two homozygotes. In practice its expression is closer to one of the latter.

(4) Recessives. Emphasis should be placed upon the fact that the study of recessive genes more and more reveals a slight expression in the heterozygote. Whether this is the result of an as yet incomplete evolutionary process, or whether this process can be regarded as complete in the sense that the heterozygote has become a wild type, to the degree necessary for the successful survival of the gene carrier, is a question the solution of which is hardly probable in a general form, and which requires a detailed study in every individual case.

Quantitatively, categories (3) and (4), taken together, form the second group of mutant genes in man.

(4 *a*) Some of the recessives are not always expressed even in the homozygote. There is reason to believe that this is true, for example, for epilepsy, schizophrenia, etc. According to Fisher these may be interpreted as follows. After rendering the heterozygote like or almost like the dominant homozygote the evolution of dominance continues with

the accumulation of modifiers of the homozygote, thus finally inhibiting its expression. However, it must be admitted that the study of this problem is as yet extremely inadequate. We still have very few data which allow us to speak convincingly about the degree to which recessive genes are expressed in the homozygote. There are theoretical and practical reasons for the accumulation of these data, both for deciding the question of whether recessive genes always remain unexpressed in the heterozygote, and of whether they are always expressed in the homozygote.

(5) Indeterminate genes. These comprise only sex-linked genes which are wholly limited to the male sex. It is impossible to distinguish dominants from recessives in this case.

Conclusions

1. The majority of pathological genes in man:

(*a*) are conditionally dominant, showing a definite expression in the heterozygote, and

(*b*) have poor penetrance. This is true both for autosomal as well as for sex-linked genes.

2. Conditionally dominant genes (in the heterozygote) tend to be expressed later in ontogeny than recessive genes (in the homozygote).

3. An explanation is given for these phenomena from the point of view of the evolution of dominance.

REFERENCES

ARDASHNIKOV, S. N. (1934). "On the genetics of leukaemia." *Proc. Maxim Gorky Med.-Biol. Res. Inst.* **3.** (Russian with an English summary.)

BAUR, FISCHER & LENZ (1927). *Menschliche Erblichkeitslehre*, **1**, III. Aufl.

BULLOCH, W. (1909). "Diabetes insipidus." *Treasury of Human Inheritance*, **1**, parts I–II.

DAVENPORT, C. B. (1930). "Sex linkage in man." *Genetics*, **15**, No. 5.

DAVIDENKOV, S. N. (1932). *Hereditary Diseases of the Nervous System.* Moscow. (Russian.)

ENGELHARD, C. F. (1915). "Eine Familie mit hereditärem Nystagmus." *Z. ges. Neurol. Psychiat.* **28.**

FISHER, R. A. (1928). "The possible modification of the response of the wild type to recurrent mutations." *Am. Nat.* **62**, 115–126.

—— (1930). *The Genetical Theory of Natural Selection.* Oxford.

—— (1931). "The evolution of dominance." *Biol. Rev.* **6**, No. 4.

FRANCESCHETTI, A. (1930). "Die Vererbung von Augenleiden." *Kurz. Handbuch d. Ophthalmologie*, **1**, herausg. von Schieck und Brückner.

FRANK-KAMENTZKY, S. O. (1927). "On a peculiar hereditary form of glaucoma in Irkutsk district." *Rus. Eug. J.*, v, No. 1. (Russian.)

FREIDBERG, S. A. (1930). "Über Nachforschungen von Blutungserscheinungen bei Frauen, die Heterozygoten des Gens Haemophilie sind." *Med.-biol. J.*, Moscow, Nos. 4–5. (Russian with a German summary.)

GASSKO, S. M. (1936). "A clinico-genetical investigation of paroxysmal tachycardia." *Proc. Maxim Gorky Med.-Gen. Res. Inst.* 4 (in the Press).

GRÖNHOLM, V. (1921). "Über die Vererbung der Megalocornea u.s.w." *Kl. Monatsbl. f. Augenh.* 67, 1.

HALDANE, J. B. S. (1930). "A note on Fisher's theory of the origin of dominance and on a correlation between dominance and linkage." *Amer. Nat.* 64, 87.

HEMMES, G. D. (1924). "Over 'hereditary nystagmus'." Diss. Utrecht.

HOGBEN, LANCELOT (1932). *Genetic Principles in Medicine and Social Science*, p. 44. New York.

—— (1933). *Nature and Nurture*, p. 51. London.

HUSKINS, C. LEONARD (1930). "On the inheritance of an anomaly of human dentition." *J. Hered.* No. 6.

KAYSER, B. (1914). "Über den Stammbaum einer Familie mit Vererbung von Megalocornea nach dem Hornerschen Vererbungstypus." *Arch. Rass.- u. GesBiol.* 11, 170.

KLEINER, WILHELM (1923). "Über den grossen schweizerischen Stammbaum, in dem sich mit Kurzsichtigkeit kombinierte Nachtblindheit forterbt." *Ibid.* 15, 1.

LENZ, F. (1932). "Über die Geschlechtsgebundheit des erblichen Augenzitterns." *Ibid.* 26.

LEVIN, A. E. & KUCHUR, B. A. (1934). "The role of hereditary factors in the etiology of ulcerous disease." *Proc. Maxim Gorky Med.-Biol. Res. Inst.* 3. (Russian with an English summary.)

LEVIT, S. G. (1929). "Genetics and pathology." *Med.-biol. J.*, Moscow, No. 5. (Russian.)

LEVIT, S. G. & PESSIKOVA, L. N. (1934). "The genetics of diabetes mellitus." *Proc. Maxim Gorky Med.-Gen. Res. Inst.* 3. (Russian with an English summary.)

—— —— (1936). "Is diabetes insipidus caused by a good dominant gene?" *Ibid.* 4 (in the Press).

LEVIT, S. G., RYVKIN, J. A., SEREISKY, M. J., VOGELSON, L. I., DORFMANN, I. A. & LIKHZIEMR, I. B. (1930). "Über die Genetik der Basedowschen Krankheit und der ihr nahestehenden pathologischen Formen." *Med.-biol. J.*, Moscow, Nos. 4–5. (Russian with a German summary.)

LIKHZIEMR, I. B., ZHISLINA, S. G. & GERCHIKOVA, J. A. (1936). "A clinico-genetical investigation of eunuchoidism" (in the Press).

McQUARRIE, MARGARET D. (1935). "Two pedigrees of hereditary blindness in man." *J. Genet.* 30, No. 1.

MALKOVA, N. N. (1936). "Clinico-genetical investigation of bronchial asthma and other allergic diseases." *Proc. Maxim Gorky Med.-Gen. Res. Inst.* 4 (in the Press).

MULLER, H. J. (1932). "Further studies on the nature and causes of gene mutations." *Proc. VI Intern. Congr. Genet.* 1.

OREL, HERBERT (1929). "Die Vererbung von Ichthyosis congenita und der Ichthyosis vulgaris." *Z. Kinderheilk.* 47, 32.

PLUNKETT, C. R. (1933). "A contribution to the theory of dominance." *Amer. Nat.* **67**, 84.

PRESNYAKOV, D. F. (1936). "Clinico-genetical investigation of pernicious anaemia." *Proc. Maxim Gorky Med.-Gen. Res. Inst.* **4** (in the Press).

ROBERTS, ELMER (1929). "The inheritance of anhydrosis associated with anadontia." *J. Amer. med. Ass.* **93**, No. 4.

RYVKIN, I. A. (1934). "On the heredity in paroxysmal tachycardia." *Proc. Maxim Gorky Med.-Biol. Res. Inst.* **3**. (Russian with an English summary.)

SCHLÖSSMANN, H. (1930). *Die Hämophilie.* Stuttgart.

SIEMENS, HERMANN WERNER (1925). "Über einen in der menschlichen Pathologie noch nicht beobachteten Vererbungsmodus: dominante geschlechtsgebundene Vererbung." *Arch. Rass.- u. GesBiol.* **17**.

—— (1929). "Die Vererbung in der Ätiologie der Hautkrankheiten." *Handb. d. Haut- und Geschlechtskrankheiten,* **3**, herausg. von J. Jadassohn.

THADANI, K. I. (1921). "A toothless type of man." *J. Hered.* **12**, 87.

—— (1934). "The toothless man of Sind." *J. Hered.* **25**, 483.

VARELMANN, HANS (1925). "Die Vererbung der Hemeralopie mit Myopie." *Arch. Augenheilk.* **96**, 385.

VERSCHUER, OTMAR (1934). *Erbpathologie.* Dresden und Leipzig.

WAARDENBURG, P. J. (1932). "Das menschliche Auge und seine Erbanlagen." *Bibliogr. Genet.* **9**.

WIELAND, MAX (1933). "Untersuchungen über Farbenschwäche bei Konduktorinen." *Graefe's Arch. Ophthal.* **130**, 144.

WRIGHT, SEWALL (1934). "Physiological and evolutionary theories of dominance." *Amer. Nat.* **68**.

7

Reprinted from *Hum. Biol.* 4:69–79 (1932)

THE RELATION OF THE MODE OF INHERITANCE TO THE SEVERITY OF AN INHERITED DISEASE

BY MADGE THURLOW MACKLIN

Department of Histology and Embryology, Medical School, London, Ontario

One often encounters the statement in textbooks of medicine that in some instances a disease, such as diabetes, for example, is familial and in other cases it is hereditary. Such a statement is unfortunate, for it at once implies that the cases which are familial, that is, that are found only in the brothers and sisters in a family, but not in the parents or in former generations, are not inherited. They are inherited just as much as are those listed as hereditary, but they exemplify a different type of inheritance. If one wishes to distinguish between those cases that appear suddenly in several of the children of a family without any previous history of a similar disease, and those cases in which parents or former generations in the direct line of descent, or even collaterals, were known to have been similarly affected, one might coin a new word to indicate the latter type of cases. They might be termed *generational* instances, while the term familial was still kept for the former variety. The term hereditary would then be applied with equal correctness to both types.

When one uses the term hereditary, one implies that the character in question has been passed on from parents to offspring via the germ plasm. It does not alter the fact that a character is inherited because it must be derived from both parents before it is manifest in the offspring, while another character, also inherited, comes from only one parent, yet is still visible in the child.

It is the first type of inheritance just mentioned, namely that in which both parents must transmit before the character becomes recognizable as such, that gives us the so-called recessive or "familial" cases. The parents need not, most often *do* not show such a character themselves.

Just as two hidden streams of water, travelling under ground unseen, unknown, suddenly meet, fuse and with the combined strength of the two, force their way to the surface as a bubbling spring, so two parents may each carry latent a character that is not manifest as such in either of them; but their germ cells meeting, fusing, produce a new individual in whom there springs forth this character in recognizable form. Just as the spring was as truly attributable to one stream as to the other, though neither were visible, so is this new character which apparently springs full grown without any obvious source, derived from both parents and so inherited.

When one studies the inheritance of disease in the human being, one is at once struck with the fact that in man, the same disease may in one family appear to be inherited through generation after generation, without a break; whereas in other families, the same disease may crop up in several children whose parents were normal; or it may appear in only one child as a sporadic example of the disease. It is this peculiarity of the human race, (for in most animals a character is more or less uniform in its mode of inheritance,) that makes the textbooks of medicine state, for example, of diabetes, that cases which are hereditary form about 15 per cent of the whole, that cases which are familial constitute about 7 to 15 per cent, while the sporadic cases make up the remaining part. Because of the erratic modes of inheritance of these diseases in man, all the cases of which cannot be classified in any one category, the tendency is to disregard the study of inheritance in man as being unprofitable.

In reviewing a large series of cases of such diseases, which in some families appear without any former history, and so are familial, and which in other families are generational, several generalizations can be made which appear to me to strengthen materially the idea that such diseases not only are inherited, but follow different modes of inheritance in different families, and that in many instances one cannot say that such and such a disease is transmitted according to any given rule, but must say that in this family it is inherited in one way while in another family it may be transmitted according to entirely different laws.

THE LOCATION OF MUTATIONS

If one conceives that in the course of evolution Nature must have made hundreds of mutations, some of which have gone in the discard, some of which have persisted as variations of the normal as we know it, it is quite comprehensible that the mutation that occurred in the germ

plasm of one person was in such a direction that all the factors necessary for the production of the character, diabetes for instance, were
located in one chromosome and if one inherited that, the disease appeared; if not, then one remained normal. In another person, the same
mutation appearing suddenly may have affected several chromosomes
so that if one got both of them one had diabetes, but if only one of
them, then one was apparently normal. If in the latter case, the factors
were located in two chromosomes, which were members of a pair, the
disease would be inherited as a recessive, whereas if the factors were
located in different chromosomes, not mates, it might have a very
sporadic type of inheritance, neither dominant nor recessive but par-
taking of the qualities of first one, then the other in different matings.

If in addition one conceives that this mutation was such that it also
determined the severity of the disease, there would be further complications. If a type of the disease, severe because it involved a great deal of
the islet tissue of the pancreas, were to arise as a mutation and also
have all its factors located in one chromosome, then such a disease would
automatically die out. It would appear early in the life of the individual,
inasmuch as there would be little normal tissue left to function and the
person would die before reaching the age of maturity, and so before he
could pass on this newly appearing mutation that was evident in him for
the first time. It would be a mutation that would rise above the
surface of the Sea of Being to be submerged again before it would
wreck the life of more than the few children in that family in whom
it might chance to appear.

But suppose that it was one which, although having the capacity to
kill early due to the severity of the disease, was also one which appeared
in the chromosomes that were mates or allelomorphs. It then would be
of such a nature that it could appear in the offspring of parents that
were themselves not affected. Under such circumstances, the patient
would die early of the disease, but that would not in the least affect
its reappearance in other children of the same family nor in subsequent
generations, for some of the children would receive, not two factors, but
merely one for the disease. They would then be like their parents,
free from it, but capable of having affected offspring if they mated
with a person like themselves who carried one factor latent for the production of this particular character. Thus there would be no direct
inheritance, but the collaterals in the same or different generations
might show the condition. The affected person would die young as in

the other case, but the stream of inheritance would flow on through others not affected.

Is there any evidence apart from the actual pedigrees, to support the idea that the same disease might be inherited in several different ways, or might one claim that rather elaborate and complicated laws of transmission governed all the cases and that were these laws known, all the pedigrees would be seen to express them perfectly? If the supposition that the disease arose in some families as a dominant and in some as a recessive, and that there were also factors linked with this that governed the severity of the disease, and hence the age at which it appeared, then it would be only in those families in which the disease was not severe that the dominant type would be propagated. If it were being inherited as a recessive, not only the mild variety, in which the person was not affected until some time past the age at which he could reproduce, but also the severe form in which he died in childhood would reappear erratically through the generations. Hence studying the age of onset of the disease in different families would be of assistance in determining whether all cases were being inherited according to the same laws.

The average age of onset of any disease which showed variations in its mode of inheritance should be uniformly higher for those families which showed the generational type of inheritance, than for those which showed the familial type; inasmuch as the former would include merely the mild cases; the latter would include the severe as well as the mild. It would explain too why the textbooks state that in the majority of cases, if the disease starts at a young age it is far more severe than if it begins later in life. Such predilection for youth is not due to the greater susceptibility of the child to these maladies, but to the mode of inheritance, whereby most of the cases in adults have been restricted to the mild variety.

VARIATION IN AGE AT ONSET WITH TYPE OF INHERITANCE

Let us examine the data to see if any support is lent to this theoretical consideration. Consider first the disease of diabetes mellitus, which we have been using as an example through the discussion. Collecting from the literature cases in which the disease was familial, and so apparently due to recessive inheritance, or even generational, but not direct in its mode of transmission, I obtained fifty-nine cases in which the age of onset was given. The average was 11.2 years. This was well below the age of reproduction. There were sixty-three cases in which the inheritance was generational and direct, so that the disease may with a fair amount of justice be considered to have arisen through a mutation acting as a dominant character. The average age of onset in these cases was 30 years; an age well above that at which the individual could have reproduced his defect.

The data on age of death would have shown the same thing, namely that the age of death from the disease was in the first class of cases below the age at which the individual would have reproduced, while in the second case it naturally would have been beyond that. So far then the actual observations bear out the theoretical deductions, namely, that diabetes is not inherited according to some more or less complex scheme which if understood would explain all the cases which are not obviously due to infection or new growths of the pancreas, etc., but which depend upon an inherent defect of the islet tissue. The comparison of the ages of onset shows that the disease in its acute form which attacks the young is inherited as a recessive, while in its mild form it may be inherited as a dominant or recessive.

If such an explanation is true, it should apply then to other diseases which do not have a uniform mode of inheritance, and which are either fatal to the patient directly, or so transform him that he no longer is to be regarded as a potential parent, his physical attributes being such that in the selection of mates he is left aside. We should find that in the familial, indirect type of inheritance the age of onset would be younger than in the cases in which the transmission is in the direct line.

Let us consider some of the diseases of the nervous system, which are examples, *par excellence,* of variation in their mode of transmission in different families. Take the progressive muscular atrophy of the adult and of the infant, the Werdnig-Hoffmann type. What do we find here? The average age of onset of twenty-five cases of the Werdnig-Hoffmann atrophy of children was at 9 months, the average age of death at 2.4 years. Obviously the parents had to be normal in these cases, none of the affected ever surviving infancy; so that the disease was inherited as a recessive. If ever a mutation occurred in the germ plasm of a person whereby this disease could be inherited as a dominant, it would be wiped out at once with the death of the tiny sufferer. Of course one might consider that these cases, or at least some of them, were just that type, namely dominant mutations which were discarded as soon as tried.

The adult form of progressive muscular atrophy was found in twenty-six cases in which the line of descent was direct, and so probably of the dominant variety, and in which the age of onset was given. This average was 43 years. Again the dominant form is found to be the milder or at least that attacking later in life.

In hereditary spastic paraplegia the two modes of inheritance are quite common. There were found in the literature 120 cases in which the disease was familial, and in which the age of onset was given. The average was 9 years. The range was from 9 months to 71 years, thus showing that there are recessive mild cases as well as dominant mild cases; but with the average at 9 years, it also shows that the majority of cases were in very young children. There were eighty-one cases in which the line of descent was direct and in these the average age of onset was 17.8 with a range from 4 to 50 years.

The same thing is true of peroneal atrophy. The recessive or familial cases in which the age of onset was given numbered thirty, and the average age was 10.5; while in the seventy-six cases in which the disease behaved as a dominant the average was 16. In muscular dystrophy there were fifty-seven familial cases with an average of 9.3 years for the age of onset, while there were twenty-four cases which showed direct inheritance and which had an average age of onset at 15.6. In Friedreich's ataxia, the familial cases were 139 in number with an average age of onset at 10.5 years, while the ninety-six cases in which the descent was direct showed an average of 16. The two corresponding figures for the eighteen familial and twenty-eight generational cases of cerebellar ataxia were 17 and 24 years.

Going outside the realm of the nervous system and viewing the cases of angioneurotic edema, we find six cases in which the disease was familial in type, the parents being normal. All these children were in the same family and all succumbed a few weeks after birth, the average duration of life being 7½ weeks. In cases showing the dominant mode of inheritance it is stated that the onset is usually between 15 and 20 years. I found ten cases in which the actual age at which the disease began was listed. The average was 15.4 years.

Polycystic disease of the kidneys is a condition that kills early or late according to the amount of tissue involved, and in that respect is like diabetes. One might get along well for years with a partial involvement, but not be able to survive with complete destruction of the secreting portion of the kidney by cysts. There are numerous reports in which the disease affects either one child in the family only, or in which more than one are affected, and in which the lesion is so extensive that the child not only cannot live after it is born but in which it cannot even be born alive. The abdomen is so distended that normal birth is impossible and the child has to be extracted piecemeal. This type of inheritance might be regarded as recessive in character, or as a dominant mutation which appears in the germ plasm of the parents, only to be snuffed out of existence immediately. In the dominant mode of transmission, there were forty-three cases with an average age of onset at 31.7 years.

Hereditary albuminuria shows somewhat the same relations, but inasmuch as it is not so necessarily fatal, even the dominant mode of transmission shows a relatively young age of onset. There were eight cases listed in the familial class with an average age of onset at 4 years. The dominant cases were nine in number with an average of 8.1 years.

Thus a review of widely different diseases which show an apparently varying mode of transmission in different families leads to the same conclusion in all, namely that the families in which the disease appears without any previous history of it, or in which the affected persons are always in collateral branches, never in the direct line of descent, show a much earlier age of onset and a more severe type of disease than do the families in which the line of descent is direct. Thus there is a rational basis for the observation that a disease (apart from the infectious diseases of childhood) which appears in either young or old, is far more apt to be of the severe variety when found in young children. The rational basis is the mode of inheritance which the disease shows in that family, and depends upon the way in which the mutation causing the departure from normal first affected the chromosomes in the ancestors of the patients investigated.

Just what name should be given to this observed generalization is hard to determine. One hesitates to call it a law, but it appears to be a relationship that runs through quite a number of inherited diseases. Its correlary is one which I have never seen stated, perhaps because so obvious, namely that if a disease or abnormality is inherited as a dominant, it is one which is not fatal, or if fatal, it is one which commences after the period of adult life is reached. The diseases which are fatal soon after their appearance and which are apt to appear in early childhood are always recessive in character, necessarily so.

CONSTANCY OF AGE OF ONSET WITHIN A FAMILY

There is another fact which I should like to emphasize. that is connected with the one just discussed, and it is an added proof to the conclusions just drawn. One finds in discussing any family in which an hereditary disease is found, that on the whole the age of onset in the various members of that family is about constant. It may differ widely from the age of onset of the next family found with the same condition, so that the range of ages in many families may run the gamut of the entire span of life, but the children in one family will all be affected at approximately the same age. That strengthens the conception that not only is there inherited the disease process itself, but also its extent or its severity, which in turn determines how long the person affected will be able to live before showing the degenerative process.

For example, in hereditary spastic paraplegia the age of onset for the familial cases was stated to be 9 years. Consider the individual families, however; the ages of onset for two children were 6 and 6; while another family showed the ages of 27, 27 and 27. Still another family had its three children affected at 6 months, 6 months and 20 months. In another the three were affected at 2 years of age, while in another they were 9 and 12; in a third family, 3½ and 5. In one family there were eight children all affected between 12 and 18 months. In another the ages were 15 and 17 years. In the dominant mode of inheritance the same rule holds; one family of children was affected at 30, 31, 35, 40 and 52.

Consider muscular dystrophy, for example. In one family, the onset was at 4 and 6 years; in another at 1 year in the two boys; in another at 19 in the two cases; in still another at 1½, 5 and 5; in another at 9 and 9; or at 14 and 15. In a very interesting family in which all thirteen patients were derived from the same two unrelated men some eight generations previously, the onset in these persons, who occurred in ten families, was at 2, 7, 5, (these last two being sisters); 15, 15, 8, 10, 12. In only two instances did the disease commence in adult life, in one case at 40, and in a second one at 24.

In Friedreich's ataxia, the same thing is noted. In one family, the age of onset was at 5½, 6½, 6½; in a second at 12, 12 and 10; or at 18 and 19; or at 18, 20, 26, 18 and 17; a group that is much later than the first group reported; or at 18 and 21; at 2 and 3; at 17 and 19. So we might go on, indefinitely, showing that although the range is fairly wide in many of these diseases, the age range in any one family is as a rule well limited. This must mean then that there is a specific factor inherited which regulates the amount of tissue that is involved in the disease process and the length of the life of the nerve cells and nerve fibres before they begin to show their atrophic condition. This we might term a virulence factor.

Still another generalization may be drawn from these cases of inherited diseases. Not all the symptoms that are possible in any one disease necessarily appear in any one individual, and it is of interest to find that if one member, in a family affected with Friedreich's ataxia for instance, complains of pains in the legs, most of the others will have the same thing; or if scoliosis be the first defect noted in a child, the subsequent members of the family to develop the disease will frequently show scoliosis before they do any other trouble. If nystagmus is present in one child, then nystagmus is apt to be present in the rest of the family who are affected, allowing time to elapse of course for the development of the various symptoms. Thus not only is the time or virulence factor fairly constant for a family, but so is the extent factor. The groups of nerve cells, the tracts of fibres that are attacked, and the sequence in which they become involved are apparently more or less the same in the children in the same family. In one family suffering with diabetes, six persons had had a leg amputated for gangrene, yet many diabetics go through life with no sign of gangrene.

THE LAW OF ANTICIPATION

A generalization noted in the study of inherited diseases was made years ago, nor am I able to say just who was responsible for it, although I myself first found it referred to in the writings of that great student of human inheritance in eye defects, Edward Nettleship. He designated it as the law of anticipation, meaning that a disease which was inherited in the direct line of descent tended to appear at an earlier age in the children than it did in the parents. For example, in progressive muscular atrophy, the age of onset in six cases in the first generation was 51 years. For ten cases in the second generation, it was 43.4; for seven cases in the third it was 43; for two cases in the fourth it was 31.5; and for the one case in the fifth it was 13. The same decrease is noted in muscular dystrophy, the average for the first being 27, for the second 14, for the third 12. When we look at the cases of polycystic kidney, we find the same thing. There were eleven cases reported which were listed as the first generation, with an average age of onset at 47. There were twenty-one cases in the second generation with an average of 29. In the third generation the average of eight cases was 25, while in the fourth the average of two cases was 14.5.

It would be assumed then that one need not worry about these diseases, that they tended to wipe themselves out by occurring at progressively earlier ages. This unfortunately is not true. If it were, we would have dispensed with these inherited defects long ere this, but we find them constantly increasing, partly because surgical skill has tended to correct those which can be modified, and modern philanthropy protects and cares for those who cannot be rehabilitated. How then reconcile the law of anticipation and the increasing amount of inherited disease? Because, although the *average* age of onset is lower, there are some among the affected children in whom it shows at a time as late or later than it did in their parents. For example, a mother who had peroneal atrophy, developed her disease at 20. Two of her three boys who became affected showed their trouble at 12, but the third son did not come into his heritage until he was 35. Thus the average for that generation was a little below the time at which the mother was affected, but one person in it became diseased so late that he started, as it were, a new cycle, thus allowing the anticipation tendency to evince itself again in his offspring.

Davenport in referring to the tendency for Huntington's chorea to appear earlier in the offspring than it did in the parents, states that this is more apparent than real. The grandparents are a selected class, so he says, selected on the basis of the late onset, otherwise they would not have become grandparents. The ones who were affected early in that generation would never get to be grandfathers and hence the age of onset for the first generation would appear to be unduly high. Such an argument is very sound in some respects, but loses its force when we leave out of consideration in any given pedigree, the one progenitor who stands as the sole representative of the first generation and consider all his children, whether they ever become parents or not, and the ages of their onset, and all the children of the third generation, etc., with their ages of onset, and determine whether the age for *all* the third generation is less than the age for *all* the second. When we do this we then overcome Davenport's objections, for we consider not only those of the second generation who lived to marry but also those who did not; thus eliminating the statement that we deal in the first generation with only a selected class. When we make this allowance we still have the average age for all in the third generation less than the average age for all in the second; and the average age for the entire fourth generation which is affected, less than the average for all those of the third, regardless of whether they become parents or not. The law of anticipation seems to be a real fact in hereditary diseases.

SUMMARY

From a survey of published cases of inherited diseases, in which the type of inheritance varies in different families, several generalizations may be made. The first is that some diseases are inherited in different families according to different laws. The families in which occurs the dominant mode of transmission of a disease which is of such nature that it is a menace to the life of the individual are usually affected either with a milder form of the disease or are affected later in life than are those families in which the disease appears without any previous history of it. Although the recessive mode of inheritance as well as the dominant may be found in the mild form, or that which appears later in life, as a rule, it is the former only which characterizes those cases which are severe and early in their onset. It is the mode of inheritance of these diseases which establishes the relation between the severity of the disease and the youth of the patient.

The second is that there is also inherited a factor which one might term the time or virulence factor, which makes the disease appear at relatively the same age in children of the same family.

The third is that in diseases of the nervous system, the sequence in which symptoms appear, is regulated by the order of the involvement of the nerve tissue, and there is a tendency for all children of the family to show the same symptoms, and in the same order. This merely means that the disease process proceeds in the same way in the children of the same family.

The fourth is, that although the law of anticipation is exemplified in many of the diseases which are inherited in the direct line, a disease does not thereby tend to eradicate itself, there usually being several members of a generation in whom the disease starts sufficiently near the age at which the parent was affected or even beyond that, to ensure his being the starting point of a new cycle.

8

Reprinted from *Am. J. Hum. Genet.* 11:360–379 (1959)

Formal Genetics of Muscular Dystrophy

N. E. MORTON AND C. S. CHUNG

Department of Medical Genetics, University of Wisconsin

SINCE ITS ORIGIN, human genetics has been subject to several sources of error, stemming both from the nature of the material and from the methods of analysis that have been applied. Two generations of geneticists have removed some of the technical limitations, but the full potentialities of genetic analysis in man have certainly not yet been realized.

These considerations led to the development of a rigorous theory of segregation analysis in man (Morton, 1958, 1959) and to the search for a trait that would test these methods to the uttermost. We required a large body of data on each of the simple modes of inheritance, made up in part of representative conventional studies, in part of the more complete data which precise analytical methods demand.

Muscular dystrophy meets these conditions admirably. Dominant, recessive, and sex-linked types are known, and a sporadic type was revealed by our investigations. A number of excellent studies have been published which contain sufficient information for genetic analysis (Sjovall, 1936; Stephens and Tyler, 1951; Tyler and Stephens, 1950; Levison, 1951; Stevenson, 1953, 1954; Walton, 1955; de Grouchy, 1953; Lamy and de Grouchy, 1954; and Becker, 1953). A small group of sibships with complete or truncate selection (Morton, 1958, 1959) and unusually good clinical and genetic data, compiled from the literature by Chung and Morton (1959), have been included as an additional sample. Casual reports from the literature, like those compiled by Julia Bell (1943), have heuristic value but, in the absence of adequate data on the number, age, and condition of relatives, mode of ascertainment, consanguinity, and other particulars, are not suitable for this kind of genetic analysis. Through the cooperation of Dr. H. A. Peters, director of the Muscular Dystrophy Clinic of the University Hospitals, we were able to collect genetic information that had never before been reported, including data on the numbers of ascertainments of probands and chemical studies on the various types of dystrophy (Chung, Morton, and Peters, in preparation). Analysis of the genetic categories by discriminant functions indicated that they could be recognized with relatively little error (Chung and Morton, 1959).

During the course of this work there was an exciting interplay between collection of data which required more refined methods of genetic analysis than were then known, and development of techniques that required data that were not then available but are included in our Wisconsin material. The result has been a stringent test of the potentialities of modern methods of analysis in human genetics.

Genetics paper number 622. This work was supported by grants from the National Institutes of Health, Muscular Dystrophy Associations of America, and the Wisconsin Alumni Research Foundation.

Segregation analysis of class 1 (facioscapulohumeral)

In a previous paper Chung and Morton (1959) defined four genetic classes which will be used here.

Class 1 consists of cases with an affected parent, excluding four families where there is a strong presumption of recessive inheritance. Most of these cases are in large kindreds, with obviously dominant inheritance. Let there be r affected, a of them probands, in a sibship of size s. The probability of only one affected, if there is at least one affected, is

$$P(r = 1 \mid r > 0) = \frac{sp\pi[x + (1 - x)q^{s-1}]}{xsp\pi + (1 - x)[1 - (1 - p\pi)^s]} \quad \text{(problem 2)}$$

where p is the expected proportion affected, $q = 1 - p$, x is the proportion of cases that are sporadic, and π is the ascertainment probability, which is unity if ascertainment was through a parent or other relative outside the sibship (Morton 1958, 1959). To facilitate reference to this distribution, its solution by maximum likelihood scores will be called problem 2; other problems will be defined and numbered presently.

For simplicity we assume that p and π are constant among sources and that $x = 0$, sibships without an affected parent having been excluded. The value of $\pi < 1$ must be estimated from the data by methods to be considered, but the value of p is determined by genetic hypothesis. For a dominant gene without appreciable viability effects before the age of diagnosis, $p = y/2$, where y is the average penetrance in the sample, or the probability of being affected if susceptible. If all carriers who live long enough are recognizably affected, y lies between $\int f(z)G(z) \, dz$ and $\int f_1(z)G(z) \, dz$, where f(z) is the frequency of age z at death or last examination among affected individuals and their normal sibs, $f_1(z)$ is this frequency with the first proband in each family excluded, and G(z) is the cumulative frequency of onset at age z among affected cases (Morton, 1958). From the ages given in class 1 sibships, the calculated value of y is .908 for affected individuals who are not probands, .895 for all affected individuals, and .707 for their normal sibs, or approximately .804 for all sibs, giving $p = .402$.

Using the same symbols, the distribution of affected individuals in sibships with more than one affected is

$$P_{(r \mid r > 1)} = \frac{\binom{s}{r} p^r q^{s-r}[1 - (1 - \pi)^r]}{1 - (1 - p\pi)^s - \pi s p q^{s-1}} \quad \text{(problem 3)}$$

The distribution of probands among affected individuals is

$$P_{(a)} = \frac{\binom{r}{a} \pi^a (1 - \pi)^{r-a}}{1 - (1 - \pi)^r} \quad \text{(problem 9)}$$

while the distribution of $t > 0$ ascertainments among r affected is

$$P_{(t)} = \frac{[-r \ln(1 - \pi)]^t (1 - \pi)^r}{t![1 - (1 - \pi)^r]} . \quad \text{(problem 6)}$$

TABLE 1. SEGREGATION ANALYSIS OF CLASS 1

Source	Prob-lem No.	p	x	π	U_p	U_π	K_{pp}	$K_{\pi\pi}$	$K_{p\pi}$
a > 0 probands	2	.402	0	.247	4.645	−0.579	98.57	1.61	−12.58
a = 0	2	.402	0	1	3.025	—	214.30	—	—
a > 0	3	.402	—	.247	−25.355	2.996	729.06	4.90	−57.11
a = 0	3	.402	—	1	70.294	—	1860.10	—	—
a > 0	9	—	—	.247	—	−20.345	—	99.18	—
Our data, a > 0	6	—	—	.247	—	−1.391	—	6.80	—
Walton, a > 0	6	—	—	.247	—	16.152	—	64.54	—
Total					52.609	−3.173	2902.03	177.03	−69.69
Sex of affected	10	$\frac{1}{2}$	—	—	48.000	—	1504.00	—	—
Sex of normal	10	$\frac{1}{2}$	—	—	−16.000	—	1896.00	—	—

	d.f.	χ^2	P
Deviation from p = .402	1	0.92	.5–.3
p among sources	3	2.85	.5–.3
p among samples	6	9.58	.2–.1
Sex of affected, p among samples	7	11.04	.2–.1

Becker (1953) distinguishes sibships with probands but not the probands themselves, hence neither problem 6 nor 9 is applicable. Walton (1955) reports ascertainments but not probands, while the remaining studies, with the exception of our Wisconsin material, indicate probands but not ascertainments. For our Wisconsin data problem 6 simplifies to the case r = 1, since we have recorded the number of ascertainments per proband. A preliminary estimate of π, made before the Wisconsin data were tabulated, was .247. The analysis by maximum likelihood scores, using p = .402, π = .247, is set out in table 1.

Assuming p = .402, an improved estimate of π is .247 + $U_\pi/K_{\pi\pi}$ = .229, where U_π is the total score for π and $K_{\pi\pi}$ is its variance. At this value the score for p by Taylor's theorem is $U_{p|\pi}$ = .247 + (.247 − .229)$K_{p\pi}$ = 52.609 + (.018)(−69.69) = 51.355, the correction to U_p being negligible. The information matrix for p and π is

$$\begin{bmatrix} K_{pp} & K_{p\pi} \\ K_{p\pi} & K_{\pi\pi} \end{bmatrix},$$

which inverts to the covariance matrix

$$\begin{bmatrix} 3.4788 & 1.3695 \\ 1.3695 & 57.0267 \end{bmatrix} 10^{-4}.$$

To test the hypothesis that p = .402 when π is estimated from the sample, we use χ_1^2 = $(51.355)^2(3.4788)10^{-4}$ = 0.92. Clearly the data satisfy this hypothesis.

The largest estimate of p is given by the pedigree of Tyler and Stephens (1950), with U_p = 84.253, K_{pp} = 1135.63. These authors eliminated "families with many children under twelve," giving $p = \frac{1}{2}$ as the expected and observed frequency of affected. There is no significant heterogeneity among samples, nor do the other samples differ significantly from $p = .402$.

The distribution of r females among s affected individuals is $\binom{s}{r}p^r q^{s-r}$, where $p = \frac{1}{2}$ on the null hypothesis of an equal sex ratio (problem 10). The data are clearly consistent with this, and the sex ratio is homogeneous among samples.

On the above evidence we conclude that almost all cases of muscular dystrophy from an affected parent (class 1) are due to one or more "dominant" genes with recognizable effects in heterozygotes of sufficient age and of either sex.

Segregation analysis of class 2 and 4(2) (limb-girdle)

Class 2 consists of cases from normal parents with at least one affected girl in the family, and class 4(2) consists of cases from normal parents with no affected girl in the family, but of the same phenotype as class 2. The probability of only one affected in a sibship of class 2, if there is at least one affected, is

$$P_{(r=1|r>0)} = \frac{sp\pi[x + (1 - x)q^{s-1}]}{xsp\pi + 2(1 - x)[1 - (1 - p\pi)^s - (\frac{1}{2}p + q)^s + (\frac{1}{2}p + q - \frac{1}{2}p\pi)^s]}$$

(problem 4)

and the distribution of affected individuals in sibships with more than one affected is

$$P_{(r|r>1)} = \frac{\binom{s}{r}p^r q^{s-r}[1 - (1 - \pi)^r][1 - (\frac{1}{2})^r]}{1 - (1 - p\pi)^s - (\frac{1}{2}p + q)^s + (\frac{1}{2}p + q - \frac{1}{2}p\pi)^s - \frac{1}{2}\pi spq^{s-1}}$$

(problem 5)

where r is the number of affected sibs including at least one affected girl and s is the sibship size (Morton, 1958, 1959). In class 4(2) with females excluded, the corresponding probabilities are the same as for problems 2 and 3.

The proportion x of sporadic cases may be estimated from the inbreeding coefficients of n isolated probands (F_I) and N familial probands (F_F) as

$$x^* = cy = c(F_F - F_I)/F_F - \alpha)$$

where α is the mean inbreeding coefficient in the general population and $c = n/(n + N)$, and y is the proportion of sporadic cases among isolated cases, estimated from the relation $F_I = y\alpha + (1 - y)F_F$. The large-sample estimate of the variance of x^* is $\sigma_{x^*}^2 = c^2\sigma_y^2 + y^2\sigma_c^2$, where $\sigma_c^2 = nN/(n + N)^3$ and

$$\sigma_y^2 = \frac{\sigma_{F_F}^2(F_I - \alpha)^2}{(F_F - \alpha)^4} + \frac{\sigma_{F_I}^2}{(F_F - \alpha)^2}$$

(Chung, Robison, and Morton, 1959). These estimates may be converted to scores for x by the transform $K_{xx} = 1/\sigma_x^2$, $U_x = (x^* - x)K_{xx}$. Weighting each sample by the number of probands, we find $\alpha = 62 \times 10^{-5}$, $F_I = 119 \times 10^{-5}$, $F_F = 804 \times 10^{-5}$, and $x^* = .500 \pm .076$.

TABLE 2. SEGREGATION ANALYSIS OF CLASS 2 AND 4(2)

Class 2

Source	Problem No.	p	x	π	U_p	U_x	U_π	K_{pp}	K_{xx}	$K_{\pi\pi}$	K_{px}	$K_{p\pi}$	$K_{x\pi}$
a > 0	4	.206	.400	.528	21.404	−18.915	−5.814	544.47	233.07	22.85	−315.00	−100.00	72.93
a > 0	5	.206	—	.528	38.593	—	−1.237	1161.40	—	4.16	—	−67.42	—
a > 0	9	—	—	.528	—	—	−58.722	—	—	260.17	—	—	—
a > 0	6	—	—	.528	—	17.375	9.051	—	173.75	86.96	—	—	—
Inbreeding	—	—	—	—	—	—	—	—	—	—	—	—	—
Total	—	—	—	—	59.997	−1.540	−56.722	1705.87	406.82	374.14	−315.00	−167.42	72.93
Sex of affected	2	½	0	1	−29.240	—	—	260.37	—	—	—	—	—
Sex of affected	3	½	—	1	−13.910	—	—	95.62	—	—	—	—	—
Sex of normal	10	½	—	—	−4.000	—	—	2184.00	—	—	—	—	—

Class 4(2) Males Only

Source	Problem No.	p	x	π	U_p	U_x	U_π	K_{pp}	K_{xx}	$K_{\pi\pi}$	K_{px}	$K_{p\pi}$	$K_{x\pi}$
♂, a > 0	2	.206	.400	.528	−3.892	3.084	0.951	378.26	77.11	8.17	−165.97	−54.68	25.05
♂, a > 0	3	.206	—	.528	2.211	—	−0.040	92.80	—	0.49	—	−6.71	—
♂, a > 0	9	—	—	.528	—	—	−10.655	—	—	61.70	—	—	—
♂, a > 0	6	—	—	.528	—	—	35.213	—	—	129.24	—	—	—
Inbreeding	—	—	—	—	—	−0.633	—	—	2.04	—	—	—	—
Total	—	—	—	—	−1.681	2.451	25.469	471.06	79.15	199.60	−165.97	−61.39	25.05

	χ^2	d.f.	P
p between 2 and 4(2)	0.55	1	.5–.3
x between 2 and 4(2)	0.08	1	.8–.7
Sex of affected, p = ½	5.23	1	.05–.02

2 + 4(2) Pooled

	χ^2	d.f.	P
Deviation from p = .206	1.59	1	.3–.2
p among sources	0.65	3	.9–.8
x between inbreeding and segregation data	2.40	1	.2–.1
p among samples	9.25	6	.2–.1
x among samples	11.14	6	.10–.05

From the ages in class 2 sibships, the average penetrance is calculated to be .794 for affected individuals who are not probands, .810 for all affected individuals, and .836 for their normal sibs, or approximately .825 for all sibs, giving p = .206 on the hypothesis of recessive inheritance for familial and chance isolated cases if there is complete penetrance in individuals of sufficient age.

The segregation analysis based on preliminary estimates of p = .206, x = .400, π = .528 is shown in table 2. The information matrix for x and π inverts to the covariance matrix

$$\begin{bmatrix} 21.3112 & -3.6394 \\ -3.6394 & 18.0510 \end{bmatrix} 10^{-4},$$

yielding

$$x = .400 + (0.911)(21.3112)10^{-4} + (-31.253)(-3.6394)10^{-4} = .413$$

$$\pi = .528 + (0.911)(-3.6394)10^{-4} + (-31.253)(18.0510)10^{-4} = .471$$

At these values the estimate of U_p by Taylor's theorem is

$$U_p = 58.316 + (.400 - .413)K_{px} + (.528 - .471)K_{p\pi} = 51.527.$$

The information matrix for p, x, and π inverts to the covariance matrix, the first element of which is $\sigma_p^2 = 5.9897 \times 10^{-4}$, giving $\chi_1^2 = (51.527)^2(5.9897)10^{-4} = 1.59$ as the test of the hypothesis that p = .206.

The data indicate that about 41 per cent of group 2 and 4(2) cases are sporadic, this fraction being homogeneous between inbreeding and segregation estimates and between classes. Estimates from different samples are consistent, indicating that all investigators have included similar proportions of sporadic cases. In segregating families, which give rise to the familial and chance isolated cases, the inheritance is monomeric recessive with complete penetrance in individuals of sufficient age. The segregation analysis gives no evidence on whether these genes in different families are allelic or at different loci. Failure of previous investigators to demonstrate the sporadic nature of many of these cases is due to the crude methods of analysis available to human geneticists in the past.

The distribution of affected girls among familial cases is a truncated binomial, which may be analysed by problems 2 and 3 with x = 0, π = 1. The scores for x are U_x = 8.667, K_{xx} = 401.86, indicating that sporadic expression of sex-linked dystrophy in females or pseudo-females is not an appreciable source of bias. Nevertheless, there is a deficiency of affected girls (χ_1^2 = 5.23), the estimated proportion of .38 being surprisingly homogeneous among samples (χ_6^2 = 0.78, P > .99). Since there is no evidence of a corresponding excess of normal girls or of sex-biased manifestation (Chung and Morton, 1959), the deficiency of affected girls may well be due to chance.

Segregation analysis of class 3 and 4(3) (Duchenne)

Class 3 consists of affected boys from normal parents with a maternal history of affected males, consistent with sex linkage, and class 4(3) consists of affected boys from normal parents with a negative maternal history, but of the same phenotype as class 3. The cases of Stephens and Tyler (1951), although not reported in detail, have all been classified as 3 or 4(3), since they are Duchenne in type and include only affected boys. Because of the transmission through carrier females to their sons, this genetic type has been considered to be due either to a recessive sex-linked gene, or to an autosomal gene that is dominant in males but recessive or not expressed in females. We shall consider the analysis under sex-linkage first, and then demonstrate that autosomal inheritance can be excluded.

Deleterious sex-linked genes are subject to stringent selection in males, one-half of the genes transmitted from carrier mothers being exposed to elimination in this way each generation. Therefore an appreciable fraction of affected cases should result from mutation in this or the preceding generation, and appear in class 4(3) as isolated or familial cases without affected relatives. If u and v are the mutation rates and P and Q the gene frequencies in eggs and X-bearing sperm, respectively, and m the coefficient of selection against the trait, the loss of genes by selection will be $m(Q + PQ)$ and the gain by mutation $2u(1 - P) + v(1 - Q)$. At equilibrium the gain and loss are equal, and if $m \gg 0$, we may neglect the product terms PQ, uP, and vQ in comparison with P, Q, u, and v, and write $Q = (2u + v)/m$. The probability that an affected male in the general population be sporadic is $x = u/Q = mu/(2u + v)$, assuming that all cases are sex-linked. Similarly, $Q = P + u$, or $P = (v + 2u - mu)/m$. The probability that a carrier female acquire the gene through an affected father is $Q(1 - m)$, and by mutation is $u + v$, and in both these cases she will have no affected brothers. But $Q(1 - m) + u + v = P$, and since the frequency of carrier females is 2P, the probability that the brother of a carrier female not be at risk is $x' = P/2P = 1/2$.

If we consider that the mother of two or more affected boys is a carrier, the probability that the grandmother was a carrier is $1 - x'$. Assuming ascertainment through nephews, the probability that none of s maternal uncles be affected is $x' + (1 + x')q^s$, counting a set of maternal uncles once for each index sibship with two or more boys affected (problem 7).

If the index case has no affected siblings, and the number of boys in the sibship is n, the probability that he is sporadic is $x/\{x + (1 - x)q^{n-1}\}$, and the probability that none of s maternal uncles be affected is

$$P(r = 0) = \{x + (1 - x)q^{n-1}[x' + (1 - x')q^s]\}/\{x + (1 - x)q^{n-1}\},$$

the complementary probability of at least one affected uncle being

$$P_{(r>0)} = \frac{(1 - x)(1 - x')q^{n-1}(1 - q^s)}{x + (1 - x)q^{n-1}} \qquad \text{(problem 8)}$$

where r is the number of affected maternal uncles. If no maternal uncle is affected, the maximum likelihood scores for x and x' are

$$u_x = \{B - C + q^{n-1}(C - AB)\}/BC \qquad (x, x' > 0)$$

$$u_{x'} = (1 - x)q^{n-1}(1 - q^s)/C$$

where

$$A = x' + (1 - x')q^s$$

$$B = x + (1 - x)q^{n-1}$$

$$C = x + (1 - x)q^{n-1}A.$$

If at least one uncle is affected,

$$u_x = -1/B(1 - x) \qquad\qquad (x, x' > 0)$$

$$u_{x'} = -1/(1 - x').$$

The variances and covariance are $k_{xx} = E(u_x^2)$, $k_{x'x'} = E(u_{x'}^2)$ and $k_{xx'} = E(u_x u_{x'})$.

Isolated and familial cases without regard to maternal uncles, maternal uncles of familial cases, and maternal uncles of isolated cases constitute three independent sources of information about x and x'. Agreement among them, and of x' with 1/2, lends support to any conclusion about the deviation of x from m/3, and hence the equality of the mutation rates in the two sexes.

Since individuals with this type of muscular dystrophy rarely reproduce, m is nearly 1 and x approaches $\frac{1}{3}$ on the hypothesis of equal mutation rates in the two sexes. From the ages of males in class 3 sibships, the average penetrance is .869 for affected individuals who are not probands, .866 for all affected individuals, and .839 for all sibs, giving p = .427 on the hypothesis of sex-linkage.

The segregation analysis, based on these values and a preliminary estimate of $\pi = .519$, is shown in table 3. An improved estimate of π is $.519 + U_\pi/K_{\pi\pi} = .492$. The information matrix for p and π inverts to the covariance matrix, the first element of which is $\sigma_p^2 = 8.8640 \times 10^{-4}$. By Taylor's theorem the score for p at $\pi = .492$ is $U_p = -3.488 + (.519 - .492)(106.60) = -0.610$, giving

$$\chi_1^2 = (-.610)^2(8.8640)10^{-4} = .0003.$$

Similarly, from the covariance matrix for x and π, the test for the hypothesis that $x = \frac{1}{3}$ is $\chi_1^2 = .216$. Since there is no covariance between x' and π, the corresponding test for the hypothesis that $x' = \frac{1}{2}$ is simply $\chi_1^2 = (U_{x'})^2/K_{x'x'} = 1.72$. The estimates of p, x, and x' are homogeneous among sources and among samples with the exception of x among samples. Walton's sample (1955) includes an unusually small frequency of sporadic cases, and the heterogeneity generated by this discrepancy is barely significant ($\chi_7^2 = 14.08$, P = .048). Considering the small amount of information in the individual sources and the approximate nature of the χ^2 test in small samples, the biological significance of this heterogeneity is doubtful. We conclude that the data are in agreement with the hypothesis that dystrophy in class 3 and 4(3) is due to a sex-linked recessive gene or genes with complete penetrance in males of sufficient age and with equal mutation rates in eggs and sperm. It is impossible to prove the assumption that all isolated cases are sex-linked since affected males rarely reproduce, but the agreement with sex-linkage is remarkable and we have shown that

TABLE 3. SEGREGATION ANALYSIS OF CLASS 3 AND 4(3)—MALES ONLY

Source	Problem No.	p	x	x'	π	U_p	U_x	$U_{x'}$	U_π	K_{pp}	K_{xx}	$K_{x'x'}$	$K_{\pi\pi}$	K_{px}	$K_{p\pi}$	$K_{x\pi}$	$K_{xx'}$
a > 0	2	.427	⅓	—	.519	5.590	−6.670	—	−1.826	311.78	306.27	—	25.66	−268.98	−80.10	88.50	—
a = 0	2	.427	0	—	1	−28.110	—	—	—	413.55	—	—	—	—	—	—	—
a > 0	3	.427	—	—	.519	.440	—	—	−.178	271.87	—	—	2.69	—	−26.50	—	—
a = 0	3	.427	—	—	1	18.592	—	—	—	150.96	—	—	—	—	—	—	—
a > 0	9	—	—	—	.519	—	—	—	−21.870	—	—	—	201.38	—	—	—	—
a > 0	6	.427	—	—	.519	—	—	—	8.377	—	—	—	338.32	—	—	—	—
Uncles	7*	.427	—	½	—	—	—	4.249	—	—	—	92.72	—	—	—	—	—
Uncles	8*	.427	⅓	½	—	—	13.646	14.550	—	—	113.42	112.97	—	—	—	—	109.80
Total						−3.488	6.976	18.799	−15.497	1148.16	419.69	205.69	568.05	−268.98	−106.60	88.50	109.80

	d.f.	χ^2	P
Deviation from p = .427	1	.0003	.99–.98
Deviation from x = ⅓	1	0.22	.7–.5
Deviation from x' = ½	1	1.72	.2–.1
p among sources	3	4.29	.3–.2
p among samples	8	11.42	.2–.1
x between sources	1	1.67	.2–.1
x among samples	7	14.08	.05
x' between sources	1	0.35	.7–.5
x' among samples	6	9.08	.2–.1

* Excluding Levison's sample, which does not report maternal uncles of class 4(3).

the frequency of misdiagnosis is small and tends to be compensated by misclassification of sex-linked cases as limb-girdle dystrophy (Chung and Morton, 1959).

Haldane (1956) first examined the question of equality of mutation rates in the two sexes. He made two assumptions that tended to overestimate x, namely that $p = \frac{1}{2}$ after an attempted age correction and that x is the same as the proportion w of families with sporadic cases among families with affected members, whereas in fact $w = xsp/\{1 - q^s - x(1 - sp - q^s)\}$ (Morton, 1959). He also made two assumptions which tended to underestimate x, namely that $\pi = 1$ and that sibships without at least one proband could be treated in the same way as index sibships. Cheeseman et al. (1958) and Smith and Kilpatrick (1958) corrected two of these sources of bias, reversing Haldane's conclusion that there was a significant deficiency of sporadic cases. They retained his assumption that $p = \frac{1}{2}$, $\pi = 1$, however. Our analysis removes these restrictions and incorporates the evidence from several other studies and from uncles of isolated and familial cases, applying the methodology of Morton (1958). Our estimate of $x = .355 \pm .050$, although with a smaller standard error, is very similar to the estimate of Smith and Kilpatrick of $.351 \pm .111$.

Several lines of evidence argue against the theory of an autosomal gene that is dominant in males and recessive or not expressed in females. On that hypothesis, half of the sons of affected males should also be affected, while none should be affected under sex-linkage, barring the highly unlikely events of mutation or mating to a carrier. In the pooled material there were 10 sons of class 3 cases and 8 sons of class 4(3) cases, all normal. (The 9 daughters of affected individuals were also normal, although several transmitted the gene to sons or grandsons, as would be expected under either hypothesis.) Walton (1957) reported a case of Turner's syndrome (ovarian agenesis) with Duchenne muscular dystrophy in a sex-linked pedigree. If the gene were autosomal, this would require a coincidence of two rare events (sexual maldevelopment and heterozygote expression or mutation). However, if the gene were sex-linked, only one unusual event is required, since many of the cases with Turner's syndrome born to a carrier mother are expected to have muscular dystrophy, being XO in chromosome type (Ford et al., 1959). Philip and Walton (1956) presented evidence consistent with linkage between this type of dystrophy and color-blindness, although far from proving such linkage. An attempt to find evidence for autosomal linkage was unsuccessful (Walton, 1955).

Linkage is a less powerful test of the hypothesis of sex-influenced dominance than the following method, which is applicable even when affected individuals fail entirely to reproduce. For such a gene, being autosomal, the gene frequencies in males and females must be the same, say P. Then the gene frequency in sperm is $P(1 - m) + u(1 - P)$, which approaches $u + P(1 - m)$, and the gene frequency in eggs is $P + v(1 - P)$, which approaches $v + P$. The genotype frequency in both sexes is $2P = u + P(1 - m) + v + P$, or $P = (u + v)/m$. The probability that the grandfather was affected and transmitted the gene to the mother is $P(1 - m)$. The probability that the grandmother was heterozygous and transmitted the gene to the mother is P. The probability that the mother is a carrier by mutation is $u + v$. Therefore the probability that the brother of a carrier female not be at risk is $x' = (u + v)/(2P - Pm + u + v) = m/2$, which is not in practice distinguishable from the relation $x' = \frac{1}{2}$ for sex-linked genes. The probability of an affected male by mutation is $u + v$, and by inheritance from the mother is P. If the father is normal, the frequency of mutations among affected cases is

$$x = (u + v)/(P + u + v) = m/(m + 1),$$

which approaches $\frac{1}{2}$ for a highly deleterious gene, whereas x approaches $\frac{1}{3}$ for a highly deleterious sex-linked gene if the mutation rates in eggs and sperm are equal. By Taylor's theorem, the score for x at $\pi = .492$ is 9.366. From the covariance matrix for x and π, $\sigma_x^2 = 24.6365 \times 10^{-4}$. The estimate of x from the data is $.333 + U_x/K_{xx} = .355$, giving $\chi_1^2 = (.500 - .355)^2/\sigma_x^2 = 8.53$, which is a highly significant deviation from the expectation under sex-influenced dominance in the direction expected under sex-linkage. We conclude that the locus for this type of dystrophy is on the X-chromosome.

Ascertainment and incidence

Although the estimates of p, x and x' are consistent among sources and problems' the estimates of π are highly heterogeneous. None of the previous studies defined the terms *propositus*, *proband*, or *ascertainment*, and it is apparent that their usages were not always consistent or correct. Stephens and Tyler (1951) reported an incidence in Utah that has never been exceeded in any other sample, indicating relatively complete ascertainment, yet except for a pair of affected twins they distinguished only one proband in each family. Walton (1955), reporting a smaller incidence of dystrophy, enumerates such a large number of ascertainments that, if they were independent, selection of cases must have been virtually exhaustive. Stevenson, after a careful study of the cases in Northern Ireland (1953), has continued to ascertain additional cases in the same population (1955, 1958). From our own experience in Wisconsin, we doubt that exhaustive ascertainment is possible for a condition not requiring institutionalization and not always correctly diagnosed.

In class 1, the estimate of π from ascertainments in Walton's data is .497, while in all other samples it is only .075. The Wisconsin material gives $\pi = .235$, which is very close to the mean for all samples. It seems likely that over-reporting of ascertainments in Walton's study is compensated by under-reporting of probands in other instances. Since the covariance between p and π is almost negligible in large pedigrees, the segregation analysis of class 1 is not greatly affected by inaccuracy in the pooled estimate of π.

The remaining classes give homogeneous estimates of π when compared within samples. However, the samples are significantly heterogeneous, even when Walton's study is excluded. For his data $\pi = .90$, for the Wisconsin material $\pi = .39$, and for the other samples $\pi = .31$. It seems likely that the mean value of π is greater than for the Wisconsin sample, since we have included only sibships from which at least one affected member underwent examination at the Muscular Dystrophy Clinic, thus excluding sibships that were represented only by patients who were unwilling or unable to make this trip. The pooled estimates of $\pi = .471$ for class 2 and 4(2) and $\pi = .492$ for class 3 and 4(3) therefore seem reasonable.

TABLE 4. SUMMARY OF THE SEGREGATION ANALYSIS

Class	Parameter	Expected	Observed
1	p	.402	.420 ± .019
2 + 4(2)	p	.206	.230 ± .024
	x	—	.413 ± .046
3 + 4(3)	p	.427	.427 ± .030
	x	.333	.355 ± .050
	x'	.500	.591 ± .070

The Wisconsin data are the most critical for estimation of π. A *proband* was defined as an affected person referred by at least one physician to the Muscular Dystrophy Clinic and examined there, independently of other cases. Many of these patients were referred without previous inquiry by the clinic, others who had been previous inpatients in the University Hospitals were re-contacted through their referring physician, and still others were ascertained through the cooperation of the State Bureau for Handicapped Children in contacting the local physician. Finally, a descriptive letter and questionnaire was sent to all members of the State Medical Society, advising them of the functions of the clinic and indicating interest in examining cases of dystrophy. No attempt was made to interview patients who were unwilling to be examined. Nearly all patients referred in these ways were extremely cooperative and have been seen several times in the clinic. Presumably the less interested patients did not consent to referral. All affected family members mentioned in a referral who subsequently came to the clinic were considered probands, whether or not they accompanied the index case on his first appointment. Siblings who became affected after examination of the index case were not considered probands.

Each referral by a different physician constituted an ascertainment. Thus if an individual was referred by Drs. X, Y, and Z, this was treated as three ascertainments regardless of the fact that Dr. X may have notified the clinic of this patient several times, through the University Hospitals, the Bureau for Handicapped Children, and in response to the letter sent to members of the State Medical Society. The high frequency of ascertainments per proband reported by Walton (1955) is probably due multiple counts of such dependent notifications, which do not follow the truncated Poisson distribution of problem 6.

Using these definitions, the scores for classes 2, 3, 4(2), and 4(3) are $U_\pi = -56.370$, $K_{\pi\pi} = 283.50$ for problem 6, and $U_\pi = 10.140$, $K_{\pi\pi} = 66.44$ for problem 9. Much the greater amount of information is supplied by the number of ascertainments (problem 6), but the two estimates are heterogeneous ($\chi_1^2 = 6.65$). We suspect that problem 6 may give too small an estimate ($\pi = .32$) because referral by one physician tends to make referral by another less likely, while problem 9 may give too large an estimate ($\pi = .67$) because two or more affected relatives may be examined and counted as probands, even when only one of them would have submitted to examination independently. For example, early cases are more likely to be brought to the clinic than terminal ones, but a terminal case may accompany an affected younger sibling to the clinic and be counted as a proband. Considering all the possible biases to which any estimate of ascertainment is exposed, the agreement between the two methods is reasonably good. We have more confidence in the pooled estimate of $\pi = .39$ than in either estimate separately.

Although π can hardly have been the same for all these studies, discrepancies among them and errors in determining the pooled value do not seem to have affected the segregation analysis, which gives internally consistent results remarkably close to expectation. In part this must be because of compensation of errors in estimating π, but also the covariances between π and the other parameters are small enough so that the analysis is rather robust, not appreciably disturbed by small departures from the ascertainment model. Since even the incomplete and crude data on ascertainment provided by these muscular dystrophy studies can apparently be tolerated, an enlightened effort to define probands and ascertainments appropriately seems all that is required to assure a valid segregation analysis by these methods.

The most serious effect of errors in estimating π is on the calculation of incidence. This has not been appreciated in human genetics because the ascertainment probability is almost never taken into account in estimating incidence, hence the recommendation by some authors of single selection, which gives no estimte of π, because it simplifies the arithmetic of segregation analysis.

Let $P = n/\pi N$ be the prevalence of a trait in the general population, where n is the number of living probands and N is the population size. To obtain I, the incidence at birth (and presumably at conception) of individuals who will develop the trait, N should be replaced by $N' = \int_z n(z)P(z)\, dz$, where n(z) is the number in the general population at age z and P(z) is the probability that a person capable of developing the trait do so by age z and not die of it before then. For practical computation P(z) may be replaced approximately by $G(z)[1 - D(z)]$, where G(z) is the cumulative frequency of onset at age z and D(z) is the cumulative frequency of death at age z among affected individuals. These distributions are given by Chung and Morton (1959); n(z) has been taken from the Demographic Yearbook for 1955.

In class 1 there are 30 living probands in the pooled sources, drawn from a total population of about 23,488,000. The pedigree of Tyler and Stephens (1950) has been omitted because the number of probands and other data were not reported. The pooled estimate of π is .229, from which the prevalence is calculated as $P = 5.6 \times 10^{-6}$. We find $N' = .606N$, from which the incidence at birth of persons who will develop the disease is $I = 9.2 \times 10^{-6}$. In the Wisconsin sample, n = 2, $\pi = .235$, N = 3,700,000, and the corresponding estimates are $P = 2.3 \times 10^{-6}$, $I = 3.8 \times 10^{-6}$.

These calculations are subject to two principal errors. Biases in n and π tend to be compensatory, but the frequency of a dominant trait that is far from lethal may be expected to vary considerably among populations, sometimes producing enormous pedigrees like the one reported by Tyler and Stephens.

In class 2 and 4(2) there are 267 living probands, $\pi = .471$, and $N' = .304N$. We find $P = 24.1 \times 10^{-6}$, $I = 79.4 \times 10^{-6}$. For Wisconsin, n = 26, $\pi = .354$, $P = 19.9 \times 10^{-6}$, $I = 65.3 \times 10^{-6}$.

In class 3 and 4(3), n = 247, $\pi = .492$, $N' = .236N$, and the number of males in the population is 11,998,000, including the sample of Stephens and Tyler (1951).

The estimate of prevalence is $P = 41.8 \times 10^{-6}$, and $I = 177.3 \times 10^{-6}$ among males. For Wisconsin males, $n = 51$, $\pi = .419$, $P = 65.8 \times 10^{-6}$, $I = 278.8 \times 10^{-6}$.

Pooling classes 1, 2, 3, and 4, the total frequency of dystrophy per million persons is $P = 50.6$, $I = 177.2$ for the pooled sample, and $P = 55.1$, $I = 208.5$ for Wisconsin.

In class 1 the gene frequency at birth is $q = I/2$, or 2×10^{-6} based on the Wisconsin sample. In class 3 and 4(3), the gene frequency in males is $q = I$, or 279×10^{-6}. By enumerating cases born during an extended period, Walton (1955), Stevenson (1958) and Stephens and Tyler (1951) obtained minimal estimates of 130, 182, and 286×10^{-6}, respectively.

If more than one locus contributes to class 1 or class 3 and 4(3), q is the sum of the gene frequencies at these loci, since I is linear in q. The problem is more delicate for class 2 and 4(2), since with recessive genes I is quadratic in q. By the theory of detrimental equivalents (Chung, Robison, and Morton, 1959), the incidence is $I = A + B\alpha$ and the mean inbreeding coefficient of probands is $\bar{F} = [A\alpha + (\sigma^2 + \alpha^2)B]/I$, where A is the incidence in a random mating population, α, σ^2 are the mean and variance of the inbreeding coefficient in the general population, and B is the mean number of detrimental equivalents per gamete. For genes with complete penetrance in homozygotes of sufficient age, the total gene frequency per gamete lies between B and $A + B$. These quantities are estimated as

$$A = I - B\alpha$$

$$B = I(\bar{F} - \alpha)/\sigma^2$$

For recessive cases we find $\bar{F} = 804 \times 10^{-5}$, $I = (.587)(65.3 \times 10^{-6}) = 38 \times 10^{-6}$, $\alpha = 62 \times 10^{-5}$, and $\sigma^2 = 35 \times 10^{-6}$, weighting each sample by the number of probands.

$$A = (33 \pm 2) \times 10^{-6}$$

$$B = .0081 \pm .0027$$

A minimum estimate of the number of loci contributing to recessive cases of dystrophy, obtained on the assumption of equal gene frequencies, is $n = B^2/A = 2.0 \pm 1.3$. Most familial and chance isolated cases of class 2 and 4(2) are evidently due to genes at a few loci, perhaps at a single locus. The average gene frequency per locus is $q = A/B = .0041 \pm .0014$. If all sporadic cases were heterozygotes for one of these genes, the penetrance in heterozygotes would be $h = (.413)(65.3 \times 10^{-6})/2B = .0017$. That is, if 17 out of 10,000 carriers developed dystrophy because of exogenous agents or genetic factors at other loci, this would account for the observed frequency of sporadic cases.

If the load represented by B were segregational (Crow, 1958), and the gene frequencies maintained by heterozygote advantage over dystrophic homozygotes, there would have to be more than $B/A = 244$ alleles at each locus capable of producing dystrophy, each allele giving a superior heterozygote with all the others! This is more than 100 times the number of genes indicated by the data. The frequency of each allele and the selection coefficient against the normal homozygote must be less than $B/244 = 3.3 \times 10^{-5}$, which is so small that virtually all homozygotes would come from consanguineous marriages. Since the postulated genes are individually very rare, the advantage of heterozygotes inter se would have to be several orders of magnitude less than this, and therefore negligible by comparison with the mutation pressure. It is incredible that so many rare alleles could be maintained by selection in a finite population. We conclude that the load represented by B is mutational, as would be expected for rare and presumably hypomorphic genes.

The argument in the above paragraph depends on the assumption that homozygotes are eliminated as dystrophic cases, and does not apply if selection is partly against some other homozygous effect. However, on the hypothesis of heterozygote advantage, the selection exerted at each locus capable of producing dystrophy is at least qm, where m is the selection coefficient against dystrophic homozygotes. Since $q = .0041$ and $m = .746$ (see below), each dystrophy locus is responsible for .003 eliminations per zygote. If we make the liberal supposition that the total number of expressed lethal equivalents is .60, so that each generation is produced by $e^{-0.6} = 55$ per cent of the previous generation, then 200 loci with mutants as detrimental as the ones that produce muscular dystrophy are sufficient to account for all selection, both genetic and phenotypic. Since the number of loci with major effects is of the order of 10,000 per gamete in Drosophila, and presumably at least as many in man, it must be true that the great majority of detrimental alleles are maintained by some other mechanism than heterozygote advantage, and the load represented by detrimental equivalents is largely mutational.

Fitness and mutation

At least three approaches have been made to the estimation of fitness for specific traits in man. The simplest is to take the ratio between the mean number of children born to affected individuals and to a suitable control sample, both of which have terminated their reproduction. Krooth (1955) has proposed a modification of this based on sib controls. Unfortunately, sib controls are neither feasible nor desirable: not feasible, because many sibs will not have completed their reproduction and their total progeny will not be known; not desirable, because information is lost about possible differences between the reproductive performance of sibs and the general population. Finally, Reed and Neel (1955) have proposed a measure of relative reproductive span, based on the assumption that fecundity up to the time of death is the same as for the general population. This method is clearly not applicable to diseases like muscular dystrophy.

To adjust for possible ascertainment bias due to association between number of sibs and number of children (Krooth's problem), let $s - 1$ be the number of sibs of an affected person and n be the number of his children. The regression of n on $s - 1$ is $n = a + b(s - 1)$. In the general population the mean value of $s - 1$ is $\mu = \Sigma s(s - 1)/\Sigma s = \bar{s} - 1 + \sigma^2/\bar{s}$, where $\bar{s}$ and σ^2 are the mean and variance of s among families with $s > 0$. Typical values of μ are 3.17 for a rural Italian population (Cavalli, unpublished) and 3.64 for an American population a generation ago (Baber and Ross, 1924). If b is not significantly different from zero, the selection coefficient against affected persons may be estimated as $m = 1 - \bar{n}\,\bar{N}$, where $\bar{n}$, $\bar{N}$ are the mean number of children from affected persons and a suitable control. If b is sig-

TABLE 5. FERTILITY OF DEAD CASES

Class	Number of Cases	Number of Children n	Adjusted Number of Children a + bμ	Selection Coefficient m
1	40	3.975	3.142	?
2 + 4(2)	62	0.694	0.642	.746
3 + 4(3)	112	0.116	.095	.962

nificantly different from zero, the selection coefficient is $m = 1 - (a + b\mu)/\bar{N}$, adjusted for possible ascertainment bias due to familial association in fertility. In the dystrophy material the deviation of b from zero is significant in class 1, and we have therefore used the adjusted estimate.

The fecundity of the general population represented by these samples is approximately twice the ratio of the population in 1950 to the population in 1920, or $\bar{N} = 2.524$. The values of ñ and m are shown in table 5, using dead cases as the closest practical approach to a representative sample of reproductive performances and taking $\mu = 3.5$. The selection coefficient is calculated to be about .746 for class 2 and 4(2) and at least .962 for class 3 and 4(3). Since several of the fertile Duchenne cases were ascertained through affected grandchildren, the selection coefficient is probably greater than this. The fertility of class 1 is actually greater than the estimated population mean, probably because many cases are drawn from previous generations. Since the mean number of sibs is 5.400, the selection coefficient against these cases is evidently less than $1-3.975/5.4 = .264$.

Assuming equilibrium between selection and mutation, the mutation rate for dominant genes is not more than $u = mq = 5.0 \times 10^{-7}$ for class 1. Therefore the contribution of dominant mutations to sporadic cases of class 2 and 4(2) is negligible. The rate for recessive genes is $u = q(\alpha' + q + h)m$, where α' is the long-term inbreeding coefficient and h is the selection coefficient against heterozygotes. The value of α' is probably about .006 (Neel et al., 1949). For $h = 0$, we find $u = 3.1 \times 10^{-5}$ per locus for the recessive genes contributing to class 2 and 4(2). If all sporadic cases are due to penetrance in heterozygotes, $h = .0017$, $u = 3.6 \times 10^{-5}$. The assumption of equal gene frequencies, which makes the estimate of the number of loci a minimum, makes this estimate of u a maximum if there is more than one locus. The mutation rate for sex-linked genes, assuming equality in eggs and sperm, is $u = mq/3 = 8.9 \times 10^{-5}$ for class 3 and 4(3). Without assuming equilibrium or equal rates in the two sexes, but using the estimated frequency of sporadic cases, the mutation rate in eggs is $u = .355q = 9.9 \times 10^{-5}$. The difference between the two estimates is trivial.

A variety of evidence indicates that, in mammals, formation of oocytes is virtually complete soon after birth (Parkes, 1956). On this hypothesis, the maternal age at birth of sex-linked mutants is critical for distinguishing two types of mutation. On one model, mutation is a failure to copy the gene correctly at cell division, in which case the mutation rate should not be related to maternal age. On the other model, mutation is caused by radiation or chemical mutagens independently of cell division and therefore accumulates with maternal age. The mean maternal age at birth of 72 isolated cases of class 4(3) is 27.736, and the mean for 147 normal sibs is 27.864, the difference being $-.128 \pm .922$ in the direction contrary to the accumulation model.

The data concur with more indirect evidence from maternal ages for autosomal dominant mutants (Penrose, 1955), in suggesting that the mutation rate in human gametes is more closely related to the number of previous cell divisions than to gonial age.

Genetic risks

In class 1, .87 of cases reported onset by age 20. Presumably nearly all carriers could be detected by a careful neurological examination at maturity. The probability that the child of a carrier be affected is $\frac{1}{2}$, and the probability that the child of a sibling who is not a carrier be affected is zero. The risk for a carrier's child who is normal at age z to develop the disease is not more than $[1 - G(z)]/[2 - G(z)]$, where $G(z)$ is given in figure 3 of Chung and Morton (1959). For example, since $G(z) = .87$ at age 20, the risk for a normal 20-year-old child of a carrier to develop the disease is not more than .115, and is probably much less than this, since $G(z)$ is based on recollection rather than clinical examination.

In case 2 and 4(2), sibships with familial cases, or an isolated case with parental consanguinity, have a risk of $\frac{1}{4}$ for dystrophy with each subsequent child. The probability that an isolated case with S normal siblings be sporadic is $Q = $

$$.413 \bigg/ \bigg\{ .413 + .587 \prod_{i=1}^{S} [1 - G(z_i)/4] \bigg\},$$ where z_i is the age of the ith sib. Since the

mean penetrance is $y = .836$ for unaffected sibs, this function may be approximated by $Q = .413/\{.413 + .587(.791)^S\}$, as shown in table 6.

The risk for dystrophy in the child of an affected homozygote is expected to be $P = F + q + h$, where F is the inbreeding coefficient, q the gene frequency, and h the penetrance in heterozygotes. The expected risk is $P/3$ for the children of normal sibs of affected homozygotes, $P/2$ for the children of carriers, and $P/4$ for the children of the sibs of carriers. The empirical risks for relatives of index cases are shown in table 7, including three families from marriages of affected homozygotes to presumptive carriers (Class 5 of Chung and Morton, 1959). The observations are consistent both with no risk for the relatives of sporadic cases and with a risk half as great as for the relatives of familial cases, as would be expected if sporadic cases are penetrant heterozygotes. The risk in relatives of familial cases is consistent with complete re-

TABLE 6. THE PROBABILITY THAT AN ISOLATED CASE OF CLASS 2 OR 4(2) BE SPORADIC, IF THERE ARE S NORMAL SIBS

S	0	2	4	6	8	12	∞
Q	.41	.53	.64	.74	.82	.92	1

TABLE 7. GENETIC RISKS FOR CLASS 2 AND 4(2) RELATIVES

	r = 1		r > 1		
	Affected	Normal	Affected	Normal	kP
Children	0	110	0	66	P
Sibs' children	0	348	1	136	P/3
Carriers' children	2	1755	3	634	P/2
Carriers' sibs' children	0	1409	0	414	P/4
P	.0014 ± .0010		.0075 ± .0037		

TABLE 8. The upper limit to the probability that a brother of an isolated case of class 4(3) be affected or that a sister be a carrier, if there are S normal brothers and s normal maternal uncles

s Uncles	S Brothers				
	0	2	4	6	∞
0	.33	.20	.09	.04	0
2	.28	.15	.06	.02	0
∞	.25	.13	.05	.02	0

cessivity (P = α + q = .0047) or with occasional penetrance in heterozygotes (P = α + q + h = .0064). All evidence indicates that the risk in noninbred relatives of class 2 and 4(2) is very small.

In class 3 and 4(3) the risk is $\frac{1}{2}$ for sons of carrier women. One-half of the daughters of carrier women and all the daughters of carrier men are carriers. The probability that an isolated case with S normal brothers and s normal maternal uncles be sporadic is at least

$$Q = x \Big/ \left\{ x + (1 - x) \prod_{i=1}^{S} [1 - G(z_i)/2] \left[x' + (1 - x') \prod_{i=1}^{S} [1 - G(z_i)/2] \right] \right\}.$$

Since x = $\frac{1}{3}$, x' = $\frac{1}{2}$, and the average penetrance in normal sibs is .839, we may write approximately Q = 1/{1 + (.58)^S[1 + ($\frac{1}{2}$)^s]}. The probability that a brother of an isolated case be affected or a sister be a carrier is not more than $(1 - Q)/2$, as given in table 8. The method for computing exact risks has been given by Binet et al. (1958), using information provided by normal male relatives more remote than uncles.

Considering the probabilities in tables 6 and 8, it is apparent that in the absence of positive identification of carriers, the separation of sporadic from chance isolated cases approaches certainty only in large families. When more is known about the biochemistry of the several genetic types of muscular dystrophy, it may be possible to determine with assurance that a given family member is, or is not, a carrier. Until then, the genetic basis, if any, of sporadic cases of class 2 and 4(2) must remain in doubt, and the specification of genetic risks be unsatisfactory.

SUMMARY

Genetic classes of muscular dystrophy delimited by pedigree information and discriminant functions were analysed by maximum likelihood scores, the theory of detrimental equivalents, and other methods.

One group (facioscapulohumeral) is due to an autosomal dominant gene or genes with complete penetrance in individuals of both sexes who survive to the age of onset. The prevalence is variable among populations, but is approximately 2 living cases per million, or an incidence of 4 persons who will develop the trait per million births. The fertility is nearly normal, and the mutation rate is not more than 5 per ten million gametes.

A second group (limb-girdle) is composed of autosomal recessive cases with penetrance in both sexes. The prevalence is 12 living cases per million, or an incidence of

38 persons who will develop the disease per million births. The fertility is 25% of normal. Most of these cases are due to a few loci, perhaps to a single locus, with a mutation rate of 31 per million loci. The frequency of heterozygous carriers, who are rarely if ever affected, is 16 per thousand persons, and this load is shown to be mutational in origin and not due to heterozygote advantage.

A third group (limb-girdle) is composed of truly sporadic cases, of unknown etiology. Fertility and other characteristics are the same as for recessive cases. The prevalence is 8 living cases per million, or an incidence of 27 persons who will develop the disease per million births. If these cases, whose etiology may well be complex, were entirely due to occasional penetrance in heterozygous carriers of recessive genes for dystrophy, the penetrance would be only 17 per ten thousand carriers. These sporadic cases are not due to dominant mutations, since all of 110 children of isolated index cases were normal.

The fourth group (Duchenne) consists of sex-linked cases, with a prevalence of 66 living cases per million males, or an incidence of 279 boys who will develop the disease per million male births. Their fertility is less than 4% of normal. One third of these cases are truly sporadic, due to new mutations. The mutation rate is 89 per million gametes, and is the same in eggs and sperm. The absence of a significantly elevated maternal age for isolated cases suggests that the mutation rate in eggs is more closely related to the number of previous cell divisions than to gonial age.

Genetic and empirical risks based on relationship, age, and the number of normal relatives are given for the different types of dystrophy.

ACKNOWLEDGMENTS

The authors wish to acknowledge the help given them by Dr. H. A. Peters and the staff of the Muscular Dystrophy Clinic in studying these families, by the Wisconsin State Medical Society and the Bureau for Handicapped Children in finding cases, and by the patients and their families whose cooperation made this study possible.

REFERENCES

BABER, R. E., AND E. A. ROSS. 1924. Changes in the size of American families in one generation. *Univ. Wisconsin Stud. in Soc. Sci. and Hist.* No. 10.

BECKER, P. E. 1933. *Dystrophia musculorum progressiva.* Thieme: Stuttgart.

BELL, JULIA. 1943. On pseudohypertrophic and allied types of progressive muscular dystrophy. *Treas. Human Inheritance* 4: 283–341.

BINET, F. E., R. J. SAWERS, AND G. S. WATSON. 1958. Heredity counselling for sex-linked recessive deficiency diseases. *Ann. Human Genet.* 22: 144–152.

CHEESEMAN, E. A., S. J. KILPATRICK, A. C. STEVENSON, AND C. A. B. SMITH. 1958. The sex ratio of mutation rates of sex-linked recessive genes in man with particular reference to Duchenne type muscular dystrophy. *Ann. Human Genet.* 22: 235–243.

CHUNG, C. S., AND N. E. MORTON. 1959. Discrimination of genetic entities in muscular dystrophy. *Am. J. Human Genet.* 11: 339–359.

CHUNG, C. S., O. W. ROBISON, AND N. E. MORTON. 1959. A note on deaf mutism. *Ann. Human Genet.* (in press).

CROW, J. F. 1958. Some possibilities for measuring selection intensities in man. *Human Biol.* 30: 1–13.

DE GROUCHY, J. 1953. *Contribution a l'étude de l'hérédité de la myopathie.* Thesis: Paris.

FORD, C. E., P. E. POLANI, J. H. BRIGGS, AND P. M. F. BISHOP. 1959. A presumptive human xxy/xx mosaic. *Nature* 183: 1030–1032.

HALDANE, J. B. S. 1956. Mutation in the sex-linked recessive type of muscular dystrophy. A possible sex difference. *Ann. Human Genet.* 20: 344–347.

KROOTH, R. S. 1955. The use of the fertilities of affected individuals and their unaffected sibs in the estimation of fitness. *Am. J. Human Genet.* 7: 325–360.

LAMY, M., AND J. DE GROUCHY. 1954. L'hérédité de la myopathie (formes basses). *J. Génét. Humaine* 3: 219–261.

LEVISON, H. 1951. Dystrophia musculorum progressiva. *Op. ex domo Biol. Hered. Human. Univ. Hafn.* vol. 26.

MORTON, N. E. 1958. Segregation analysis in human genetics. *Science* 127: 79–80.

MORTON, N. E. 1959. Genetic tests under incomplete ascertainment. *Am. J. Human Genet.* 11: 1–16.

NEEL, J. V., M. KODANI, R. BREWER, AND R. C. ANDERSON. 1949. The incidence of consanguineous matings in Japan. *Am. J. Human Genet.* 1: 156–178.

PARKES, A. S. 1956. *Marshall's physiology of reproduction.* v. 1, pt. 1. 688 pp. N. Y.: Longmans.

PENROSE, L. S. 1955. Parental age and mutation. *Lancet.* 2: 312–313.

PHILIP, URSULA, AND J. N. WALTON. 1956. Colour blindness and the Duchenne-type muscular dystrophy. *Ann. Human Genet.* 21: 155–158.

REED, T. E., AND J. V. NEEL. 1955. A genetic study of multiple polyposis of the colon (with an appendix deriving a method of estimating relative fitness). *Am. J. Human Genet.* 7: 236–263.

SJOVALL, B. 1936. Dystrophia musculorum progressiva. *Acta Psychiat. neur.,* suppl. 10.

SMITH, C. A. B., AND S. J. KILPATRICK. 1958. Estimates of the sex ratio of mutation rates in sex-linked conditions by the method of maximum likelihood. *Ann. Human Genet.* 22: 244–249.

STEPHENS, F. E., AND F. H. TYLER. 1951. Studies in disorders of muscle. V. The inheritance of childhood progressive muscular dystrophy in 33 kindreds. *Am. J. Human Genet.* 3: 111–125.

STEVENSON, A. C. 1953. Muscular dystrophy in Northern Ireland. I. An account of 51 families. *Ann. Eugen.,* Lond., 18: 50–93.

STEVENSON, A. C. 1955. Muscular dystrophy in Northern Ireland. II. An Account of nine additional families. *Ann. Human Genet.* 19: 159–164.

STEVENSON, A. C. 1958. Muscular dystrophy in Northern Ireland. IV. Some additional data. *Ann. Human Genet.* 22: 231–234.

TYLER, F. H., AND F. E. STEPHENS. 1950. Studies in disorders of muscle. II. Clinical manifestations and inheritance of facioscapulohumeral dystrophy in a large family. *Ann. Int. M.* 32: 640–660.

WALTON, J. N. 1955. On the inheritance of muscular dystrophy. *Ann. Human Genet.* 20: 1–38.

9

Reprinted from *Acta Genet.* **10**:63–70 (1960)

THE SIMULATION OF MENDELISM

By J. H. EDWARDS

Mendelism, in its original sense, refers to the segregation of characters in such a manner that their inheritance may be explained in terms of their being the consequence of a single factor, or of a pair of single factors, and these concepts have been successfully applied to a large number of rare diseases in man.

Familial concentration is a feature common to all diseases in man; in some, as in silicosis, malnutrition and schizophrenia, the familial concentration is so striking that it is necessary to postulate some innate or acquired peculiarity common to such families. An innate peculiarity may be variously conditioned; in addition to single factor inheritance a large number of characteristics are now known to be dependent upon the additive effects of a multiplicity of genes, and such characteristics may appear discontinous or, in Grüneberg's terminology, quasi-continuous.

The threshold or line or dichotomy may be due to developmental variations, as in the case of the foramen ovale of the mouse acetabulum, in which it appears that the genetic determinants related to the formation of this hole, if present at less than a certain intensity, are inadequate to lead to its development (*Grüneberg*, 1951).

In man such developmental abnormalities as cleft palate, hare lip, anencephaly, spina bifida and most heart defects, which are probably related to the unpunctual fusion of the margins of various grooves and holes, may be conditioned by continuous variations, innate and acquired, in developmental punctuality.

In other conditions the threshold is biochemical, the response to some substance showing some sudden discontinuity, as in the various hormonal switch mechanisms, or in the crystallization, incorporation or overflow which develops at above a certain level, as in gouty deposits, atheroma, or glycosuria. In other cases the discontinuity may be mechanical, as in the bursting of an artery in apoplexy or sub-arachnoid haemorrhage being predisposed to by high internal pressures and thin or diseased walls.

The discontinuity is not necessarily explicit in the data but may be entirely imposed by the observer for semantic economy, as in such words as thin, neurotic and hypertensive, or for administrative convenience, as in recording the eleven-plus examination or in the specification of prematurity.

In many cases of disease it may be self-perseverating, so that, once a certain level of abnormality is present, further damage occurs of such a nature that it excites the mechanism causing damage. Malignant hypertension, in which a vicious circle becomes established through pressure damage to the arteries increasing their resistance, is a simple example. Others include the social isolation induced by eccentricity, the intellectual deprivation of the institutionalized defective, and the physical inactivity of the very fat. In diabetes it is usually maintained that adequate control reduces the rate of increase of insulin requirements, and this suggests that a vicious circle may be established at an earlier stage in the disease. In focal epilepsy there is extensive clinical evidence suggesting that the influence of the focus increases with each fit. All such vicious circles will have the effect of causing a break-away at one end of any continuous distribution to which they are related, and in such cases an overt bimodality may overlie a basic continuity.

The concept of quasi-continuous variation in relation to human disease was clearly defined in the Hippocratic doctrine of diathesis, a diathesis being an inherited continuum extremes of which are peculiarly predisposed to the development of various diseases. The concept was explicitly related to congenital malformations by Darwin, who regarded these as the likely price of an extreme variability, while more recently *Lerner* (1953) has advanced a similar hypothesis with special reference to the increase in phenotypic variability resulting from an increased homozygosity.

When the variates underlying these discontinuities show familial correlations, the discontinuities will show familial aggregations, and, when the correlations are high and the disorder common, these familial aggregations may be so high that they simulate the pattern of inheritance expected in single factor inheritance. If the qualification of necessity is removed by the concept of penetrance, the hypotheses may be indistinguishable on even extensive observational data.

As nearly all disorders develop from a stage of apparent normality, there are times when the manifestation of disease is only present in a proportion of disposed cases, and, in such cases, this proportion is termed the penetrance. In other conditions the variability of other developmental influences is such that manifestation may occasionally not occur as in the common forms of polydactyly, or may only occur under extreme conditions, as in haemophilia, or osteogenesis imperfecta, which may only become manifest under considerable trauma. In other conditions, as Huntington's chorea or polycystic kidneys, death from other causes commonly precedes the usual age of manifestation. In such relatively rare conditions, whose incidence does not exceed 1 per thousand or so, and is usually less than 1 in 10,000, and in which our observational powers do not allow the genotype to be specified by the phenotype, the concept of penetrance appears not merely valuable but inescapable.

This concept is, however, often used in relation to common disorders when it is a refuge for an ignorance which it is the business of the investigator to disperse rather than to concentrate. Like the epicycle of Ptolemy it revolves round nothing and, like the epicycle, should only be invoked if alternative explanations, even if of greater analytic complexity, prove less adequate. Further, the assumption of the constancy of penetrance in different individuals assumes an independence of both the inherited constitution and the environment, neither of which is likely. If the penetrance is assumed to vary in relation to either the genetic background, or the environment, or both, then it is difficult to see how this can be manifest other than in an effectively quasi-continuous manner. In this case the difference between single gene concepts with penetrance and quasi-continuous variation is the difference between a single gene combined with multifactorial inheritance and multifactorial inheritance alone. The former hypothesis is more complex and shades into the latter as the single gene becomes of diminishing effectiveness or "lower penetrance".

If we consider the simple cases of multifactorial inheritance, and of an abrupt threshold such that a proportion p of the population lies beyond it, then the intensity of familial aggregation shows a simple approximate relationship to p. In practice the threshold is not abrupt, which has the effect of reducing the genotype intraclass correlations. On the other hand the environmental features common to families, and, in the case of siblings, the common uterus and similar post-natal handling, will have the effect of increasing the phenotypic correlations. In the case of first degree relatives the correlations of the genotypes will be approximately $\frac{1}{2}$ and, more generally, in the n the degree relations $(\frac{1}{2})^n$.

In the case above we may ask; firstly, what is the incidence in the n th degree relatives of a propositus? Secondly, to what extent does this incidence exceed the population incidence? The former question is of primary interest to the geneticist who examines formal data; the latter to the clinician who tends to base his opinion on the familial tendencies of diseases on the extent of the excess risk of relatives. *Penrose* (1953) has discussed this increase in liability, which he terms k, in relation to dominant, intermediate and recessive inheritance in common diseases.

If we consider any relative of a propositus, and the relationship is such that the intraclass correlation is r, then we may specify the above question with reference to the quadrants of the bivariate normal surface in figure 1.

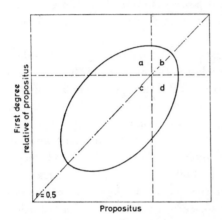

Figure 1

Diagram showing the incidence of affected first degree relatives of propositi in quasi-continuous inheritance. The ellipse represents an arbitrary contour on the bivariate normal surface of correlation r = 0.5. The lines represent the dichotomizing planes defining the level of discontinuity or the threshold of specification: as these are equal in the propositus and their relatives the diagram is symmetrical about the main diagonal.

The bivariate normal surface defines the expected distribution of a mass of points, each representing the relevant parameters for each pair of relatives. The dichotomies which divide the continuum into two discontinuous segments, and the surface into four quadrants, represent the thresholds which will be equal. If the numbers a, b, c, d represent proportions, and a + b + c + d = 1, then if p is the population incidence of the disorder

$$p = b + d$$

the incidence in specified relatives of the propositus is

$$\frac{b}{b+d}$$

and the ratio of this incidence to the incidence in the general population is

$$\frac{b}{(b+d)^2}$$

The proportions a, b, c, d exhibit the approximate relationship

$$\ln \frac{bc}{ad} \simeq \frac{8}{\pi} z \qquad \text{where } z = \tanh^{-1} r$$

and this is largely invariant to the levels of dichotomy (*Edwards*, 1957). In the symmetric case when the planes of dichotomy are equidistant from the mean the approximate relationship

$$b \simeq (b + d)^t$$

where t is a constant is also largely invariant to the levels of dichotomy (*Sandon*, 1957). It may be shown by substitution that the satisfaction of these approximate identities implies the approximation

$$t \simeq 1 + \ln (1 + e^{-8z/\pi}) / \ln 2$$

which varies between 2 and 1 as *r* varies between 0 and 1. The approximation becomes progressively less exact as the distance of the dichotomies from the centre increases. When r = 0.5, the value $t \simeq 1\frac{1}{2}$ is simple and adequate to a first approximation, and, in this case, the incidence in first degree relatives

$$= (b + d)^{(t - 1)} \simeq p^{\frac{1}{2}}$$

and the ratio of this incidence to the population incidence

$$= (b + d)^{(t - 2)} \simeq p^{-\frac{1}{2}}$$

Figure 2 shows this approximation, and compares it to exact values calculated from the tetrachoric tables of Karl Pearson. It is seen that when *p* is less than about 16% the incidence in first degree relatives exceeds $p^{\frac{1}{2}}$.

It is interesting to note that in the range of incidence between 0.1% and 1.0%, within which lie epilepsy, schizophrenia, diabetes, spina bifida, anencephaly, pyloric stenosis and mental deficiency of unknown cause, the incidence in first degree relations expected on a multifactorial hypothesis is very similar to that found. In the common conditions of schizophrenia and diabetes the incidence in first degree relatives is such that hypotheses of recessive inheritance have been authoritatively entertained. In these two conditions the frequent association with failure to marry and procreate respectively, and the fact that families often contain more children than parents, have given further irrelevant support to such hypotheses.

Some additional information relevant to these hypotheses may be obtained from the incidence in cousins, aunts and uncles, or other more distant relatives. Such additional information is however of very little help

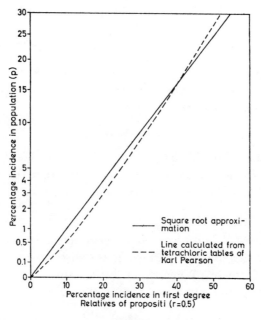

Figure 2

as the operational similarity of the two hypotheses persists. This is hardly surprising as in both hypotheses the more distant the relationship the lower the incidence. We may express the incidence and relative incidence in various relatives on the three hypotheses of dominant, recessive and multifactorial inheritance (tables 1 and 2).

There are, however, two observable consequences of multifactorial inheritance which differ from single factor inheritance. Firstly, in the former the risk to the unborn increases with the number of relatives affected. If, for example, two sibs are affected in one two-child family, and one in

Table 1

Incidence in relatives of propositi

	Dominant	Recessive	Multifactorial
Sibs	$^1/_2$	$^1/_4$	$\simeq p^{1/2}$
Parent or child	$^1/_2$	$p^{1/2}$	$\simeq p^{1/2}$
First cousin	$^1/_8$	$^1/_4 p^{1/2}$	$\simeq p^{4/5}$

Table 2

Relative incidence in relatives of propositi

	Dominant	Recessive	Multifactorial
Sibs	$\frac{1}{2}p^{-1}$	$\frac{1}{4}p^{-1}$	$\simeq p^{-1/2}$
Parent or child	$\frac{1}{2}p^{-1}$	$p^{-1/2}$	$\simeq p^{-1/2}$
First cousin	$\frac{1}{8}p^{-1}$	$\frac{1}{4}p^{-1/2}$	$\simeq p^{-1/5}$

another, the risks to any further sib will be higher in the former family on a multifactorial hypothesis but equal on a single factor hypothesis. The expected extent of this excess risk must await tabulation of the partitions of the symmetrically dichotomized trivariate normal hypersurface. Secondly, when the arbitrarily dichotomized variate is measurable, as in height, gout or feeble mindedness, there should, on the multifactorial hypothesis, be no bimodality in the scores, unless some self-perpetuating mechanism can reasonably be postulated as in hypertension or glaucoma; further, if dominant inheritance is simulated, the hypothetical non-carrier parent should tend to have above-average scores.

Attempts to discover so-called carriers by imposing special stresses, as in giving cortisone to the relatives of diabetics, or flashing lights at the children of epileptics, cannot necessarily provide any distinction between single gene and multiple gene hypotheses, although they may appear to confirm the former as it will often be possible to find a stress which will lead to the desired segregation ratio. To take an analogy, if we arbitrarily defined tallness in women as being more than 5′ 8″ tall, it would be possible, by trying out various types of high-heeled shoes, to find brands which would provide a segregation ratio appropriate to the required genetical hypothesis, and to define women on whom tallness was thus imposed as exhibiting latent tallness.

Although these two types of inheritance may lead to similar patterns of familial aggregation, their distinction is far from trivial in importance. If single genes of impaired penetrance are involved then an increase in the mutation rate will lead to an increase in incidence of the disorders to which they predispose. If, on the other hand, the manifestation is the result of multifactorial inheritance, no prediction can be made of the influence of an increased mutation rate without assuming values for several unknown parameters.

Summary

The operational distinction between quasi-continuous variation and the effects of major genes of low penetrance is discussed.

It is shown that, in quasi-continuous variation, a trait with an incidence p would be expected to have an incidence of the order of $\sqrt{p}$ in the first degree relatives of propositi.

Zusammenfassung

Die methodische Unterscheidung zwischen quasi-kontinuierlicher Variation und den Wirkungen von Hauptgenen mit niedriger Penetranz wird diskutiert.

Es wird gezeigt, daß bei quasi-kontinuierlicher Variation ein Merkmal mit der Wahrscheinlichkeit p bei den Verwandten 1. Grades des Probanden eher eine Wahrscheinlichkeit $\sqrt{p}$ haben muß.

Résumé

L'auteur discute la différence entre les variations presque continues et les effets des gènes principaux ayant une faible pénétrance. Il montre que lorsque dans une variation presque continue la fréquence d'un caractère est p, la fréquence probable de ce caractère chez les parents du 1er degré du proband est $\sqrt{p}$.

REFERENCES

Edwards, J. H.: A Note on the practical interpretation of 2×2 Tables. Brit. J. prev. soc. Med. *11:* 73 (1957).

Grüneberg, H.: Genetical studies on the skeleton of the mouse IV. Quasi-continuous variations. J. Genet. *51:* 95 (1952).

Lerner, I. M.: Genetic homeostasis (Oliver and Boyd, Edinburgh 1953).

Penrose, L. S.: The genetical background of common diseases. Acta genet. *4:* 257 (1953).

Sandon, F.: The relative numbers and abilities of some ten-year-old twins. J. R. Statist. Soc. A. *120:* 440 (1957).

10

Reprinted from *Proc. Roy. Soc. London,* B, **153**:147–170 (1948)

The formal genetics of man

By J. B. S. HALDANE, F.R.S.

(*Delivered* 7 *November* 1946—*Received* 18 *December* 1947)

Man has obvious disadvantages as an object of genetical study. The advantages are that very large populations are available, and that many serological differences and congenital abnormalities have been intensively investigated.

Some characters are found to obey Mendel's laws with great exactitude. In others the deviations are such as to suggest the existence of a considerable selective mortality, perhaps prenatal. In yet other cases the observations are biased because we only know that we are investigating the progeny of two heterozygotes when the family includes at least one recessive. Statistical methods which eliminate this bias are described.

Still more complex methods are needed for the detection and estimation of linkage. Several such cases have been detected with greater or less certainty, and the frequency of recombination between the loci of the genes for colour-blindness and haemophilia is now estimated at $10 \pm 4 \%$. If the theory of partial sex-linkage be accepted, it is possible to make a provisional map of a segment of the human sex chromosome.

When a gene is sublethal, as are those for haemophilia and achondroplasic dwarfism, its elimination by natural selection is in approximate equilibrium with its appearance by mutation, and the frequency of the latter process can be estimated. The mutation rates at five human gene loci lie between 4×10^{-5} and 4×10^{-6} per locus per generation. These are the only estimates available for vertebrates. The rates per generation are rather higher than those in *Drosophila*, but those per day are so small that much, or even all, human mutation may be due to natural radiations and particles of high energy.

In 1906 Bateson delivered a lecture to the Neurological Society on 'Mendelian heredity and its application to man' in which he described the genetics of brachydactyly and congenital cataract, which are dominant to the normal (the word is used loosely, since the abnormal homozygote is unknown). He suggested that albinism and alcaptonuria were recessive, and he described the laws of inheritance of haemophilia and colour-blindness, though he did not, of course, give the explanation of these laws which is now accepted.

In the ensuing 40 years a very large number of pedigrees have been collected, unfortunately with very variable standards of accuracy. These show that more than a hundred different human abnormalities are certainly due to single gene substitutions, and that several hundred more are probably so. For example, Cockayne (1933) listed eighty abnormalities of the skin, hair, nails, and teeth which are probably due to dominant autosomal genes, eighteen to autosomal recessives, and thirteen to sex-linked genes. In over half the cases·the evidence is adequate.

On the other hand the genetical analysis of the normal polymorphism of a race such as our own for colour, size, and shape has not gone far. The genetics of eye colour, for example, are far more complex than was originally thought, and stature is undoubtedly determined by a large number of genes, as well as by environmental influences. Still less progress has been made in the analysis of the genetics of those differences in skin colour and hair shape which exist between the major human races. However, immunology has revealed a polymorphism existing in all races which was wholly unexpected when Bateson wrote. Its genetical basis is exceedingly simple; perhaps because antigens are direct products of gene action, while pigments are the end products of complex chains of metabolic processes in which many, if not all, of the steps are controlled by different genes; and the processes of morphogenesis are even more complicated. Meanwhile, genetics have developed along many lines, of which three are especially important. It has been shown that genes are material structures located at definite points in the chromosomes. If we can homologize the genes of organisms which conjugate and the 'transforming principles' of bacteria, which can apparently transfer them to one another without conjugation, just as they carry out a communal metabolism, the work of Avery, MacLeod & MacCarty (1944) suggests that genes, at least in some phases of their life cycles, may consist wholly of desoxy-ribose nucleic acid.

We have learned a good deal about the causal chain between a gene and its manifestation. Goldschmidt was a pioneer in this work. You, Mr President, played an important part in the analysis of the genetical control of anthocyanin production in flowers. Beadle, Tatum and others were able to specify the stages in the production of arginine and other essential metabolites controlled by different genes in *Neurospora*. In this country, Grüneberg, in the mouse, and Waddington, in *Drosophila*, investigated the genic control of morphological development.

Finally, a number of workers, notably Dobzhansky, Dubinin, Fisher, Haldane, Teissier & l'Heritier, Tsetverikov, and Wright have investigated the genetics of populations both practically and theoretically, and they and others have discussed the bearing of their results on the problem of evolution.

All these methods are applicable to our own species. There is, however, a widespread belief that what I may call formal genetics, that is to say the study of heredity and variation, based on the description and counting of individuals, has ceased to be important, and that in future genetics will consist mainly of the study of biochemical and morphogenetic processes controlled by genes, and of evolutionary changes in populations; while the mere enumeration of the results obtained from various matings, and deductions drawn from such enumeration, are no longer of great interest. As I propose to devote this lecture to the pure or formal genetics of man, I may perhaps be pardoned if I state what seem to me to be the legitimate aims of human genetics, and so to justify what some will regard as a reactionary standpoint.

The final aim, perhaps asymptotic, should be the enumeration and location of all the genes found in normal human beings, the function of each being deduced from the variations occurring when the said gene is altered by mutation, or when several allelomorphs of it exist in normal men and women. In addition, information would be gathered on the effect of changes involving sections of chromosomes, such as inversions, translocations, deficiencies, and duplications.

The number of genes in a human nucleus almost certainly runs into thousands, possibly tens of thousands. Each has, so far as one can judge, a highly specific biochemical function. The end result of such a genetical study as I have adumbrated would be an anatomy and physiology of the human nucleus, which would be incomparably more detailed than the anatomy and physiology of the whole body as known at present. This end will perhaps be achieved in part by non-genetical methods, such as ultramicroscopic operations on the nuclei of human cells in tissue culture.

No doubt one result of such a study will be the possibility of a scientific eugenics, which may bear the same relation to the practices now or recently in vogue in certain countries as chemotherapy bears to the bleedings and purgations of early medicine. But other results may be more important. A knowledge of the human nucleus may give us the same powers for good or evil over ourselves as the knowledge of the atomic nucleus has given us over parts of the external world.

In this lecture I shall be largely concerned with the localization of genes in human chromosomes. A simple example will show why this is important. One of the common causes of blindness is retinitis pigmentosa. Ten years ago it could be said that in some pedigrees this disease was transmitted as a dominant, in others as a recessive of the ordinary type, occasionally as a sex-linked recessive. In 1936 I argued that some pedigrees showed partial sex-linkage, a phenomenon which I shall describe later. We can now say tentatively that one of the genes, the abnormality of which causes this condition, is carried in that segment of the sex chromosomes of which women possess two, and men only one; another, which may give dominant or recessive mutants, in that segment of the same chromosomes of which both sexes possess two; while other such genes, how many we do not know, are carried in the other chromosomes. It is reasonably sure that they control different processes. And this is borne out by the fact that the partially sex-linked recessive type is never associated with deafness, while one of the autosomal recessives is so associated.

Pathologists will have to work out the aetiology of the different genetical types. They can hardly hope to do so until they are distinguished, if, as seems probable, each gene controls a different process. And just as the methods for the cure of bacillary and amoebic dysentery are very different, so it is unlikely that the same therapeutic measures will succeed against diseases, however similar in their symptoms, which are due to different genes. They certainly do not do so in *Drosophila melanogaster*. In that species at least four different recessives give eye colours which are scarlet because they lack a yellow pigment found in the normal eye, which is a derivative of tryptophane. The eye colour of the mutant *vermilion* can be made normal by injecting the larvae with kynurenine, for the gene present in normal flies but inactive in *vermilion* flies is concerned in the oxidation of tryptophane to kynurenine. But *cinnabar* flies are not cured, because they cannot catalyze a further stage in the pigment formation; and *cardinal* and *scarlet* flies are not cured because their eye rudiments cannot take up the pigment precursor (Ephrussi 1942). The four genes in question are carried at different loci in three chromosomes.

The first step in formal genetics is to establish that certain characters are inherited in accordance with Mendel's laws, and in particular that segregation occurs in Mendelian ratios.

This is certainly true in many cases where large numbers have been studied. Thus according to theory a member of blood group AB produces equal numbers of A and B gametes. Table 1* shows that this is the case, the deviation from theory being less than the standard error of sampling. In the mating $A \times AB$ the A children are derived from A gametes of the AB parent, the B and AB children from B gametes; and so on. Such an agreement implies not only that the two types of gamete are

TABLE 1

	children in group				
parents	O	A	B	AB	total
$O \times AB$	8 !	633	646	3 !	1290
$A \times AB$	0	533	247	312	1092
$B \times AB$	2 !	183	406	232	823
$AB \times AB$	0	28	36	65	129

total A gametes 1609⎱
total B gametes 1647⎰ $(d-1)\, n^{-\frac{1}{2}} = 0.648$.
total homozygotes from $AB \times AB$ 64.
total heterozygotes from $AB \times AB$ 65.

* From Wiener (1943), p. 190. This includes all data published since 1931. Before this date only groups of over 250 children are included. This criterion omits three children of AB mother assigned to group O by two workers who had tested seven and nine children of such mothers, and whose findings have perhaps received undue attention.

formed in equal numbers, but that there is no marked selective mortality of either type of zygote. Thirteen children occur in unexpected groups. These represent the combined effects of illegitimacy, technical errors, and conceivably mutation or abnormal segregation of chromosomes. Clearly these causes combined will produce results smaller than sampling error.

In the case of the pair of allelomorphic genes which respectively produce the M and N agglutinogens, there is at least prima facie evidence for very abnormal segregation. Where such cases have been investigated in animals, they have so far almost always been found to be due to selective mortality, or in fact natural selection, one genotype having a higher deathrate in the early stages of life than another. Taylor & Prior (1939a) found that the progeny from the marriages of heterozygotes which they had examined included a considerable excess of heterozygotes. They were convinced that this was not due to technical errors, and found a similar excess in the pooled data of other workers, giving 54·85 % of heterozygous children, the excess being 3·08 times its standard error. Wiener (1943) believes the excess to be an artefact due to the use of incompletely absorbed testing fluids, and that no such excess occurs where this error is avoided. Now if the sera used give too many MN children they should also give too many MN parents. Taylor & Prior (1939b) showed that in most series of data the square of the number of MN individuals approximates to four times the product of the numbers of M and N, while a few show a great excess. I have therefore applied their test to all the series in Wiener's table 46, and eliminated cases where χ^2 exceeds four for the parents.[†] I have also omitted two small series containing 40 children between them, and two Japanese and one German series which were not available in English libraries, and have inserted one German series. The results are shown in table 2. The values of χ are calculated by Taylor & Prior's method, a positive value denoting an excess of MN over expectation among the parents. It will be seen that there are 54·85 % of heterozygotes. The excess is 3·16 times its standard error. The probability of so large an excess or defect by sampling error is 0·0016. If it were due to the systematic use of incompletely absorbed testing fluids, we should expect to find a lower percentage of heterozygotes in those series where χ is negative. The value found is 56·77 %, which is slightly, but not significantly higher. The excess of M over N children is just below twice its standard sampling error, and not statistically significant.

If the numbers of heterozygotes (MN) and homozygotes ($M + N$) in the different groups are compared, we find $\chi^2 = 17·21$ for 11 degrees of freedom. Thus $P = 0·10$, and the data are not significantly heterogeneous, as they would probably be if some workers had used faulty methods.

Table 3 shows the totals found for all types of marriage by the twelve authors cited in table 2. Illegitimacy, technical errors, and mutation clearly account for

[†] Wiener gives the reasons for excluding Lattes & Garrasi's (1932) data. The other relevant values of χ are: Dahr (1940) +4·51, Hirzfeld & Kostuch (1938) +2·16, Landsteiner & Wiener (1941) +2·002, Wiener & Sonn (1943) −2·20. The last three values may well all be due to sampling error.

very few unexpected classifications. In the case of the $MN \times M$ and $MN \times N$ marriages, the differences between the classes where equality is expected are less than twice their standard errors. It will be seen that the mean number of children examined per $MN \times MN$ marriage was decidedly less than that in the other groups. The figure 2·73 is not of course the mean fertility per marriage, as sterile marriages were excluded, it was not always possible to examine all the living children, and

TABLE 2

authors	number of families	χ	children		
			M	MN	N
Landsteiner & Levine	11	+ 0·017	17	31	7
Wiener & Vaisberg	25	− 0·006	29	58	29
Schiff	33	+ 1·808	18	48	22
Crome	9	+ 0·083	4	10	4
Clausen	70	− 0·777	38	74	28
Blaurock	23	+ 1·002	25	40	25
Moureau	53	+ 0·444	45	102	41
Hyman	32	+ 0·906	10	41	16
Matta*	$\begin{cases} 20 \\ 20 \end{cases}$	$\begin{array}{l} − 0·781 \\ − 1·741 \end{array} \Big\}$	9	45	10
Dahr & Bussmann	30	− 0·514	38	70	18
Taylor & Prior (a)	56	− 0·349	10	38	8
Holford	34	+ 1·185	24	37	14
total	416	—	267	594	222

* One group in Egypt, one in Glasgow.

TABLE 3. PROGENY OF DIFFERENT MARRIAGES INVOLVING M AND N

parents	number of families	M	MN	N	total	mean children per family
$M \times M$	147	425	3 !	0	428	2·91
$M \times N$	151	1 !	477	2 !	480	3·18
$N \times N$	74	0	0	232	232	3·13
$MN \times M$	397	597	662	4 !·	1263	3·18
$MN \times N$	292	2 !	428	483	913	3·13
$MN \times MN$	396	267	594	222	1083	2·73
total	1457	—	—	—	4399	3·02

investigators tend to choose large families. Nevertheless, it suggests that such marriages are less fertile than the average. The shortage of total children and of homozygotes can both be explained if homozygotes have a higher deathrate (probably prenatal) than heterozygotes. The prenatal and infantile fitness of the homozygotes is about 82 % of that of the heterozygotes, so at least 18 % of them must die at an early stage. If there had been 594 homozygotes instead of 489, the mean family size would have been 3·00. The hypothesis of selective death implies that 105, or 2·3 % of a group of 4504, human zygotes were eliminated, probably before birth. If the $MN \times MN$ marriages had been as fertile as the rest, we should have expected 155 more children from them, making a total of 4659, of whom 3·4 % were eliminated.

This is a substantial fraction of all conceptions, and it would seem that if a scientific study of the problem of human population is to be undertaken, it would be desirable to investigate a group of say 5000 married couples (including sterile couples) serologically, in order to discover whether certain types are less fertile than others, and whether certain human genotypes are eliminated prenatally. It would be essential, in such a study, to tabulate the results of reciprocal unions such as $MN♀ \times M♂$, and $M♀ \times MN♂$, separately. Unfortunately, many of the authors cited did not do so. It is of course possible that Wiener's hypothesis is correct. Nevertheless, the matter seems sufficiently important to warrant further study.

In the case of the *Rh* group of genes it is known that certain classes of offspring are killed off because they immunize their mother, and their blood corpuscles are destroyed by her antibodies. Such a mechanism will not explain the results found with M and N. Moreover, the elimination of homozygous offspring of two heterozygous parents would make the equilibrium between the two genes unstable, whereas in fact their frequencies in different peoples are much less variable than those of other genes responsible for serological differences.

Whatever may be the final answer to these questions, I hope I have shown that the exact investigation of the segregation of common genes is not a matter of merely academic interest.

I must pass on to the methods which are used in the investigation of the segregation of rare genes. When the rare gene is a dominant there are no statistical difficulties provided the gene manifests itself in all heterozygotes, and early in life. We cannot possibly expect to find Mendelian ratios for such a character as Huntington's chorea, whose mean age of appearance is about 35 years. We should expect to find good results in the case of hereditary skin diseases, which are easily and accurately diagnosed, and mostly manifested at an early age. Table 4 shows the children from unions of affected and normal persons in the sixteen diseases inherited as dominants of which Cockayne (1933) in his classical treatise was able to collect records of over 100 such children. A few of my numbers differ slightly from his totals through the exclusion of doubtful pedigrees. As a result of sampling error we should expect a normal distribution about zero with unit variance of the values of $(d-1)\,n^{-\frac{1}{2}}$, when d is the number of affected minus that of normal, and n is their sum. There are three aberrant values. The low incidence of neurofibromatosis may possibly be accounted for by its variable age of onset and sublethal character. Some individuals carrying the gene may have died prenatally, others may not yet have developed it when observed. Hypoplasia of the enamel is due to genes in at least two different chromosomes (Haldane 1937) and therefore presents complications. Tylosis, which is an abnormal thickening of the skin of the palms and soles, generally develops in the first year of life. It seems to present a definite exception to the usual rules, and demands further investigation. The similar anomalous cases which occur in the literature of dominant abnormalities of other organs are easier to explain by faulty diagnosis. There seems no reason to doubt that the segregation of most human dominant abnormalities follows Mendel's laws.

The ratios in which a gene pair segregates cannot be obtained so simply when one allelomorph is fully recessive. This is due to the fact that the compilation of a pedigree introduced a certain bias. The bias may be of a very simple kind. Birch (1937) in Chicago and Andreassen (1943) in Copenhagen collected 146 pedigrees of haemophilia, a sex-linked recessive condition, transmitted to and from males

TABLE 4. PROGENY OF INDIVIDUALS AFFECTED WITH DOMINANT ABNORMALITIES OF THE SKIN, HAIR, NAILS, AND TEETH

abnormality	affected	normal	$(d-1) n^{-\frac{1}{2}}$
piebaldness	133	118	+0·88
cutaneous xanthomatosis	98	111	−0·83
telangiectasis	320	302	+0·43
epidermolysis bullosa simplex	193	163	+1·54
epidermolysis bullosa dystrophica	147	181	−1·82
monilethrix	92	89	+0·15
porokeratosis	70	91	−1·58
tylosis plantaris et palmaris	594	483	+3·35
ichthyosis vulgaris	86	98	−0·81
alopecia congenita	130	118	+0·70
onychogryphosis	242	253	−0·45
hypoplasia of enamel	84	50	+2·81
neurofibromatosis	115	160	−2·61
naevus aplasticus	53	61	−0·66
fistula auris	63	60	+0·18
angioneurotic oedema	182	206	−1·17
total	2602	2544	+0·79

TABLE 5. SONS OF HAEMOPHILICS' DAUGHTERS

	families	normal sons	haemophilic sons	
mothers of patients	17	11	26	(−17)
other mothers	26	25	23	+1?
total	43	36	49	+1?(−17)
				=32+1?

through females. Each pedigree began with a patient whose relatives were then traced. In order to verify that the condition is due to a single gene we must show, among other things, that the daughters of haemophilics bear equal numbers of normal and haemophilic sons. If we study the daughters of haemophilics in the pedigrees we find a considerable excess of haemophilic sons. However, a further analysis (table 5) shows that this excess is confined to the mothers of the patients from whom the compilation of the pedigree started.

The reason is simple. The mothers of patients were investigated because the patient was discovered to be haemophilic. Hence at least one of their sons must have been haemophilic. The other daughters of haemophilics may have borne no haemophilic sons, indeed one was fortunate enough to bear three normal sons and no haemophilic. We can allow for this bias by subtracting one haemophilic from each family including a patient. The total then becomes 36 normal sons, with 32 haemophilics and one doubtful, a very good approximation to equality. A similar but more complicated analysis shows that about half the sisters of haemophilics transmit the disease. A neglect of this elementary point has led to the most remarkable conclusions as to the fertility of human stocks afflicted with hereditary disease. For if a character is passed on to half the children of an afflicted person, it will not be recognized as hereditary unless at least one child possesses it. We shall thus exclude all families of no children, half the families with only one child, a quarter of those with two, and so on, thus giving a wholly false impression of the fertility of such stocks. Where, on an average, the character only appears in one-quarter of the children, the exaggeration is still greater.

Unfortunately, the very simple type of correction which was applicable to the pedigrees of haemophilia cannot always be applied.

Consider a recessive character such as albinism or amaurotic idiocy which, by analogy with animals, is to be expected in one-quarter of the progeny of unions between two heterozygotes of normal appearance. We have in general no evidence that a pair of parents is heterozygous, except that they have produced at least one recessive. We cannot therefore study the progeny of a number of pairs of known heterozygotes, as we can in animal experiments. We can only study the progeny of those pairs which have produced at least one recessive.

Clearly the frequency of recessives in such sibships* is greater that the expected quarter. For it is 100 % in sibship of one, and over 50 % in sibships of two. The method for assessing the frequency p which would be found in a very large sibship from data on small ones depends on how the data are collected.

Let a_{rs} be the number of sibships of s members, of which r are abnormal.

Let $t_s = \sum_r a_{rs}$, i.e. the total number of sibships of s members.

Let $N = \sum_{r,s} a_{rs}$, i.e. the total number of sibships.

Let $R = \sum_{r,s} (ra_{rs})$, i.e. the total number of abnormals.

Let $S = \sum_{r,s} (sa_{rs})$, i.e. the total number of sibs.

Let $q = 1 - p$.

Now consider two ideal cases. In the first case a whole population is surveyed, and all sibships containing at least one abnormal are tabulated. This is possible in a small European country. Thus Sjögren (1931) probably tabulated over 90 % of the Swedish families in which a case of juvenile amaurotic idiocy had occurred in the twentieth century. In this case the estimate of p is given by:

$$\frac{R}{p} = \sum_s \left[\frac{st_s}{1 - q^s} \right],$$

* The word sibship means a set of siblings, that is to say brothers and/or sisters.

and its standard error is given by:

$$\sigma_p^{-2} = \frac{R}{p^2 q} - \frac{1}{q^2} \sum_s \left[\frac{s q^s t_s}{(1-q^s)^2} \right].$$

Unfortunately, this cumbrous equation, due to Haldane (1938), can be shown to yield a result with a smaller standard error than a simpler one due to Weinberg. Perhaps a quicker but equally efficient method may be devised.

In the second ideal case all children leaving school in a certain year, or better all children born in a certain 6 months, are examined. The sibs of all abnormal children are tabulated. Clearly if a sibship contains three abnormals, it is three times as likely to be tabulated as if it contains only one, and so on, apart from an obvious correction for twins. In this case the estimate of p is

$$p = \frac{R-N}{T-N}, \quad \text{and} \quad \sigma_p^2 = \frac{(T-R)(R-N)}{(T-N)^3},$$

an elegant result due to Weinberg (1927).

Applying these methods to Pearson, Nettleship & Usher's (1913) collection of 411 sibships from normal parents including at least one albino, 864 out of 2435, or 35·48 %, were albinos.

Applying the first correction

$$p_1 = 0\cdot3082 \pm 0\cdot0107,$$

applying the second $\qquad\qquad p_2 = 0\cdot2238 \pm 0\cdot0092.$

The Mendelian value of $\frac{1}{4}$ lies between these two estimates, and there is reason to think that if an exact correction were possible, the Mendelian ratio would be found. A simple example will show the need for other corrections. According to Andreassen (1943), in the year 1941 there were 1,820,000 males in Denmark, of whom 81 were haemophilic. Almost all their families were investigated, and valuable results were obtained. However, haemophilics have a much shorter average life than ordinary males (about 18 years in Denmark). So a large fraction, probably the majority, of the haemophilics born between 1910 and 1930 were dead by 1942. A family into which three haemophilics were born in that time was more likely to contain a living haemophilic in 1942 than one into which only one was born. But it was not exactly three times as likely. If half the haemophilics had died, and there was no correlation in the age of death between haemophilic sibs, it was 1·75 times as likely. Hence the true value of p would lie somewhere between the two extreme estimates. Special methods could and should be developed for such cases.

They are important because they offer a possibility of investigating selective prenatal death, of verifying the general applicability of genetical principles to man, and of developing, in comparatively simple cases, the quite peculiar statistical methods which are required when the genotypes of parents must be deduced from the phenotype of the children, with an accuracy which increases with the size of the sibship. These methods may be said to play the same part in human genetics that standard culture methods play in animal and plant genetics. Without them qualitative conclusions may be drawn, but quantitative work is impossible in the case of many genes.

We now pass to the methods for the location of genes on the human chromosomes. A serious beginning has been made with the mapping of two sections. One is the segment of the *X* chromosome which is responsible for sex determination, in which the ordinary sex-linked genes are located. The other is the segment common to the *X* and *Y* chromosomes. With regard to the remaining twenty-three there is fairly good, but never conclusive, evidence, for the compresence in one chromosome of two genes. For each type of location appropriate statistical methods have been developed.

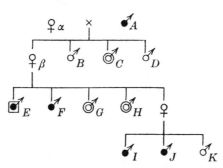

FIGURE 1. ♂, normal male; ♂, deuteranopic male, not haemophilic; ♂, haemophilic male, vision untested; ♂, haemophilic trichromatic male.

Let us begin with the differential segment of the *X*. A large number of sex-linked recessive genes have been located there. The mode of inheritance of the characters determined by them is highly characteristic. The abnormality determined by any one of them is far commoner in men than women. It is not transmitted from a father to his son. But it occurs in about half the sons of heterozygous women, who include the daughters of affected men. The *X* chromosomes of *Drosophila* species have been mapped by studying the segregation of genes in the progeny of mothers who are heterozygous for two or more pairs of sex-linked allelomorphs.

Cytological studies have shown that the maps so obtained depict real material structures, as X-ray diffraction and reflexion have shown that the structural formulae of the organic chemist depict real objects.

At one locus in the human *X* chromosome abnormalities are very common. About 8 % of all men are colour-blind or anomalous colour-matchers. Hence if we wish to estimate the percentage of recombination between the loci of haemophilia and colour-blindness we must search for colour-blindness among haemophilics and their brothers. Seventeen pedigrees are known in which both abnormalities are found. Of the total information available from them, about a third was collected by Bell & Haldane (1936), another third by the Dutch physician Hoogvliet (1942) and the remainder by five others.

The method employed can be illustrated by two simple examples. Figure 1 shows a pedigree in which A was a haemophilic, while α carried the gene for colour-blindness in one of her two X chromosomes. She also gave it to β, who had two colour-blind sons. So β had the genes for these two defects in different chromosomes (in the *trans* position, if we like a metaphor from organic chemistry, or in repulsion in Bateson & Punnett's terminology). Of β's three surviving sons, one was a haemophilic trichromat; two were colour-blind, but had normal blood. Now if x is the frequency of recombination between the two loci concerned, the probability of β producing just these three sons is $(1-x)^3$, given that she had one haemophilic and two non-haemophilics. If one of the non-haemophilics had had normal vision it would have been $x(1-x)^2$, and so on. On this pedigree taken alone the best estimate of x is clearly zero.

Figure 2, which is part of Bell & Haldane's (1937) pedigree A, of 98 members, raises a rather more subtle problem.

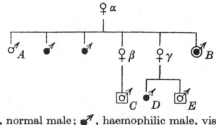

FIGURE 2. ♂, normal male; ♂, haemophilic male, vision untested;
♂, trichromatic male, not haemophilic; ♂, haemophilic deuteranopic male.

This was solved approximately by Bell & Haldane (1936), but my colleague Dr Smith has produced a more accurate method (Haldane & Smith 1947). We ask what is the probability that seven persons related in the manner shown should have just these phenotypes. If p_c and p_h are the frequencies of the genes **c** and **h** for colour-blindness and haemophilia, we ask what is the probability that α should have been heterozygous for both of them. Clearly it is $4p_c(1-p_c)p_h(1-p_h)$. If α was doubly heterozygous and her husband normal, there are eight possible sets of events in the formation of ova by α, β, and γ which could have given the observed results. They are shown in figure 3 and the probability of each is given, putting $y = 1-x$. It is the product of five factors representing the probabilities of the formation of five different eggs. These are shown in each case. Since γ had a haemophilic son who died in infancy we know that she received the gene **h** from her mother. This excludes sixteen other possibilities whose probability is zero. It does not exhaust the possibilities. For though we can be sure that neither α nor her husband was haemophilic, we cannot be sure that one or both of them was not colour-blind. So the total probability has two more terms, each containing p_c^2. It is

$$\tfrac{1}{8}p_c p_h[3(1-p_c)(1-2x+4x^2-4x^3+x^4)+p_c(x^2-x^4)].$$

Since p_c is of the order of 0·01 it can in practice be neglected in this case, but not in all cases.

The corresponding expression for the pedigree of figure 1 is

$$2^{-11}p_c p_h (1-x)^3.$$

These probabilities are of course small, partly because the genes in question are rare, partly because the particular pattern of segregation found is one of a vast number which are possible, like the 635,013,559,600 equiprobable bridge hands. However, each is maximal for some value of x between 0 and 1 inclusive; in the cases considered, for $x = 0$. In other pedigrees a cross-over has occurred, i.e. the genes **c** and **h** have entered a woman in one gamete and left her in different ones or vice versa. In these the polynomial is maximal for some other value of x. The product of the seventeen polynomials derived from the different pedigrees is maximal when $x = 0.098$, and we may estimate the frequency of recombination between the two loci as $9.8 \pm 4.2 \%$.

White (1940) found $64.8 \pm 12.7 \%$ of recombination between the loci of colour-blindness and of myopia with nystagmus, so the genetical map of the human X chromosome is likely to be as long as that of *Drosophila melanogaster*, though probably shorter than that of *Gallus domesticus*.

Ten years ago I accounted for the peculiar inheritance of certain characters on the hypothesis that the genes concerned in their determination were located in that segment of the sex chromosomes which is common to the X and Y, and may be exchanged between them (Haldane 1936). Such genes are said to be partially sex-linked.

At that time I put forward the hypothesis with considerable misgiving, but it has been generally accepted, and I shall therefore state it with comparative confidence. Penrose (1946b) has recently given an alternative explanation for some cases which appear to conform to it, but he does not think that this will explain all the cases. First, consider a dominant gene in this segment. If a woman has it, necessarily in an X chromosome, she will transmit it, on an average, to half her children, regardless of sex. If a man has it in his Y, he has inherited it from his father, and will probably transmit it to most of his sons but a few of his daughters. If he has it in his X he has inherited it from his mother, and will transmit it to most of his daughters but a few of his sons. Thus the sex of the affected children of affected males will generally be the same as that of the affected paternal grandparent. The fewer the exceptions, the nearer the locus to the differential segments containing the sex determiners, and (in the X) the loci of such genes as haemophilia.

Only one such partially sex-linked dominant is known, namely retinitis pigmentosa in some pedigrees. Penrose (1946b) has, I think disproved Pipkin & Pipkin's (1945) claim to have found a second such, zygodactyly, or webbing of the toes. It is possible that my own claim in the case of retinitis pigmentosa will equally be disproved.

The location of partially sex-linked recessives is not so simple, but I think some of its results are more certain. Where the parents are first cousins, the sex of the affected offspring is usually the same as that of the paternal grandparent through whom the parents are related. For if this grandparent was a male the father carries the gene in his Y chromosome, and will transmit it predominantly to his sons, as in figure 4. If she was a female they will mostly be daughters. If the parents are not known to be related, we can only say that in any particular sibship the affected members will be predominantly of one sex, though in all sibships together no such predominance is to be expected, except in the case of very rare conditions where the parents are mostly consanguineous.

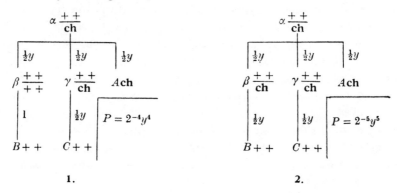

1.   2.

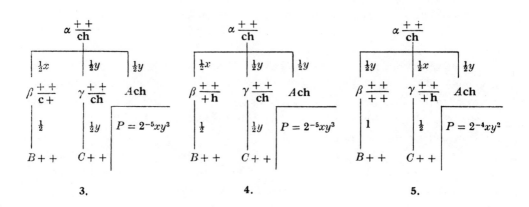

3.   4.   5.

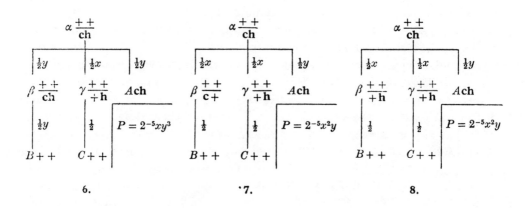

6.   ·7.   8.

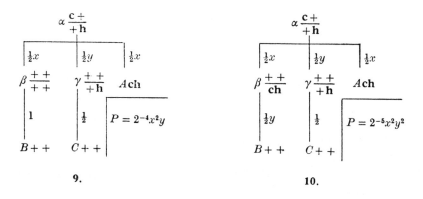

9.   10.

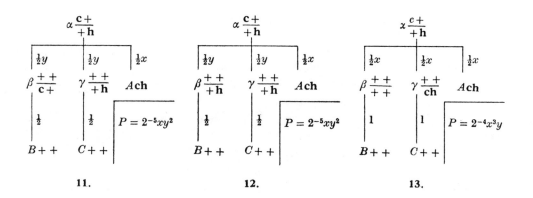

11.   12.   13.

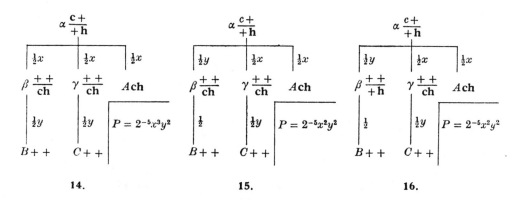

14.   15.   16.

FIGURE 3. Sixteen possible explanations of the pedigree of figure 2. In each case the probability of the five different steps is given, together with their product. The overall probability is the sum of these products, multiplied by the probability that **c** and **h** should be compresent in α.

Fisher (1936) developed a most elegant method for the detection of partial sex-linkage in such cases. Suppose a sibship consists of N normal females, n normal males, A affected females, a affected males, and that

$$u = (N-n-3A+3a)^2 - (N+n+9A+9a),$$

then in the absence of partial sex-linkage the expected value of u is zero, in its presence it is $\frac{1}{9}k(1-2x)^2$, where $k = (N+n+9A+9a)^2 - (N+n+81A+81a)$, and x is the frequency of recombination. Thus if the sum of a large number of u values is significantly positive partial sex-linkage can be inferred, and its intensity estimated.

I have since shown (Haldane 1948) that if we calculate a polynomial for each sibship on the lines developed for the investigation of linkage between sex-linked genes, its logarithm can be expressed as a series in ascending powers of $(1-2x)^2$. The coefficient of $(1-2x)^2$ is Fisher's u. Thus the sum of Fisher's u scores gives a perfect test for the presence of linkage, though not a quite unbiased estimate of its intensity. There is, however, a further complication. In the absence of linkage the sampling distribution of Σu is not normal, but positively skew. So a high positive value gives a rather exaggerated estimate of the significance of the evidence for linkage. When allowance is made for this (Haldane 1946) most, but not quite all, of the data formerly regarded as significant remain so.

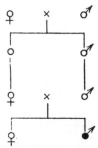

FIGURE 4. ♂, male homozygous for a partially sex-linked recessive. The sex of his maternal grandparent is irrelevant.

The Croonian Lecture was originally intended to be on 'local motion', and I shall therefore illustrate recessive partial sex-linkage by discussing spastic paraplegia, a disease in which the tonus of the limb muscles gradually increases until walking becomes impossible. Bell (1939) collected forty-four pedigrees in which one or more children of normal parents were affected with this disease.

Applying Fisher's method we have $\Sigma u = 1256 \pm 231$, and the estimated frequency of recombination is 17·5 %. The significance is not as high as it appears, since the sampling distribution in the absence of linkage is very skew positively. But it is not in doubt. It is wholly possible that while most of the families are segregating for a partially sex-linked gene, others are segregating for an autosomal one. To determine whether this is so, about five times the present number of families would be required, and it would be necessary to devise new statistical methods. There is also a suggestion, both from the results of the direct method applied to the progeny of cousin marriages, and the indirect method based on Fisher's 'u' scores, that a few cases diagnosed as spastic ataxia, and perhaps even as Friedreich's ataxia, may be due to partially sex-linked genes. The large majority are not (Haldane 1941 a).

On the basis of such statistical work I located seven genes on this segment. The standard errors of their distances are so large that I do not think that a map is worth publishing. However, the loci of dominant and recessive genes for retinitis pigmentosa, probably allelomorphs, lie about 30 units from the sex-determining or differential segment, while the loci of five other genes, namely those for achromatopsia, epidermolysis bullosa, xeroderma pigmentosum, spastic paraplegia, and Oguchi's disease, seem to lie between 21 and 44 units from it. In addition a lethal gene for convulsive seizures with mental deterioration is probably located in this segment (Snyder & Palmer 1943) while a gene concerned in some cases of hare-lip and cleft-palate may be so (Philip & Mather 1940).

The next step in this investigation will perhaps be the discovery of families in which colour-blindness is segregating along with a partially sex-linked gene such as those for spastic paraplegia or xeroderma pigmentosum. In the latter case in particular we should be able to detect linkage in doubly heterozygous females between genes in the two segments of the X, and thus to produce a unified map. Even more valuable would be the discovery of a gene as common as those for colour-blindness and anomalous matching, in either section of the X chromosome. The most hopeful field for such a discovery is among the antigens.

I have deliberately restricted my own work on human linkage to the sex chromosomes, because in every satisfactory pedigree sex as well as abnormality is recorded. Other workers more industrious than myself have looked for linkage between autosomal genes. Since man has twenty-three pairs of autosomes, the probability that a particular pair of loci will lie in the same chromosome is of the order of $\frac{1}{23}$, though rather more, because the chromosomes are of unequal size. On the other hand a recombination value of over 25 % is unlikely to be detected until data accumulate in very considerable quantity.

Two genes are very possibly linked with those for the blood-group antigens. These are the recessive gene for phenylketonuria (Penrose 1946a) and the partially dominant gene for allergy (Finney 1940; see also Finney 1941, 1942). A phenylketonuric apparently lacks some enzyme concerned in the metabolism of phenylalanine, and consequently excretes up to 1 g./day of phenylpyruvic acid. As a further consequence there is a shortage of material for melanin formation, and the hair is of a lighter colour than that of other members of the family. Far more important, thought is impossible. Phenylketonurics are usually idiots or imbeciles, at best feeble-minded.

Penrose (1946a) has developed a statistical method for dealing with such cases, when the children in a family can be examined, but the parents cannot. Consider a series of sibships in which some members are normal, others are phenylketonurics, some have the B antigen, i.e. belong to group B or AB, others belong to other groups. Consider what is to be expected if the loci concerned are on the same chromosome. In some sibships the recessive genes for phenylketonuria and absence of B antigen have been in the same chromosome in both parents. In these, the sibs who are phenylketonurics will probably not possess antigen B, or more accurately, will be less likely than normals to possess it. If the genes are in different chromosomes in one or both parents, those who are phenylketonurics will probably possess antigen B. Over a group of sibships there will be no association between the two characters, but in any particular sibship they will be positively or negatively associated.

Now if we observe any pair of sibs, they must fall into one of the nine categories of table 6.

TABLE 6. THE NINE POSSIBLE TYPES OF SIB-PAIRS IN A SIBSHIP SEGREGATING FOR
THE B ANTIGEN AND PHENYLKETONURIA. EXPLANATION IN TEXT

	BB	Bb	bb	total
NN	$a_{11}+$	$a_{12}-$	$a_{13}+$	c_1
NP	$a_{21}-$	$a_{22}+$	$a_{23}-$	c_2
PP	$a_{31}+$	$a_{32}-$	$a_{33}+$	c_3
total	b_1	b_2	b_3	1

Here N and P denote normality and phenylketonuria B and b the presence and absence of the B antigen. Thus a pair of sibs of whom one is normal and one phenyl-ketonuric, but both belong to group O, falls into the category $NP.bb$. If the two gene pairs are located in different chromosomes the expected number E_{23} of sib-pairs in this category will be the product of the numbers of NP pairs and bb pairs, divided by the total, that is to say $b_3 c_2/s$. If they are in the same chromosome it will be less. In general, linkage will increase the numbers in the categories labelled $+$ in table 6, and diminish it in the remainder. Penrose (1946a) found the figures of table 7.

TABLE 7. NUMBERS OF SIB-PAIR TYPES FOUND BY PENROSE, WITH
EXPECTATIONS IN THE ABSENCE OF LINKAGE

	BB		Bb		bb		total
NN	17	(9·125)	11	(16·947)	89	(90·928)	117
NP	10	(15·365)	34	(28·535)	153	(153·100)	197
PP	1	(3·510)	7	(6·518)	37	(34·972)	45
total	28		52		279		359

In this table the bracketed numbers are the expectations in the absence of linkage. Thus $9{\cdot}125 = \dfrac{117 \times 28}{359}$. In the absence of linkage, Penrose finds that

$$\xi = \left[\frac{a_{11}}{E_{11}} + \frac{a_{13}}{E_{13}} + \frac{a_{31}}{E_{31}} + \frac{a_{33}}{E_{33}} + \frac{4a_{22}}{E_{22}} - \frac{2a_{12}}{E_{12}} - \frac{2a_{21}}{E_{21}} - \frac{2a_{23}}{E_{23}} - \frac{2a_{32}}{E_{32}} \right]$$
$$\div \left[s \left(\frac{1}{b_1} + \frac{4}{b_2} + \frac{1}{b_3} \right) \left(\frac{1}{c_1} + \frac{4}{c_2} + \frac{1}{c_3} \right) \right]^{\frac{1}{2}}$$

is normally distributed with mean zero and unit variance. A positive value indicates linkage. In this case the value is $+1{\cdot}51$ which is not in itself significant. But Penrose informed me that when the O antigen and the two forms of the A antigen are also taken into consideration, the value rises to $1{\cdot}78$. Since this is in the direction expected on theoretical grounds, the probability of obtaining so large a value by chance is $0{\cdot}046$ or one in twenty-two. Such a value is usually taken as on the borderline of significance. A few more families may well establish this linkage conclusively.

Finney (1940) used modifications of Fisher's u statistics, and concluded from Zieve, Wiener & Fries' (1936) data that the gene for allergy was linked with those for the blood groups. However, it must be emphasized that the genetics of allergy are not so simple as those of phenylketonuria. The probability of obtaining so large a deviation in the expected direction is 0·04, after correcting for skewness. His result and Penrose's are weaker evidence for linkage than appears at first sight because man has twenty-three pairs of autosomes, so the *a priori* probability of linkage is only about 0·04. This means that considerably stronger evidence for linkage is required than in *Drosophila melanogaster* where the *a priori* probability is about 0·5.

Burks (1939) published preliminary results which very strongly suggest linkage between genes for hair colour and defective teeth. Unfortunately only a statistical summary was given, and as I have pointed out elsewhere (Haldane 1941 b) other explanations besides linkage are possible, though perhaps not very likely. She also obtained evidence of linkage between genes for myopia and eye colour. Penrose (1935) and Rife (1941) obtained suggestions of linkage between blood groups and hair colour, and interdigital pattern and left-handedness respectively.

Finally, Kloepfer (1946) has made a most comprehensive study involving nineteen characters, and obtained evidence suggesting a number of linkages. The most impressive are those between eye colour and flare (or projection) of the ears, and between ability to taste phenyl-thio-urea and ear size. Unfortunately nothing is yet known as to the genetics of ear size and structure.

Finally, there is massive negative evidence showing that various genes, notably those for blood groups, blood types (M and N), and ability to taste phenyl-thio-urea, are not linked with one another nor with sex. Such work is inevitably tedious, but it is striking how long a time elapsed before linkage was discovered in poultry or peas, and how rapidly knowledge accumulated once the first linkages were discovered.

Up till now we have considered the behaviour of genes in so far as they reproduce their like (or perhaps better, are copied exactly) at each nuclear division. When this does not happen, a new type of gene arises which generally, but by no means always, reduces the fitness of the organism either (a) at once, if it is a dominant; (b) when it appears in a male, if it is a sex-linked recessive; or (c) when two genes of the new type are contributed by different parents to the same zygote, if it is an autosomal recessive. This process of change is called mutation. Clearly it may be due to a failure of the copying process, or to a change induced in the model between copyings by physical means such as X-rays or chemical means such as $\beta\beta'$-dichlorethyl disulphide. Mutation occurs spontaneously, that is to say under normal conditions, in all organisms so far studied; but as it is a rare process, it can only be measured when vast numbers are available. The rate was first measured in *Drosophila melanogaster*, then in *Zea Mays*, and finally by Gunther & Penrose (1935) and Haldane (1935) in man. More exact estimates, fully confirming these figures, have been made in Denmark in the last 6 years.

The rate can be measured directly, as was done by Mørch (1941) for achondroplasia, or chondrodystrophy. This is the condition found in the familiar short-legged type of dwarf. In 1938 there were eighty-six such dwarfs in Denmark among 3,800,000 people. They have a very low fertility, but when they breed, about half their off-spring are similar dwarfs. The large majority of dwarfs, however, are the offspring of normal parents with no dwarfs in their families. It is clear that the gene for dwarfism arises sporadically by mutation. Out of 132,761 children born of normal parents in hospitals in Copenhagen and Lund over a period of 21 years, eleven were dwarfs of this type. This gives a mutation rate of $4 \cdot 1 \times 10^{-5} \pm 1 \cdot 2 \times 10^{-5}$ per normal gene per generation, or about $1 \cdot 2 \times 10^{-6}$ per year, since the mean age of normal parents is 35 years. The probability that the true value should be as low as 10^{-5} per generation is $0 \cdot 0011$, the probability that it should be as high as 10^{-4} is $0 \cdot 0001$, so the order of magnitude is certain.

Mørch also estimated the mutation frequency indirectly. Most such dwarfs die at or within 2 days of birth, and a number more in the first year of life, but after this their expectation of life is normal. If 80 % die in the first year, which is his estimate, there would be 415 such dwarfs in Denmark but for this mortality, or a frequency of $1 \cdot 09 \times 10^{-4}$. Now 108 dwarfs had 27 children, and their 457 normal sibs had 582, thus their fitness from a Darwinian point of view is $\frac{27}{108} \times \frac{457}{582}$, or $f = 0 \cdot 204$. That is to say in each generation natural selection effectively eliminates 80 % of the dominant genes, and but for mutation there would be no dwarfs left on earth within seven generations, or say two centuries, if Danish figures are typical. However, the two processes are in approximate equilibrium. So if x is the frequency at birth, the mutation rate $\mu = \frac{1}{2}(1-f)x = 4 \cdot 3 \times 10^{-5}$. The factor $\frac{1}{2}$ arises because we are dealing with a population of chromosomes equal to twice the population of human beings. The two estimates agree very well, but the second is much less accurate, since it depends on the figure for the infant mortality. Mørch, using a rather different argument, gets $4 \cdot 8 \times 10^{-5}$.

Professor Penrose has pointed out to me that Mørch's data are open to three criticisms. In some pedigrees, though not in any of those which he collected, there is evidence that the gene for achondroplasia can fail to manifest itself, as in Richsbieth's (1912) pedigree 608. The cases where two normal parents had more than one achondroplasic child may be due to this cause or to a mutation at an early stage in the development of a gonad. A correction for this possibility makes very little difference to the estimate of the mutation rate, since the gene is detected on its first appearance, even if this be occasionally delayed. Secondly, Mørch did not personally examine all the infants, and it is possible that some may have been wrongly diagnosed. This is plausible, since he himself failed to confirm the diagnosis of achondroplasia made by another worker in a Norwegian family. Finally, the frequency increases with parental age to an extent inexplicable if all the dwarfs born of normal parents are mutants, but explicable if some of them are due to bad prenatal conditions, as with mongoloid imbeciles. It may therefore well be that Mørch's figure is too high. But the true value is almost certainly above 10^{-5}.

For most diseases only the indirect method is available. Andreassen (1943) has applied it to haemophilia in Denmark. However, I believe (Haldane 1947b) that his calculations give rather too low a result. Haemophilia is due to a sex-linked recessive gene. Hence only about one-third of the genes for haemophilia in a population are exposed to natural selection at any moment. More accurately, if μ and ν are the mutation rate in the female and male sexes respectively,

$$2\mu + \nu = (1-f)\,x,$$

where x is the frequency in males at birth. Now there were just eighty-one haemophilics alive in Denmark in 1941, and their mean life is 18 years compared with 55 for Danes in general. So $x = 1\cdot 33 \times 10^{-4}$. The fitness f, that is to say the mean number of progeny, compared with that of the population in general, appears to be $0\cdot 28$. I have criticized Andreassen's much higher figure, $f = 0\cdot 59$. It follows that $2\mu + \nu = 9\cdot 6 \times 10^{-4}$, or a mean mutation rate of $3\cdot 2 \times 10^{-4}$. Now if the mutation rates were equal in the two sexes, i.e. $\mu = \nu$, nearly a third of all haemophilics would be single cases due to mutation in homozygous mothers. Andreassen has shown that the gene for haemophilia is not completely recessive. Heterozygous women sometimes bleed abnormally, but always have an abnormally long coagulation time, by which they can be detected. Using this technique he has not yet found a case where the mother of a haemophilic was homozygous. Doubtless such a case will be found. But it can be concluded that ν is much larger than μ, very likely ten times as large. If this is correct we should have, very roughly, $\nu = 8 \times 10^{-5}$, and $\mu = 8 \times 10^{-6}$.

Similar estimates are available for three other conditions. Gunther & Penrose (1935) found 4–8×10^{-6} for epiloia, Philip & Sorsby (unpublished) found $1\cdot 4 \times 10^{-5}$ for retinoblastoma, and Møllenbach (according to Kemp 1944) finds 5–10×10^{-6} for aniridia (Kemp's figure of double this value appears to be the mutation rate per zygote, not per gene). The median rate is about 10^{-5}. Unfortunately, this method cannot be applied to autosomal recessive conditions.

Pätau & Nachtsheim (1946) have estimated the mutation rate of the autosomal dominant gene *Pg* which is responsible for the Pelger anomaly, a failure of segmentation of the nuclei of polymorphonuclear leucocytes. *Pg*/+ individuals are thought to be less resistant to disease than +/+, whilst by analogy with rabbits, it is suggested that *Pg*/*Pg* is a lethal genotype. However, too little is known as to the viability of heterozygotes to allow an indirect estimate of the mutation frequency. The authors estimated the frequency of the condition as $0\cdot 001$, and found that out of twelve persons showing the Pelger anomaly, and both of whose parents could be examined, one parent was affected in ten cases, neither in two cases. This gives a mutation rate of $\frac{1}{2} \times 10^{-3} \times \frac{2}{12}$, or 8×10^{-5}. Even if the frequency were accurately known, and if 120 cases had been examined instead of twelve, this estimate would be somewhat high, simply because the cases with both parents living are a selected group, and include a higher fraction of cases with normal parents than of those with one affected. However, the order of magnitude agrees well with the figures given above, and is unlikely to be incorrect by a power of ten.

It is certain that these figures are not representative. Consider a well-known and unmistakable dominant such as lobster claw. Five families with this gene are known in England. They are quite fertile, but presumably their fitness, or net fertility, is a little below the average, or the condition would be commoner. Lewis & Embleton's (1908) pedigree goes back to a son of allegedly normal parents born in 1793. In the case of such a conspicuous abnormality mutation is a far likelier explanation than adultery. But it is extremely doubtful whether the mutation occurs in Britain once in 10 years. Five isolated cases were described in Britain between 1895 and 1918. That is to say its mutation frequency is of the order of 10^{-7} per generation. This is probably a much more representative figure than those of 10^{-5} or over. Unfortunately, the indirect method becomes quite unreliable when, as in this case, the fitness is near unity. Finally, we have such cases as that of the 'porcupine men' of the Lambert family (literature, see Cockayne 1933), a most striking dominant mutation, perhaps a translocation, carried by the Y chromosome. This has only been recorded once, and would have stood a good chance of being recorded in any civilized country, in the last 2000 years. It was twice described in the *Philosophical Transactions* of this Society. The mutation rate is probably below 10^{-10}.

A man or woman consists of about 2^{48} cells, that is to say a representative cell is separated from the fertilized ovum by about fifty mitoses. The primordial ova are all formed at birth, and do not undergo further mitoses. A man may produce 2^{40} spermatozoa in a lifetime, so the mean number of mitoses is somewhat greater in the male than the female germ-line, but probably not over 100 in the former.

Thus a mutation rate even of the order of 10^{-4} means that the gene-copying process, at worst, goes wrong about once in a million times, whether as the result of a failure of copying, or of a change in between two copying processes. A similar degree of accuracy in crystal growth would give a crystal with under ten flaws per millimetre, and 10^{10} successive flawless layers would give a perfect crystal several metres in length. The living substance of our bodies is clearly far more accurately copied than the successive layers of a crystal.

In *Drosophila* the natural mutation rate is of the order of 10^{-6} to 10^{-5} per generation for the more mutable loci, such as that whose mutation produces a white eye, and considerably lower for the stabler genes. Natural mutation is increased about threefold by a rise of $10°$ C, and is therefore largely due to a chemical reaction. As a generation in *Drosophila melanogaster* takes about 10 days and a fly contains about 2^{23} cells, while the mutation rate of the more labile genes is about one-fifth of that of man per generation, it follows that human mutation rates are about twice those of *Drosophila* per nuclear division, and about one two-hundredth of those of *Drosophila* per day, though the body temperature is about $13°$ higher. It has been calculated that natural radiations and particles of high energy will account for only $0·001$ of the mutations in *Drosophila*. It is clear that if so they may account for about a fifth of those in man, and in view of the uncertainty of our knowledge as to the efficiency of particles from K^{40} and cosmic rays in producing mutations, and the different radiosensitivity of different genes, it is quite possible that radiation may account for most human mutation. Mørch found that the rate of mutation to achondroplasia increased with age, but it was not clear whether maternal or paternal age was most important. If this finding is confirmed it suggests a cumulative effect, either of radiation, or of successive nuclear divisions, during a lifetime. The apparently higher rate in males suggests that the number of nuclear divisions may be an important factor in human mutation. To sum up, there are three possible known causes of mutation, a chemical reaction with a temperature coefficient, radiation, and imperfections of copying, which might have a positive or negative temperature coefficient. The first predominates in *Drosophila*, the second or third probably does so in man. There must be about a thousand achondroplasic dwarfs in Britain. If the ages of their parents at their births were determined, it would be possible to decide between these alternatives, since the egg of a woman of 45 has undergone no more nuclear divisions than that of a girl of 15. It is worth remarking that it is quite practicable to obtain data of this kind on populations of 40 million human beings, and wholly impracticable to do so on 40 million of any other mammal.

The mutation rate is probably more or less adaptive. Too high a mutation rate would flood a species with undesirable mutations, too low a one would probably slow down evolution. Man and *Drosophila melanogaster* have about the same rate per generation, and this could not be increased ten times without a very great loss of fitness (Haldane 1937). Other species such as five species of *Sciara* (Metz 1938) have far lower rates per generation though not necessarily less than the human rate per day. But it is doubtful whether the human rate could be lowered much further, since a substantial fraction of it is due to natural radiation. In fact a very great prolongation of human life, or at any rate of the reproductive period, might be incompatible with the survival of the human species.

I hope that, in this brief survey, I have shown that human genetics has reached the stage when it can claim to be a branch of biology with its own peculiar problems and methods. I have only dealt with a few of them. This lecture could equally well have been devoted to the human antigenic structure, to human prenatal physiology, or to variation in human sensory and intellectual capacity, all of which a human geneticist must study. If I have confined myself to the more quantitative aspects, my excuse must be that in dealing with a branch of science where erroneous views may have important political consequences, in such a lecture as this it is desirable to concentrate on those problems where political or social bias is least likely to be effective, and where we may hope to raise a solid theoretical structure by methods like those which have been fruitful in the other branches of science.

REFERENCES

Andreassen, M. 1943 Haemofili i Danmark. *Opera ex domo biologiae hereditariae humanae Universitatis Hafniensis*, **6** (Copenhagen).
Avery, O. T., Macleod, C. M. & MacCarty, M. 1944 *J. Exp. Med.* **79**, 137–157.
Bateson, W. 1906 *Brain*, pt. 2, pp. 157–179.
Bell, J. & Haldane, J. B. S. 1936 *Proc. Roy. Soc. B*, **123**, 119–150.
Bell, J. 1939 *Treas. Hum. Inher.* **4**, pt. 3.
Birch, C. L. 1937 *Illinois Med. Dent. Monogr.* **1**, 4.
Burks, B. 1939 *Proc. Nat. Acad. Sci., Wash.*, **24**, 512–514.
Blaurock, G. 1932 *Münch. Med. Wschr.* **74**, 1552–1556.
Clausen, J. 1932 *Hospitalstidende*, **75**, 196–206.

Cockayne, F. A. 1933 *Inherited abnormalities of the skin and its appendages.* Oxford: University Press.

Crome, W. 1933 *Dtsch. Z. ges. gerichtl. Med.* 21, 435–450.

Dahr, P. & Bussmann, R. 1938 *Dtsch. Med. Wschr.* 64, 818–821.

Dahr, P. 1940 *Z. ImmunForsch.* 97, 168–188.

Ephrussi, B. 1942 *Cold Spring Harbor Symp.* 10, 40–48.

Finney, D. J. 1940 *Ann. Eugen.* 10, 171–214.

Finney, D. J. 1941 *Ann. Eugen.* 11, 10–30, 115–135.

Finney, D. J. 1942 *Ann. Eugen.* 11, 224–244.

Fisher, R. A. 1936 *Ann. Eugen.* 7, 87–104.

Gunther, E. R. & Penrose, L. S. 1935 *J. Genet.* 31, 413–430.

Haldane, J. B. S. 1935 *J. Genet.* 31, 317–326.

Haldane, J. B. S. 1936 *Ann. Eugen.* 7, 28–57.

Haldane, J. B. S. 1937*a J. Hered.* 28, 58–60.

Haldane, J. B. S. 1937*b Amer. Nat.* 71, 337–348.

Haldane, J. B. S. 1938 *Ann. Eugen.* 8, 255–262.

Haldane, J. B. S. 1941*a J. Genet.* 41, 141–144.

Haldane, J. B. S. 1941*b New paths in genetics.* London: Allen and Unwin.

Haldane, J. B. S. 1946 *Ann. Eugen.* 13, 122–134.

Haldane, J. B. S. 1947 *Ann. Eugen.* 13, 262–271.

Haldane, J. B. S. 1948 Unpublished.

Haldane, J. B. S. & Smith, C. A. B. 1947 *Ann. Eugen.* 14, 10–31.

Holford, F. F. 1938 *J. Infect. Dis.* 63, 287–297.

Hirszfeld L. & Kostuch Z. 1938 *Schweiz. Z. Path. u. Bakt.* 1, 23.

Hoogvliet, B. 1942 *Genetica,* 23, 94.

Hyman, H. S. 1935 *J. Immunol.* 29, no. 3.

Kemp, T. 1944 *Acta Path. microbiol. Scand.* suppl. LIV.

Kloepfer, H. W. 1946 *Ann. Eugen.* 13, 35–71.

Landsteiner, K. & Levine, P. 1928 *J. Exp. Med.* 47, 757–775.

Landsteiner, K. & Wienin, A. S. 1941 *J. Exp. Med.* 74, 309–320.

Lattes & Garrasi 1932 *Atti IV Congr. Naz. Microbiol.* p. 146.

Lewis, T. & Embleton 1908 *Biometrika,* 6, 26.

Mather, K. & Philip, U. 1940 *Ann. Eugen.* 10, 403–416.

Matta, D. 1937 *Faculty Med. Publ. Egypt. Univ. Cairo,* no. 11.

Metz, C. W. 1938 Cooperation in research, *Carn. Inst. Wash. Pub.* 501, 275–294.

Mørch, T. 1941 *Chondrodystrophic dwarfs in Denmark.* Copenhagen: Eijnar Munksgaard.

Moureau, P. 1935 *Rev. Belg. Sci. Med.* 7, 541–588.

Pätau, K. & Nachtsheim, H. 1946 *Z. Naturforsch.* 1, 345.

Pearson, K., Nettleship & Usher 1913 *Drap. Co. Res. Mem. Biom.,* Series IX.

Penrose, L. S. 1935 *Ann. Eugen.* 6, 133–138.

Penrose, L. S. 1946*a Ann. Eugen.* 13, 25.

Penrose, L. S. 1946*b J. Hered.* 37, 285.

Philip, U. & Mather, K. 1940 *Ann. Eugen.* 10, 403–416.

Pipkin, A. C. & Pipkin, S. 1945 *J. Hered.* 36, 313.

Richsbieth, H. 1912 *Treas. Hum. Inher.* 1, 355–553.

Rife, D. C. 1941 *Science,* 94, 187.

Schiff, F. 1933 *Dtsch. Z. ges. gerichtl. Med.* 21, 404–434.

Sjögren, T. 1931 *Hereditas,* 14, 197–425.

Snyder, L. H. & Palmer, D. M. 1943 *J. Hered.* 34, 207–212.

Taylor, G. L. & Prior, A. M. 1939*a Ann. Eugen.* 9, 18–44.

Taylor, G. L. & Prior, A. M. 1939*b Ann. Eugen.* 9, 97–108.

Weinberg, W. 1927 *Z. indukt. Abstamm.- u. VererbLehre,* 48, 179–228.

White, M. 1940 *J. Genet.* 40, 403–438.

Wiener, A. S. 1943 *Blood groups and transfusion.* Springfield: Charles C. Thomas.

Wiener, A. S. & Sonn, E. B. 1943 *Genetics,* 28, 157–161.

Wiener, A. S. & Vaisberg, M. 1931 *J. Immunol.* 20, 371–388.

Zieve, M. A., Wiener, A. S. & Fries, J. 1936 *Ann. Eugen.* 3, 163–178.

Part II
GENES, CHROMOSOMES, AND DISEASE

Editors' Comments
on Papers 11 Through 20

Earlier in this century when genetics was still young, it was common to conceive of the relationship of genes to chromosomes as that of beads on a string. This conception emerged from a number of lines of evidence, two of which were especially prominent, namely, the linear order of genes suggested by the phenomenon of linkage and the cytological studies of meiosis and mitosis. The fact that we now know this analogy to be less than apt, however, does not denigrate the ingenuity revealed by those early investigators who provided the first compelling evidence that genes and chromosomes were intimately related one to another. While Montgomery (1901) seems to have apprehended much of the evidence, Sutton (1902, 1903) was the first to set forth clearly in all its significance the cytological explanation of the Mendelian phenomenon that is offerred by the behavior of the chromosomes. It was another quarter of a century, though, before Creighton and McClintock (1931) and Curt Stern (1931) were to show almost simultaneously that chromatid exchange was the mechanism of recombination.

Throughout these years human cytogenetics floundered. It is arguable that human tissues are more fragile, more sensitive to autolysis if not rapidly fixed and that the early missteps reflect the intractableness of the tissues to the cytological techniques then current. Be this as it may, it was long unclear whether human beings had forty-seven chromosomes (Winiwarter 1912), or forty-eight (Painter 1923), or both (Oguma and Kihara 1923). This uncertainty was resolved when Tjio and Levan (1956) demonstrated unequivocally that the number was forty-six. Contemporary cytogeneticists are inclined to regard the flourishing of human cytogenetics that has followed this demonstration as due largely to four developments: (1) the study of peripheral leucocytes rather than solid tissues and the uses of (2) hypotonic saline solution to promote separation of the chromosomes upon squashing of the cell, (3) colchicine or its analog, methyldeacetylcolchicine, to suppress spindle formation, and (4) phytohemagglutinin to enhance the number of dividing cells.

Interestingly, the cytological value of all of these techniques save phytohemagglutinin had been known for years prior to 1956, but for some reason they had not been brought to bear simultaneously on the study of human chromosomes. Tage Kemp (1929) had shown that the "buffy coat" of peripheral blood was relatively rich in cells undergoing division, and Chrustschoff (1935) and Andres and Navashin (1935; see also Andres and Jiv 1936) parlayed this observation into a formidable series of drawings of human chromosomes to be seen in one of the early volumes of *Doklady Acad. Sci. USSR*, before the Maxim Gorky Institute and Russian genetics was reduced to a shambles. Yet all of these drawings were in error, at least in the number of chromosomes the investigators purported to see. Why? How was it possible for this error to have been perpetuated for almost four decades (as we have indicated, Winiwarter as early as 1912 had suggested that human beings had forty-eight chromosomes)?

We have selected two papers to represent the ferment that followed the developments just cited. In 1959, a mere three years after Tjio and Levan had established the normal diploid number, three chromosomal abnormalities with obvious clinical manifestations—namely, Down's, Turner's and Klinefelter's syndromes—were described by Lejeune, Gauthier, and Turpin (Paper 11), Ford et al. (1959), and Jacobs and Strong (1959), respectively. Much of the early impetus given to human cytogenetics by the developments just cited was expended on a search for other chromosomally determined disorders. As a result, many are now known. Most are rare, but they include two different syndromes resulting from a supernumerary autosome (trisomy 13, trisomy 18) in addition to Down's syndrome (trisomy 21), of course. Although no viable and generally accepted abnormality associated with the loss of an entire autosome has yet been reported, at least three syndromes resulting from the partial deletion of an autosome (partial deletion of short arm of 5, short arm of 18, long arm of 18) are well established. It is also known that some 20 percent of spontaneously aborted fetuses have abnormal chromosome numbers, and among the latter triploidy is surprisingly common, as indicated by Carr (Paper 12; but see also Edwards, Yuncken, Rushton, Richards, and Mittwoch 1967; WHO 1966). It has been suggested, in fact, that triploidy may be one of the commonest causes of natural death between conception and puberty.

Impressive as the list may be, an even greater number of abnormalities of the X and Y chromosomes exists. Presumably such departures from normality are less life-threatening, although not without clinical consequence. Here, variation in number of X chromosomes from 1 to 5, and in Y chromosomes from 0 to 3 have been documented. Of these, none has received more attention than the XYY male. First described by Hauschka and his colleagues (1962), it was not until Jacobs and her associates (1965) reported persons of this genetic constitution to be larger than average in stature and of a violent and aggressive temperament that interest in the XYY male heightened. Further attention has been centered on this defect by several court trials where it has been maintained that the defendant was not accountable for his actions because of this chromosomal abnormality. Courts have held quite differently in Paris, Chicago, and Sydney, and much legal concern has been expressed over the implications of rulings favorable to the defendants.

Another frequent cytologic abnormality is the existence within one person of two or more cell lines with differing numbers of chromosomes or differences in the structure of one or more chromosomes. Such persons are said to be mosaics—that is, numerical in the first instance and structural in the second. Especially common in this regard are differences in numbers of sex chromosomes or mosaics with at least one cell line containing a structurally abnormal X-chromosome. Much less common are mosaics involving autosomal chromosomes. Ford (1969), for example, has reported that no less than 28 of 163 cases of Turner's syndrome in four large studies proved to be structural mosaics of various kinds.

Of late, interest has moved away from these gross, naturally occurring abnormalities and converged on a search for minor variations among individuals and groups in chromosome morphology; agents, physical and chemical, that disturb or alter chromosome structure generally through "breaking" chromosomes or chromatids; and finally, a better understanding of chromosomal fine structure and duplication. Among the minor variations recorded between individuals and groups are the number and size of secondary constructions and satellites as well as the length of chromosomes and their degree of coiling or condensation. A fairly extensive literature on these variations and efforts to associate them with disease now exists (see Hamerton 1971 for a recent review), but much of the evidence that supports the existence of the variants as well as their purported association with disease is fairly shaky. Typical of these minor variants are differences in the length of the Y chromosome that Cohen, Shaw, and MacCluer (1966) have shown to vary significantly among races and the inherited abnormality known as "uncoiler" (Cooper and Hernits 1963). Persons who possess this defect have first chromosomes that are markedly longer than normal. The increase appears to stem from an alteration in the coiling structure of the portion of the long arm adjacent to the centromere. The recognition of this abnormality has already proven a boon to investigators interested in human linkage relationships. The genes responsible for the Duffy blood group system were localized on the first chromosome because of the "uncoiler" (Donahue, Bias, Renwick, and McKusick 1968).

Although it has long been known that a variety of chemicals as well as ionizing radiation can produce both chromatid and chromosome aberrations in many plants and animals, the search for environmental chromosome-damaging agents in man had to await the technical developments previously described. Once it was possible, however, to culture leukocytes conveniently, the hunt was on. Not surprisingly, the first damaging agent to be described was ionizing radiation, which was shown to induce aberrations in cultured cells as well as *in vivo*. Bender and Gooch (1962) demonstrated chromosome abnormalities in cells cultured from persons irradiated in the course of the Y-12 criticality accident at Oak Ridge, and in the same year, Buckton et al. (1962) discovered chromosomal abnormalities among persons with ankylosing spondylitis irradiated as much as twenty years before the cytogenetic investigation. We now know that the survivors of the atomic bombings of Hiroshima and Nagasaki (Bloom, Neriishi, Kamada, Iseki, and Keehn 1966) as well as many persons who experienced procedures such as intravenous pyelograms or barium examination of the gastro-intestinal tract (Bloom and Tjio 1964) exhibit an increased frequency of cells with chromosomal aberrations.

Many geneticists see as more threatening, however, the disclosure that an increasing frequency of aberrations is to be seen in persons following exposure to measles or after innoculation with yellow fever vaccine (Harnden 1964). Numerous chemical agents ranging from industrially encountered ones—ambient benzene (Tough and Court Brown 1965)—to antibiotics—streptonigrin (Cohen, Shaw, and Craig 1963; Cohen 1963)—are recognized to produce chromosomal damage. Concern over these cellular consequences has mounted steadily as the list of drugs and chemicals that damage chromosomes and to which man is exposed has grown. The day rapidly approaches when guidelines to such exposure will be set just as they have been set in the case of ionizing radiation. Indeed, current guidelines of the Environmental Protection Agency and the Food and Drug Administration that govern licensing of drugs, food additives, pesticides, and the like already constitute a major step in this direction.

Interest in chromosome fine structure and duplication has been spurred by still other new developments. Significant among these are autoradiography (for a review, see German 1967), new staining techniques that reveal chromosomal substructure not previously recognized (see Dutrillaux and Lejeune 1975), and methods to anneal DNA.

Almost two decades ago, Barski, Sorieul, and Cornefert (1960) called attention to the occurrence *in vitro* of cell fusion or hybridization. This phenomenon, though infrequent, patently presented almost limitless possibilities for the genetic analysis of somatic cells. Evidence of cell hybridization in the earlier work was all of a karyotypic nature, but Littlefield (1964) demonstrated that biochemically marked cell lines could be used to select for hybrid cells. It has since been shown that interspecific hybridizations involving many species could occur and would, in fact, yield populations of hybrid cells capable of indefinite serial propagation. The first such interspecific hybrid involving human cells was reported in 1967 by Weiss and Green (Paper 13). It resulted from crossing human diploid fibroblasts with an established mouse fibroblast line. The hybrid line underwent a noncoordinate and gradual, but preferential loss of chromosomes. The human characteristics of the hybrid's phenotype could be related to the number of chromosomes of the human complement retained. More specifically, the human surface membrane antigens detectable in the hybrid cell membrane depended on the number of human chromosomes in the hybrid cell.

These investigators then showed that when the human diploid strain was crossed with a thymidine kinase deficient mouse fibroblast strain, the resulting hybrid cell line would grow on a medium containing aminopterin whereas the deficient mouse cell line would not. It was possible, however, through the use of a medium containing 5-bromo-deoxyuridine to select variants in the hybrid cell line that lacked the capacity to grow on aminopterin. These variants had invariably lost a chromosome of a particular group (every case of 34 analyzed), and Weiss and Green concluded that it was this chromosome that contained the thymidine kinase gene. This line of inquiry has already borne much additional fruit. The use of human-mouse hybrids has markedly accelerated the assignment of human genes to linkage groups and to identifiable chromosomes, a process that had previously been inordinately slow. A fine recent review of the mapping of the human chromosome can be found in McKusick and Ruddle (1977).

Another area of human genetics that has profited greatly from postwar technological developments is human somatic cell genetics. Although efforts at cell culture are decades old, the modern era may be said to have begun shortly after World War II with the work of Eagle and his colleagues. This work more clearly defined the nutrient requirements of cultured cells and ultimately, in concert with the work of others, standardized and simplified the growth of mammalian cells *in vitro*. Added stimulus to research in this area came from the belief—a fairly widespread one—that the techniques so spectacularly successful in microbial genetics should be applicable to cultured mammalian cells. Clearly, if true, the somatic cell could serve as a substitute for the whole organism in many genetic analyses. The aims of somatic cell genetics have been said to be fourfold: to describe the biology and biochemistry of the cultured mammalial cell; to map the human genome through analyses of crosses between the cultured somatic cells of different individuals; to determine how the mammalian genome controls the properties of the cell; and finally, to determine the circumstances under which the cell will change genetically (Krooth, Davenport, and Velasquez 1968).

One of the first genetic abnormalities known to confer a distinct and biochemically recognizable phenotype on the cultured mammalian cell—galactosemia—was described by Krooth and Weinberg (Paper 14). Galatosemia, an autosomally and recessively inherited defect, results in demonstrably different levels of cellular UDP galactose transferase. Today, several dozen different inherited abnormalities have been shown to have biochemical effects discernible at the cellular level. Some seven or so of these are associated with the X-chromosome—that is, are sex-linked. Collectively, these abnormalities include disorders of carbohydrate, amino acid, nucleotide, and lipid metabolism. In at least two instances, Hurler's syndrome and orotic aciduria, the characteristic biochemical phenotype of the mutant cell can be reversed by experimental manipulation. Seegmiller, Rosenbloom, and Kelley (Paper 15) have shown that a deficiency of the enzyme hypoxanthine-guanine phosphoribosyl transferase (HPRT) is the principal cause for the primary defect in the X-linked recessive neurological disorder described by Lesch and Nyhan (1964).

No account of these advances in somatic cell genetics would be complete without mention, however briefly, of the so-called Lyon hypothesis. It has been known for a quarter of a century that some 20 to 60 percent of cells of normal women have a darkly staining chromatin mass next to the inner surface of the nuclear membrane. This mass can be seen in few, if any, cells of normal men. We now know this mass to be an X chromosome—one that replicates late and is genetically inert. Mary Lyon (1962) was the first to propose that in mammalian cells only one chromosome is active and that the chromatin body was merely the inactive X chromosome that was heteropycnotic. This propposition was an attractive one for it would account for the absence of any evidence of genic imbalance between males and females in sex-linked genes; moreover it was testable. Evidence to support it came from observations on coat color in mice, the cytologic findings cited above, and biochemical studies of females known to be heterozygous for X-linked genes. Beutler, Yeh, and Fairbanks (1962), for example, were able to show from studies of enzyme activities that red cells of an individual heterozygous for the enzyme, glucose-6-phosphate dehydrogenase, are a mixture of two types of cells, each containing only one form of the enzyme. Other studies of both this enzyme as well as others associated with the X chromosome have confirmed and extended their findings.

Our perceptions of how genes act to give rise to inherited diseases have undergone momentous changes in this century. As we have previously indicated, in 1902 A. E. Garrod suggested that alcaptonuria was inherited as a simple recessive characteristic, and in so doing, he was the first to interpret an inherited abnormality in man in Mendelian terms. Out of this early work grew his notions of "inborn errors of metabolism" (Paper 16), which was first set forth in a series of papers that appeared in *Lancet* in 1908 and subsequently in a book with the title *Inborn Errors of Metabolism* (1923). These notions had a profound influence on subsequent work by George Beadle and Edward Tatum, out of which grew the hypothesis "one gene-one enzyme," that is, genes effect their role in development through the control of the structure and rates of syntheses of enzymes. This hypothesis was subsequently broadened to recognize that genes might control the structures and rates of syntheses of nonenzymic proteins as well.

In an elegant set of experiments, Pauling, Itano, Singer, and Wells (Paper 17) demonstrated in 1949 that sickle cell anemia was, in fact, as they put it, a "molecular disease." They were able to demonstrate that two electrophoretically different hemoglobins existed—each related to a different gene—and that persons might possess one or the other, or both, of these hemoglobins in their red blood cells. The combination of specific hemoglobins determined whether they were normal, had sickle cell anemia, or merely sickle cell trait. Beet (1949) and Neel (1949), on clinical and hematologic grounds, had earlier reached a similar conclusion, namely, that persons with sickle cell anemia were homozygous for a gene that in single dose gave rise to sickle cell trait. See also Valentine and Neel (1944) for a still earlier, analogous suggestion of the mode of transmission of thalassemia. The line of investigation that Pauling and his colleagues initiated was expanded greatly in the following decades. Soon other hemoglobins were recognized, and before the 1950s were to end, the sequence of amino acids that characterize not only normal but sickle hemoglobin were defined. Today, literally dozens of inherited hemoglobin variants are known. Most are rare, but a few reach polymorphic proportions in some populations.

Concurrently with these developments, another line of research was being exploited. In 1952 the Coris (Paper 18) succeeded in demonstrating that one form of glycogen storage disease—that is, von Gierke's disease—arose from a deficiency of a specific enzyme, glucose-6-phosphatase, which they succeeded in characterizing. Today, at least six other differently inherited glycogen storage diseases are known, and countless tens of other disorders that stem from the deficiency of specific enzymes have been described (see Stanbury, Wyngaarden, and Frederickson 1972 for a description of many).

Our final two selections for Part II fit none of the areas we have previously described well. In 1941 Levine, Burnham, Katzin, and Vogel's demonstration (Paper 19) that erythroblastosis fetalis stems from maternal-fetal isoimmunization revealed still another important role played by red cell antigens and solved the etiologic puzzle associated with a perplexing problem. Numerous other maternal-fetal incompatibilities have since been shown to produce hemolytic disease of the newborn, as this anemia of the newborn is now more commonly called. Our final selection in Part II (Paper 20) represents an effort to recognize that while the rare and simply inherited diseases, many of which have biochemical manifestations, have preoccupied human geneticists for much of this century, these disorders do not impose the burden upon contemporary society imposed by the more common disorders such as diabetes, gout, hypertension, and the like. Most of the latter are undoubtedly multifactorial in origin—that is, they result from the interplay of a number of genes at different genetic loci.

153

Generally, in the past investigators been content to attempt to estimate the relative contributions of genetic and environmental factors to the variability within these traits. They have usually assumed that there existed some scale on which genetic and environmental factors acted additively, and the number of such factors and the manner in which they interrelated could, of course, be elaborated on almost indefinitely. Available from which to estimate these contributions have been such quantities as the means and variances of particular groups associated with specifiable classes of pairs of individuals. As is increasingly evident to all, however, this general approach has not been particularly profitable, and a new method of inquiry is required that will permit the concurrent testing of genetic, environmental (ecological), medical, and even sociopsychological hypotheses. New statistical approaches are being advanced, some of which hold promise of better control or delineation of many of the sources of extraneous variation that have clouded past efforts, and old approaches (path analyses) are being resurrected. A major focus of many of these statistical approaches is the identification of "major genes"—that is, loci that contribute disproportionately to the quantitative variability that is observed. Still other investigators are seeking biochemical ways of partitioning out major contributors. The most successful such effort thus far has been Goldstein and Brown's (1973) study of hypercholesterolemia. They have shown that in some families the abnormality of lipid metabolism, commonly termed "essential hypercholesterolemia," stems from a membrane receptor defect.

REFERENCES

Andres, A. H., and B. V. Jiv. 1936. Somatic chromosome complex of the human embryo. *Cytologia* 7:371–88.

Andres, A. H., and M. S. Navashin. 1935. A morphological analysis of chromosomes in man. *Doklady Acad. Sci. USSR* 3:309–12.

Andres, A. H., and M. S. Navashin. 1936. Morphological analysis of human chromosomes. *Proc. Maxim Gorky Medicogenetical Research Inst.* 4:506–24.

Barski, G., S. Sorieul, and F. R. Cornefert. 1960. Production dans des cultures in vitro de deux sonches cellulaires en association de cellules de charactere "hybrid." *C.R. Acad. Sci.* 251:1825–27.

Beet, E. A. 1949. The genetics of sickle cell trait in a Bantu tribe. *Ann. Eugen.* 14: 279–84.

Bender, M. A., and P. C. Gooch. 1962. Persistent chromosome aberrations in irradiated human subjects. *Rad. Res.* 16:44–53.

Beutler, E., M. Yeh, and V. F. Fairbanks. 1962. The normal human female as a mosaic of X-chromosome activity: Studies using the gene for G-6-PD deficiency as a marker. *Proc. Nat. Acad. Sci.* 48:9–16.

Bloom, A. D., and J. H. Tjio. 1964. In vivo effects of diagnostic X-irradiation on human chromosomes. *New Engl. J. Med.* 270:1341–44.

Bloom, A. D., S. Neriishi, N. Kamada, T. Iseki, and R. J. Keehn. 1966. Cytogenetic investigation of survivors of the atomic bombings of Hiroshima and Nagasaki. *Lancet* ii:672-74.

Buckton, K. E., P. A. Jacobs, W. M. Court Brown, and R. Doll. 1962. A study of the chromosome damage persisting after X-ray therapy for ankylosing spondylitis. *Lancet* ii:676-82.

Chrustschoff, G. K. 1935. Cytological investigations on cultures of normal human blood. *J. Genet.* 31:243-61.

Cohen, M. M., M. W. Shaw, and A. P. Craig. 1963. The effect of streptonigrin on cultured human leukocytes. *Proc. Nat. Acad. Sci.* **50**:16-24.

Cohen, M. M., M. W. Shaw, and J. W. MacCluer. 1966. Racial differences in the length of the human Y chromosome. *Cytogenetics* **5**:34-52.

Cooper, H. L., and R. Hernits. 1963. A familial chromosome variant in a subject with anomalous sex differentiation. *Am. J. Hum. Genet.* **15**:465-75.

Creighton, H. B., and B. McClintock. 1931. A correlation of cytological and genetical crossing over in Zea mays. *Proc. Nat. Acad. Sci.* **17**: 492-97.

Donahue, R. P., W. B. Bias, J. H. Renwick, and V. A. McKusick 1968. Probable assignment of the Duffy blood group locus to chromosome 1 in man. *Proc. Nat. Acad. Sci.* **61**:949-55.

Dutrillaux, B., and J. Lejeune. 1975. New techniques in the study of human chromosomes: Methods and applications. *Adv. Hum. Genet.* **5**:119-56.

Edwards, J. H., C. Yuncken, D. I. Rushton, S. Richard, and U. Mittwoch. 1967. Three cases of triploidy in man. *Cytogenetics* **6**:81-104.

Ford, C. E. 1969. Mosaics and chimaeras. *Brit. Med. Bull.* **25**:104-09.

Ford, C. E., and J. L. Hamerton. 1956. The chromosomes of man. *Nature* **178**:1020-23.

Ford, C. E., K. W. Jones, P. E. Polani, J. D. de Almeida, and J. H. Briggs. 1959. A sex-chromosome anomaly in a case of gonadal dysgenesis (Turner's syndrome). *Lancet* i:711-13.

Garrod, A. E. 1923. *Inborn errors of metabolism.* Frowde, and Hodder and Stoughton, London.

German, J. L. 1967. Autoradiographic studies of human chromosomes. I: A review. In J. F. Crow and J. V. Neel (eds.), *Proc. III Int. Cong. Hum. Genet.* Johns Hopkins Press, Baltimore, pp. 123-36.

Goldstein, J. L., and M. S. Brown. 1973. Familial hypercholsterolemia: Identification of a defect in regulation of 3-hydroxy-3-methylglutaryl coenzyme reductase activity associated with overproduction of cholesterol. *Proc. Nat. Acad. Sci.* **70**:2804-08.

Hamerton, J. L. 1971. *Human Cytogenetics.* Vol. 1, General Cytogenetics. Academic Press, New York.

Hamerton, J. L. 1971. *Human Cytogenetics.* Vol. 2, Clinical Cytogenetics. Academic Press, New York.

Harnden, D. G. 1964. Cytogenetic studies on patients with virus infections and subjects vaccinated against yellow fever. *Am. J. Hum. Genet.* **16**:204-13.

Hauschka, T. S., J. E. Hasson, M. N. Goldstein, G. F. Koepf, and A. A. Sandberg. 1962. An XYY man with progeny indicating familial tendency to non-disjunction. *Am. J. Hum. Genet.* **14**:22-30.

Jacobs, P. A., and J. A. Strong. 1959. A case of human intersexuality having a possible XXY sex-determining mechanism. *Nature* **183**:302.

Jacobs, P. A., M. Brunton, M. M. Melville, R. P. Brittain, and W. F. McClermont. 1965. Aggressive behaviour, mental subnormality and the XYY male. *Nature* **208**:1351-52.

155

Kemp, T. 1929. Das Verhalten der Chromosomen in den Somatischen Zellen der Menschen. *Zeits. fur Mikro. u. Anat. Forschung.* **16**:1–20.

Krooth, R. S., G. A. Darlington, and A. A. Velazquez. 1968. The genetics of cultured mammalian cells. *Ann. Rev. Genet.* **2**:141–64.

Lesch, M., and W. L. Nyhan. 1964. A familial disorder of uric acid metabolism and central nervous system function. *Am. J. Med.* **36**:561–70.

Littlefield, J. W. 1964. Selection of hybrids from matings of fibroblasts in vitro and their presumed recombinants. *Science* **145**:709–10.

Lyon, M. F. 1962. Sex chromatin and gene action in the mammalian X-chromosome. *Am. J. Hum. Genet.* **14**:135–48.

McKusick, V. A., and F. H. Ruddle. 1977. The status of the gene map of the human chromosomes. *Science* **196**:390–404.

Montgomery, T. H. 1901. A study of the chromosomes of the germ cells of metazoa. *Trans. Am. Phil. Soc.* **20**:154–236.

Montgomery, T. H. 1901. Further studies of the chromosomes of the hemiptera and heteroptera. *Proc. Acad. Nat. Sci. Phila.* **53**:261–71.

Neel, J. V. 1949. The inheritance of sickle cell anemia. *Science* **110**:64–66.

Oguma, K., and H. Kihara. 1923. Études des chromosomes chez l'homme. *Arch. de Biol.* **33**:493–514.

Painter, T. S. 1923. Studies in mammalian spermatogenesis. II: The spermatogenesis of man. *J. Exp. Zool.* **37**:291–338.

Stanbury, J. B., J. B. Wyngaarden, and D. S. Frederickson. 1972. *The Metabolic Basis of Inherited Disease,* 3rd ed. McGraw-Hill, New York.

Stern, C. 1931. Zytologische-genetische untersuchungen als Beweise fur die morgansche theorie des faktorenaustauschs. *Biol. Zentralbl.* **51**:547–87.

Stern, C. 1936. Somatic crossing over and segregation in Drosophila melanogaster. *Genetics* **21**:625–730.

Sutton, W. S. 1902–03. Chromosomes in heredity. *Biol. Bull.* **4**:231–248.

Tjio, J. H., and A. Levan. 1956. The chromosome number of man. *Hereditas* **41**: 1–6.

Tough, I. M., and W. M. Court Brown. 1965. Chromosome aberrations and exposure to ambient benzene. *Lancet* i:684.

Valentine, W. N., and J. V. Neel. 1944. Hematologic and genetic study of transmission of thalassemia (Cooley's anemia: Mediterranean anemia). *Arch. Int. Med.* **74**:185–96.

WHO. 1966. Standardization of procedures for chromosome studies in abortion. *Bull. WHO* **54**:765–81.

Winiwarter, H. 1912. Études sur la spermatogenise humaine: Cellule de Sertoli, heterochromosome et mitoses de l'epithelium seminal. *Arch. de Biol.* **26**:91–189.

11

Reprinted from *Compt. Rend. Acad. Sci.* **248**:1721–1722 (1959)

ÉTUDE DES CHROMOSOMES SOMATIQUES DE NEUF ENFANTS MONGOLIENS

Jérome Lejeune, Marthe Gautier, and Raymond Turpin

La culture de fibroblastes de neuf enfants mongoliens révèle la présence de 47 chromosomes, le chromosome surnuméraire étant un petit télocentrique. L'hypothèse du déterminisme chromosomique du mongolisme est envisagée.

Chez neuf enfants mongoliens l'étude des mitoses de fibroblastes en culture récente ([1]) nous a permis de constater régulièrement la présence de 47 chromosomes. Les observations faites dans ces neuf cas (cinq garçons et quatre filles) sont consignées dans le tableau ci-après.

Le nombre de cellules comptées dans chaque cas peut sembler relativement faible. Ceci tient au fait que seules ont été retenues dans ce tableau les images ne prêtant qu'à un minimum d'interprétation.

L'apparente variation du nombre chromosomique dans les cellules « douteuses », c'est-à-dire celles dont chaque chromosome ne peut être individualisé avec une absolue certitude est signalée par de nombreux auteurs ([2]). Ce phénomène ne nous semble pas correspondre à une réalité cytologique, mais reflète simplement les difficultés d'une technique délicate.

Il nous semble donc logique de préférer un petit nombre de dénombrements absolument certains (cellules « parfaites » du tableau) à une accumulation d'observations douteuses dont la variance statistique ne dépend que de l'imprécision des observations.

L'analyse de la garniture chromosomique des cellules « parfaites » révèle chez les garçons mongoliens la présence de 6 petits télocentriques (au lieu de 5 chez l'Homme normal) et chez les filles mongoliennes de 5 petits télocentriques (au lieu de 4 chez la Femme normale).

Les cellules « parfaites » d'individus non-mongoliens, ne présentant jamais ces caractéristiques (¹), il nous semble légitime de conclure qu'il existe chez les mongoliens un petit chromosome télocentrique surnuméraire, rendant compte du chiffre anormal de 47.

Nombre de cellules examinées dans chaque cas.

Nombre de chromosomes...	Cellules diploïdes						Cellules tétraploïdes			
	Cellules « douteuses ».			Cellules « parfaites ».			Cellules « parfaites ».			
	46.	47.	48.	46.	47.	48.	–	94.	–	
Garçons. Mg 1	6	10	2	–	11	–	–	1	–	30
Mg 2	–	2	1	–	9	–	–	–	–	12
Mg 3	–	1	1	–	7	–	–	2	–	11
Mg 4	–	3	–	–	1	–	–	–	–	1
Mg 5 (*)	–	–	–	–	8	–	–	–	–	8
Filles. Mg A	1	6	1	–	5	–	–	–	–	13
Mg B	1	2	–	–	8	–	–	–	–	11
Mg C	1	2	1	–	4	–	–	–	–	8
Mg D	1	1	2	–	4	–	–	–	–	8
	10	27	8		57			3		105

(*) Cet enfant est issu d'une grossesse gémellaire. Son cojumau normal, examiné parallèlement possède 46 chromosomes dont 5 petits télocentriques.

Discussion. — Pour expliquer l'ensemble de ces observations, l'hypothèse de la non-disjonction lors de la méiose, d'une paire de petits télocentriques pourrait être envisagée. Comme on sait que chez la Drosophile la non-disjonction est fortement influencée par le vieillissement maternel, un tel mécanisme rendrait compte de l'accroissement de fréquence du mongolisme en fonction de l'âge avancé de la mère.

Il n'est cependant pas possible d'affirmer que le petit télocentrique surnuméraire soit certainement un chromosome normal et l'on ne peut écarter à l'heure actuelle la possibilité qu'il s'agisse d'un fragment résultant d'un autre type d'aberration.

(¹) J. Lejeune, M. Gautier et R. Turpin, *Comptes rendus*, 248, 1959, p. 602.
(²) P. A. Jacobs et J. A. Strong, *Nature*, 183, 1959, p. 302-303.

(*Chaire d'Hygiène et Clinique 1ʳᵉ Enfance, Institut de Progénèse, Paris.*)

Reprinted from *Obst. Gynec.* 26:308–326 (1965)

Chromosome Studies in Spontaneous Abortions

DAVID H. CARR, M.B., Ch.B.

I T WAS only speculation that lethal chromosome abnormalities would be found in spontaneous abortions until 2 specimens with 69 chromosomes (triploidy) were described from the Galton Laboratory.[21, 66] In a preliminary communication of the present work, it was reported that chromosome abnormalities were found in 12 of 53 previable human abortions, an incidence of 22.5%.[9] Since then, other unselected series have confirmed the high incidence of chromosome abnormalities in spontaneous abortions.[16, 31, 36]

MATERIAL AND METHODS

With the cooperation of many physicians in London and other cities in southwestern Ontario, over 400 abortuses were collected between January 1963 and December 1964. The specimens, consisting of whole or parts of sacs, embryos, or fetuses, were collected in normal saline without addition of preservatives or antibiotics. About half way through the study, it became evident that the small number of stillborn infants were selected by physicians for the presence of abnormalities. It was decided to restrict the study to abortions earlier than 154 days after the first day of the last menstrual period, excluding fetuses weighing over 500 gm. There were, in addition, 11 specimens with a gestational age over 154 days but which weighed less than 500 gm. Many of these specimens were missed abortions. Javert presents the argument that some infants weighing little more than 500 gm. survive, and one is put in the difficult position of calling them "living abortions." It was for this reason that a gestational age under 154 days and weight under 500 gm. were selected as criteria of previable abortions.

Tissue was selected according to its availability and included amnion, chorion, umbilical cord, parts of embryos, especially limbs and eyes, and muscle sheath from larger fetuses. Care was taken to avoid any possibility of culturing maternal tissue. The selected tissue was grown in culture by a technic described by Lejeune *et al.* The final preparation of the chromosomes for analysis was based on a method used in this laboratory for leukocyte culture.[12]

After final preparation, cells were studied under an oil-immersion lens, and the best cells were photographed for chromosome analysis. The chromosomes were then cut

From the Department of Anatomy, University of Western Ontario, Faculty of Medicine, London, Ontario, Canada.

This study was supported by grants from the Medical Research Council of Canada and the D. H. McDermid Medical Research Fund.

Professor Murray L. Barr has given constant encouragement during this study and helpful suggestions in the preparation of this paper. Through Professor R. A. H. Kinch, I wish to extend sincere thanks to all the obstetricians and general practitioners, too numerous to mention individually, whose cooperation in collecting aborted specimens, made this project possible.

This publication is dedicated to Professor S. Makino on the occasion of his sixtieth birthday.

Submitted for publication Apr. 29, 1965.

from prints and paired according to an international convention agreed upon in Denver in 1960[37] and modified at a meeting in London in 1963.[51]

According to these conventions, the chromosomes are arranged in descending order of size and separated into 7 easily identifiable groups sometimes given the letters A to G.

Of all specimens received, 54% produced adequate cell growth in culture for chromosome analysis. In addition to the study of chromosomes, other data were collected. These consisted of sex-chromatin study of cells in culture or in a whole mount of amnion, recording the phenotypic sex of larger fetuses, and a study of the obstetric history including maternal age, date of last menstrual period, date of abortion, and number of previous pregnancies and their results. When case histories were reviewed, an attempt was made to exclude all abortions resulting from external interference.

RESULTS

This study is based on the results of chromosome analysis of 200 successful cultures of tissue from previable spontaneous abortions. Among these 200 specimens, 44 were found to have chromosome abnormalities.

There were 11 specimens in which a sex chromosome was missing, thus reducing the count to 45. This is known as the XO sex-chromosome anomaly (or X monosomy*) and is the same anomaly as is found in the majority of subjects with Turner's syndrome.

The second commonest anomaly was the presence of 3 instead of 2 chromosomal sets resulting in a chromosome count of 69. This anomaly, well-known in plants and lower animals, is called triploidy. Examples

---

* Although the normal male is monosomic for the X chromosome, the term X monosomy will be used as synonymous with XO throughout this paper.

of each of the 3 possible sex-chromosome complexes were found among the 9 triploid specimens. There were 6 with an XXY, 2 with an XXX, and 1 with an XYY sex-chromosome complex. A karyotype from the XYY triploid, which has not been described previously, is shown in Fig. 1.

There were 22 specimens, from which cells grew, which had one extra chromosome. If the extra chromosome is the same as one in the normal complement, it is evident that one of the chromosomes will be represented 3 times instead of twice. This anomaly is known as trisomy. Only when an extra normal chromosome is a sex chromosome in the male does it not give rise to trisomy. Such an abnormality would then produce an XXY or XYY sex-chromosome complex.

It must be emphasized that when a specimen produces cells which are said to be trisomic for a certain chromosome, this conclusion is based only on the appearance of the extra chromosome. The possibility must not be excluded that the extra element may be similar to a member of the normal complement while being an abnormally constituted chromosome, resulting from a structural abnormality such as a translocation or deletion.

Of the 22 trisomic specimens, 7 were found to have an extra chromosome in Group E. In 6 of these, the extra member could not be distinguished from the normal No. 16 chromosome, while in the other, the extra element resembled No. 17 or 18. Chromosomes 17 and 18 are not readily distinguishable, and the difficulty of identifying the extra element in this specimen has been discussed elsewhere.[9]

In 6 trisomic specimens the extra chromosome was in the D group. The 3 pairs of chromosomes which make up this group cannot be distinguished from one another. The extra chromosome in all 6 D trisomic specimens was indistinguishable from the normal members of this group.

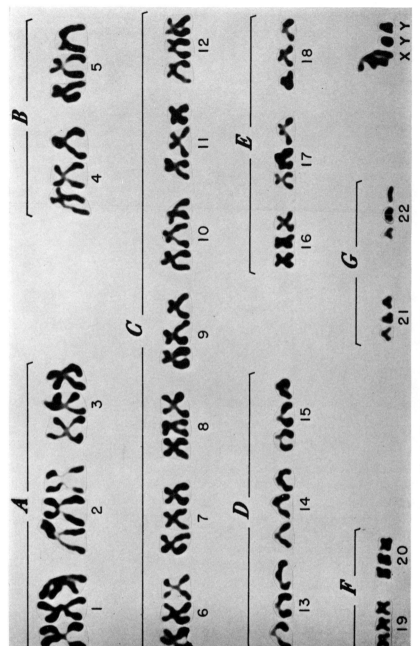

Fig. 1. Triploid karyotype with XYY sex-chromosome complex.

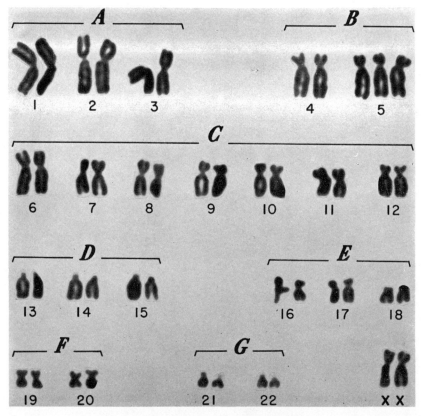

Fig. 2. Karyotype showing B group trisomy.

There was an extra Group G chromosome in 5 specimens, and in 4 of these the extra element could not be distinguished from the normal members of this group. However, in 1 specimen the extra chromosome, though resembling the normal members of Group G, had no short arm. In none of the 5 specimens with Group G trisomy did the extra chromosome show features of the normal Y chromosome.

In 2 of the 22 specimens with 1 extra chromosome, the element resembled the normal members of Group C. The X chromosome falls in this group in addition to 7 pairs of nonsex chromosomes or autosomes. It was concluded on the basis of sex-chromatin studies that the extra element was an autosome in each specimen.

The other 2 trisomies were found in 1 specimen each. One had an extra chromo-some in Group B, while the other had an extra member indistinguishable from chromosome No. 3. These 2 trisomies have not been previously described in man, and their karyotypes are shown in Fig. 2 and 3.

Finally, there were 2 specimens which consistently produced cells with 92 chromosomes. This anomaly is known as tetraploidy and has 2 possible sex-chromosome complexes, XXXX and XXYY. One specimen with each type of sex-chromosome complex was found, and their karyotypes are depicted in Fig. 4 and 5.

The cytogenetic features of the abnormal specimens are summarized in Table 1.

Among the 156 specimens which produced cells with normal chromosome counts, there were 5 with a small number of cells with excess chromosomes. In 1 of these, the abnormal cells probably arose in culture,

162

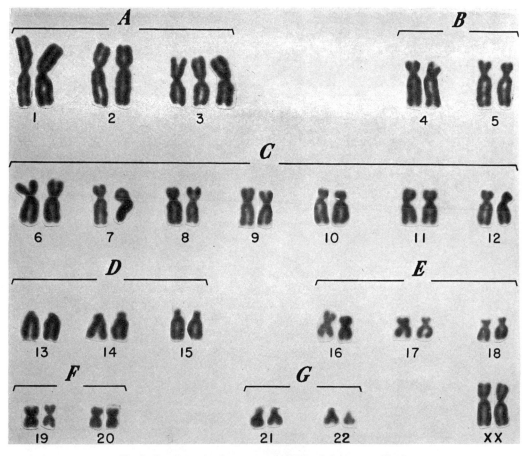

Fig. 3. Karyotype showing apparent trisomy of chromosome No. 3.

whereas in the other 4, the reason for the cells with extra chromosomes is not known. In none of these 4 specimens was there sufficient evidence confidently to diagnose mosaicism—that is, the presence of more than 1 cell type in the tissue cultured.

In order to assess the sex ratio, the polyploid specimens were assigned as male or female on the basis of the presence or absence of a Y chromosome. Thus, there were 7 triploid and 1 tetraploid specimens with 1 or more Y chromosomes, while the other 3 polyploid specimens lacked a Y chromosome and were regarded as female. The over-all sex ratio was 104 female:85 male, the 11 XO specimens being excluded.

The mean maternal age at abortion for all abnormal specimens was 28.9 ($\pm$ 1.19) years. This was not significantly different from the mean maternal age of 151 women producing chromosomally normal abortuses, which was 27.1 (± 0.52) years. However, it is known that trisomy in neonates becomes increasingly common with increasing maternal age.[65, 85] For the 22 trisomic specimens in this series, the mean maternal age was 31.3 ($\pm$ 1.88) years. This was significantly different from the mean maternal age for women producing abortions with normal chromosomes ($p < 0.01$).

The mean duration of pregnancy for the 44 specimens with chromosome anomalies was 85.9 ($\pm$ 3.21) days and for 144 normal

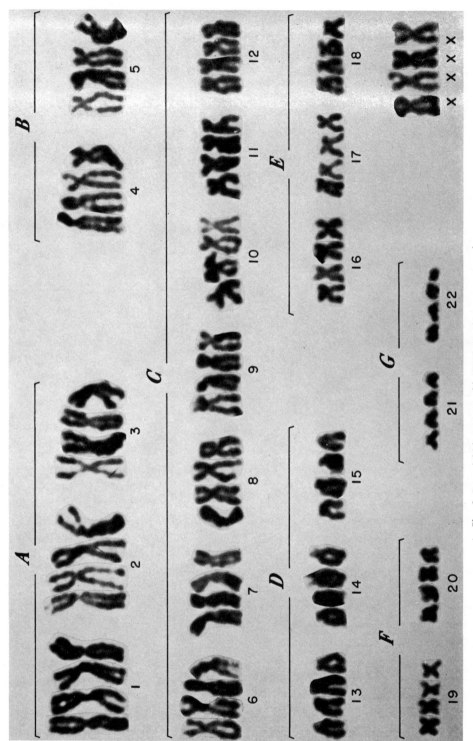

Fig. 4. Tetraploid karyotype with XXXX sex-chromosome complex.

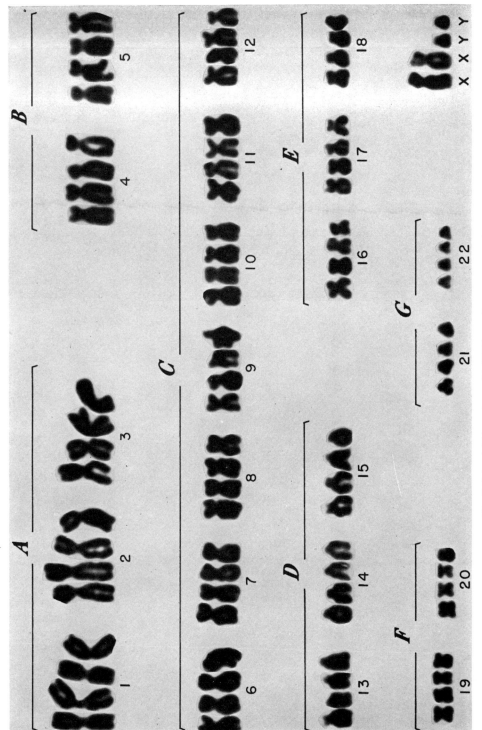

Fig. 5. Tetraploid karyotype with XXYY sex-chromosome complex.

TABLE 1. CYTOGENETIC FEATURES OF CHROMOSOMALLY ABNORMAL SPECIMENS

Specimen No.	First cells studied (days)	Cells counted	Distribution					Exact or near triploid	Exact or near tetraploid	No. of karyotypes	Anomaly
			<45	45	46	47	>47				
1	12	31	5	25	1	0	0	0	0	9	XO
2	7	55	0	1	2	49	3*	0	1*	25	Trisomy 17–18 ♀
3	7	32	1	2	2	27	0	0	0	9	Trisomy D ♀
4	6	30	2	0	0	28	0	0	0	13	Trisomy D ♀
5	14	42	8	34	0	0	0	0	0	13	XO
6	6	31	1	3	1	25	1*	0	1*	11	Trisomy G ♀
7	8	60†	1	0	0	0	59*	59*	0	10	Triploid XXY
8	6	32	0	1	4	26	1	0	0	9	Trisomy D ♀
9	9	44	3	38	1	2	0	0	0	7	XO
10	8	42†	0	0	0	0	42*	40*	0	5	Triploid XXY
11	7	88†	1	0	0	0	87*	0	87*	3	Tetraploid XXXX
12	9	41	2	1	4	34	0	0	0	16	Trisomy 16 ♂
13	11	38	2	0	4	31	1	0	0	19	Trisomy C ♀
14	7	34†	0	0	0	0	34*	34*	0	6	Triploid XXY
15	8	53†	0	0	0	0	53*	52*	0	10	Triploid XYY
16	7	35	2	1	4	28	0	0	0	4	Trisomy D ♂
17	6	23	2	0	2	18	1*	0	1*	10	Trisomy 16 ♂
18	7	61	2	1	4	52	1*	0	1*	32	Trisomy G ♂
19	6	58	1	7	0	48	2	0	0	18	Trisomy C ♂
20	9	39	0	1	4	34	0	0	0	11	Trisomy 16 ♀
21	15	18	3	14	1	0	0	0	0	7	XO
22	8	31	2	28	1	0	0	0	0	6	XO
23	6	32	2	29	1	0	0	0	0	5	XO
24	7	42	4	2	3	32	1*	0	1*	15	Trisomy B ♀
25	14	75†	0	0	0	0	75*	75*	0	2	Triploid XXY
26	4	37	1	1	2	32	1	0	0	8	Trisomy G ♀
27	8	34	3	31	0	0	0	0	0	6	XO
28	6	20	0	0	0	20	0	0	0	7	Trisomy 16 ♀
29	17	22	1	0	1	19	1	0	0	5	Trisomy D ♂
30	5	157†	1	0	0	0	156*	156*	0	5	Triploid XXY
31	9	30	5	24	1	0	0	0	0	3	XO
32	10	104†	0	0	0	0	104*	104*	0	7	Triploid XXX
33	7	36	2	1	3	30	0	0	0	10	Trisomy G ♀
34	7	156†	0	0	0	0	156*	156*	0	6	Triploid XXY
35	8	120†	0	0	0	0	120*	0	120*	7	Tetraploid XXYY
36	11	41	1	1	2	36	1*	0	1*	19	Trisomy 16 ♀
37	8	106†	0	0	0	0	106*	106*	0	3	Triploid XXX
38	6	39	4	35	0	0	0	0	0	5	XO
39	13	24	3	0	2	18	1	0	0	7	Trisomy D ♀
40	19	19	1	0	1	17	0	0	0	4	Trisomy G ♀
41	5	30	2	28	0	0	0	0	0	3	XO
42	7	30	3	26	0	0	1*	0	1*	4	XO
43	20	12	0	0	1	11	0	0	0	5	Trisomy 16 ♀
44	6	53	2	1	0	49	1*	0	1*	16	Trisomy 3 ♀

* Some or all cells counted in both columns.

† Includes recognizable polyploid cells which were not counted.

TABLE 2.   OBSTETRIC HISTORY RELEVANT TO THE 44 CHROMOSOMALLY ABNORMAL SPECIMENS

Specimen No.	Last menstrual period	Date of abortion	Length of pregnancy (days)	Maternal age	Previous abortions, stillbirths, and ectopic pregnancies	Live births
1	10/16/62	1/4/63	80	35	0	3
2	10/19/62	2/4/63	108	33	1	3
3	12/15/62	2/11/63	58	47	1	7
4	1/4/63	3/15/63	70	34	1	3
5	1/10/63	3/21/63	70	24	0	0
6	1/10/63	4/9/63	89	37	2	3
7	2/20/63	5/16/63	85	35	1	5
8	2/20/63	5/16/63	85	35	1	5
9	3/23/63	6/13/63	82	28	0	2
10	5/5/63	7/2/63	58	33	0	1
11	3/15/63	6/20/63	97	21	0	1
12	3/17/63	7/9/63	114	36	1	0
13	6/25/63	9/5/63	72	46	2	4
14	7/22/63	10/17/63	87	21	0	1
15	8/3/63	11/6/63	95	22	0	1
16	8/9/63	11/3/63	86	18	0	0
17	11/6/63	1/15/64	70	24	1	1
18	11/8/63	1/18/64	71	34	1	2
19	11/22/63	1/26/64	65	36	3	2
20	11/26/63	2/15/64	81	16	0	0
21	11/13/63	1/30/64	78	29	0	4
22	12/12/63	3/15/64	94	27	0	3
23	10/20/63	3/20/64	152	17	0	1
24	12/25/63	2/28/64	65	29	2	2
25	1/3/64	3/30/64	87	37	1	3
26	1/11/64	4/12/64	92	39	1	2
27	3/20/64	6/26/64	98	25	1	3
28	2/1/64	4/10/64	69	17	0	0
29	1/23/64	4/17/64	85	34	0	3
30	11/20/63	3/5/64	106	23	0	0
31	12/2/63	5/9/64	158	33	0	1
32	2/20/64	6/3/64	103	37	0	1
33	3/28/64	6/26/64	90	27	2	2
34	4/4/64	7/23/64	110	26	1	3
35	5/8/64	7/27/64	80	26	0	1
36	4/26/64	8/6/64	102	32	0	1
37	6/15/64	8/10/64	56	18	0	1
38	6/15/64	9/24/64	101	18	0	1
39	6/30/64	9/8/64	70	42	1	4
40	6/25/64	8/28/64	64	28	1	3
41	7/28/64	10/22/64	86	23	0	0
42	7/15/64	10/6/64	83	26	1	1
43	8/4/64	10/8/64	65	20	0	1
44	9/21/64	11/24/64	64	25	1	0
TOTALS					27	85

specimens for which data were available, 106.7 ($\pm$ 8.97) days. The difference is highly significant statistically ($p<0.001$).

The results of previous pregnancies did not differ significantly among the women who produced abortions with or without chromosome abnormalities. This finding was unchanged when the 22 women who aborted trisomic specimens were considered separately.

Obstetric histories relevant to the 44 abnormal specimens are summarized in Table 2.

DISCUSSION

It is considered unlikely that chromosome anomalies of the types described could arise as technical artifacts. It is also believed that amnion or chorion reflects the chromosome constitution of the embryo or fetus.[9]

Before the incidence and types of chromosome abnormality, individually and collectively, are discussed, it is necessary to establish the validity of certain related data.

The incidence of spontaneous abortion has been studied repeatedly, but in order to include early abortions, figures based on personal interview by conscientious investigators are the most reliable. Two recent studies on the frequency of spontaneous abortion, based on personal interview, reported an almost identical figure averaging 15%.[68, 91]

There was no conscious selection of material for this study. However, it is possible that some selection could have occurred

accidentally. In order to test for this, the present series was compared with other series of unselected spontaneous abortions in which the date of abortion was given by the week. Abortions in these series occurring after 22 weeks of pregnancy have been omitted from this comparison, which is summarized in Table 3. The figures for the present series correspond closely to those of Peckham. However, when they are compared with 3 more recent studies,[40, 77, 82] there is an excess of specimens from abortions after the sixteenth week of gestation. As a corollary, there are fewer abortions from the earlier gestational period—those in which most of the specimens with chromosome anomalies are found. It follows, then, that any accidental selection entering into the present series was against the earlier abortions which were responsible for the highest number of chromosome anomalies.

If the incidences of Down's syndrome,[65] the sex-chromosome anomalies,[53] and the 17–18 and D_1 trisomy syndromes are taken as 1 in 600, 1 in 540, 1 in 2000, and 1 in 4000 respectively, this gives an over-all incidence of 1 in 240 for chromosome anomalies at birth. Support for this figure comes from a study of 4412 newborn babies in whom the incidence of chromosome anomalies was 1 in 245.[57] The over-all incidence of chromosome anomalies in this series of abortions was 22%, or more than 50 times the incidence at birth. In contrast, among nearly 500 cultures grown from tissue collected from unselected human material from

TABLE 3. DATA IN THIS AND 4 OTHER SERIES, REGARDING TIME OF ABORTION IN POSTMENSTRUAL DAYS

Reference	Total abortions	Aborted between day				
		29–56 (%)	57–84 (%)	85–112 (%)	113–140 (%)	141–154 (%)
Peckham[64]	576	10	30	25	24	11
Javert[40]	1112	25	47	17	8	3
Stevenson et al.[82]	1073	19	51.5	21.5	6	2
Shapiro et al.[77]	808	17	50	22	8	3
Present series	175	5	34	32	23	6

the prenatal period to old age, no numerical chromosome anomalies were found.[38, 54]

Monosomy for the X is the commonest chromosome anomaly in Turner's syndrome and gonadal dysgenesis.[15, 19, 39, 50] Maclean et al. found that X monosomy had an incidence of 1 in 5000 among newborn infants, while its incidence in this series of abortions was 5.5%, a figure supported by data from other reports.[16, 31, 84, 86] If the incidence of spontaneous abortion is 15%, it appears from these data that only 1 in 10 XO zygotes survives to term. In addition, there is evidence that the anomaly is associated with a raised perinatal mortality.[19, 29] The XXY anomaly, which is the commonest chromosomal defect in Klinefelter's syndrome, is about 5 times as common as the XO anomaly in newborn infants. One suggested reason for this difference, supported by the evidence presented here, was a high prenatal mortality of the XO embryo.[6, 88] At the time of fertilization of the zygote, the XO anomaly apparently is commoner than XXY. This finding is in accord with theoretical calculations, if one assumes that both anomalies arise by the same mechanism (nondisjunction).[6] However, although the XXY anomaly was not found in abortions in this series, it may be associated with an increased mortality in late pregnancy and in the perinatal period.[53]

The XO anomaly has been described in the mouse with a spontaneous incidence of 0.1–1.7%, according to stock.[69] The average calculated incidence, 0.7%,[71] is very close to that for the probable incidence at fertilization in man (0.83%). However, the XO mouse is a normal fertile female with no evidence of a raised perinatal lethality.[14] In matings which should lead to equal occurrence of XX, XY, XO mice, the actual incidence of the XO anomaly is only about one-third that expected. There is no evidence that this discrepancy is due to death during embryogenesis, and other explanations are possible.[14]

The high lethality of the XO anomaly in man requires a reference to Lyon's hypothesis that one X chromosome in female cells becomes contracted and genetically inactive and that this heteropyknotic X forms the sex chromatin in interphase nuclei. This inactivation randomly affects the paternally or maternally derived X chromosome in any one cell, and after this "time of decision," early in embryonic life, the inactivation remains fixed. McKusick has reviewed the evidence on Lyon's hypothesis in a monograph on the X chromosome. There were certain items of genetic evidence which were against the hypothesis, and this, and other evidence, has recently been discussed by Russell.[70] She suggests a modification of the hypothesis, which involves the idea that "inactivation may emanate from a certain point in the X chromosome and may spread along gradients." This view, then, proposed that the second X chromosome in female cells is only partially inactivated.

Whether or not the inactivation of 1 X chromosome is total or not, it is clear that normal females are not genetically the same as XO females for 2 reasons. Firstly, the normal female has X^M (maternally derived) active in half her cells and X^P (paternally derived) in the other half. Secondly, there is a period, early in embryonic development, when neither X chromosome is contracted and both are presumably active.[62] The possible effect of this period of full activity of both X chromosomes on future cell generations cannot be ignored.

These differences between the XX and XO do not apply to the normal male (XY) and XO female since, in both instances, the single X chromosome is active in all cells. It is not disputed that the Y chromosome in man is strongly male-determining, but if this is its only function, why is the XO anomaly so lethal? It has been shown that the Y chromosome can suppress the expression of X-linked recessive, lethal genes in Drosophila,[49] but whether this explanation can be

169

applied to man is unknown. It is clear that the Y chromosome in man is in some way responsible for protecting a zygote with only 1 X chromosome against intrauterine death. The mechanism by which this is achieved is not yet clear.

Triploidy, the presence of 3 chromosome sets instead of 2, was found in 9, or 4.5%, of the previable abortions in this series. In addition to the 2 triploidy specimens reported from the Galton Laboratory,[21, 66] the finding of triploidy in spontaneous abortions has now been repeatedly confirmed.[1, 2, 36, 55, 84, 86]

There have been 3 reports of triploidy in live-born individuals. These subjects were all studied because of mental retardation and/or congenital anomalies. In each instance there was a normal cell line in culture in addition to that with 69 chromosomes.[8, 26, 27]

Previous sex-chromatin findings in triploidy have been variable,[60] and the results in this series are therefore of interest. They are summarized in Table 4. The triploid specimens with an XXY sex-chromosome complex had either chromatin-positive or chromatin-negative nuclei. The 2 specimens with an XXX sex-chromosome complex were both chromatin-positive, and 1 contained a number of cells with duplicated sex chromatin. In the case of the culture which produced only cells with a triploid-

XYY karyotype, the sex-chromatin findings in uncultured cells are in conflict with the sex-chromosome complex. Cells in culture were chromatin-negative, as were those in sections of chorionic villi. On the other hand, cells in a whole mount of amnion were undoubtedly chromatin-positive. One is driven to the inevitable conclusion that the specimen was a chromosome mosaic, in spite of the fact that only triploid-XYY cells were found in culture.

The most likely mosaicism is a mixture of normal female (XX) cells and a triploid-XYY cell line. There are several ways in which such a zygote could arise.[4, 7] The occurrence of an XYY sex-chromosome complex, however, means that the triploid cell line resulted from double fertilization with 2 Y-bearing sperm or an abnormal (diploid) sperm containing 2 Y chromosomes. The specimen itself presented features which have become important. Though there was no frank hydatidiform mole, chorionic villi were swollen on gross examination. An exactly similar appearance has been described by Makino *et al.* in 3 early abortions.[55] They found, in each of these specimens, triploid cells with the commoner XXY sex-chromosome complex.

Of the 7 specimens with an extra element in the E group, only 1 had an extra chromosome resembling No. 17 or 18, as found in cases of the well known E trisomy syn-

TABLE 4.   SEX-CHROMATIN FINDINGS IN THE CELLS OF TRIPLOID SPECIMENS

Specimen No.	Chromatin-positive cells/ cells studied in culture	Chromatin-positive cells/ cells studied in whole mount of amnion	Chromatin-positive cells/ cells studied in sections of chorion
7	1/200	—	—
10	0/200	—	—
14	1/200	9/200	—
15	1/200	53/100	0/50
25	—	28/100	—
30	56/200	—	—
32	80+7 duplicated/200	66+24 duplicated/100	—
34	4/100	25/200	—
37	—	43+4 duplicated/200	—

drome.[23] The clinical and pathologic features of this syndrome have recently been reviewed.[47, 85, 92, 93] Trisomy 17–18 has been found in a spontaneously aborted 25-week fetus[24] and in an embryo delivered by hysterotomy.[30] Though there is considerable difficulty in distinguishing between Chromosomes 17 and 18,[9] No. 16 is a readily recognized member of the complement. An extra chromosome indistinguishable from No. 16 was found in 6 of the 7 specimens with trisomy in Group E. The occurrence of trisomy 16 in aborted specimens is supported by other workers,[16, 84, 86] but it has been reported in all the cells of only 1 living subject.[48] In addition, a boy with metaphysial dysostosis was found to have an extra element indistinguishable from Chromosome 16 in about 10% of cells in repeated cultures.[72]

The 6 specimens with D trisomy must be considered together, since no consistent way has been found to separate the 3 chromosome pairs in this group. The first example of an extra chromosome in the D group was reported in a child with multiple congenital anomalies.[63] The syndrome associated with this trisomy has recently been reviewed, and its high mortality in the neonatal period was emphasized.[44, 58, 85]

Let us suppose that the extra element in all 6 cases of D trisomy reported here is the same D chromosome as that found in the clinical syndrome. If the incidence of this syndrome is 1 in 4000 live births[85] and the incidence of the anomaly in aborted material is 3%, then the prenatal lethality of this syndrome would be 120:1, if it is assumed that 15% of all pregnancies end in abortion. On the other hand, none of the aborted specimens may have an extra D_1 chromosome (the one involved in the known clinical syndrome). This would of course suggest a low prenatal but a high perinatal mortality in association with D_1 trisomy.

Among the 5 specimens with Group G trisomy, the extra element could not be distinguished from the normal 21 and 22 chromosomes in 4 of them. One specimen produced cells with an extra element which resembled chromosomes in Group G but had no short arms. The history of 2 previous abortions and 3 normal children suggested the possibility of the parents carrying a balanced translocation. However, leukocyte cultures from both parents showed normal chromosomes.

It is not known whether the extra chromosome in Down's syndrome is No. 21 or 22, though it is traditionally known as No. 21.[45] In 4 of the aborted specimens, therefore, the extra chromosome may have been the one found in Down's syndrome. If the latter proposal is true, then the incidence of "trisomy 21" in this series is 2%. If it is assumed that spontaneous abortion occurs in 15% of pregnancies, the number of 21-trisomy zygotes lost as abortions is 1 in 333. As the incidence of Down's syndrome is about 1 in 600 births,[65] the figure for the abortions suggests that neonates with Down's syndrome represent only one-third of all 21-trisomy zygotes. On the other hand the prenatal lethality of Down's syndrome may be zero if the extra element found in the G trisomy abortuses is not "Chromosome 21."

There were 2 specimens in which the extra chromosome was found to lie in the C group. As the X chromosome is indistinguishable from the autosomes in this group, it was necessary to decide whether the extra element in the abortuses was an autosome or an X chromosome. An extra X chromosome would be expected to produce an extra sex-chromatin mass in interphase nuclei.[5] In these 2 aborted specimens, the chromosomes were apparently autosomes, since the sex-chromatin findings corresponded with normal male and normal female sex-chromosome complexes. The possibility that an extra X chromosome may fail to produce an extra sex-chromatin mass has been discussed but discounted as unlikely.[10]

Extra Group C autosomes have been reported in 7 living individuals, though in each instance there were also cells with a normal chromosome complement.[25, 35, 39, 67, 74, 78, 83]

The single specimen with trisomy in Group B was aborted by a woman with 2 normal children and a history of 2 previous abortions. Such a history suggests the possibility of a balanced translocation in a parent. The chromosomes in blood cultures of the parents were, however, found to be normal. No living individual has ever been found to have trisomy in Group B. Shaw *et al.* recently described an infant with serious anomalies who was effectively trisomic for the long arms of a Group B chromosome as a result of a parental translocation. They have also reviewed the literature on anomalies involving chromosomes in Group B.

The remaining specimen with a trisomy appeared to have an extra chromosome No. 3. There are other possible sources for a chromosome resembling No. 3 but actually having a different origin, including translocations[22, 90] and isochromosomes.[11, 28] In an attempt to define the extra element resembling No. 3, a study of the parents' chromosomes was made. Chromosomes seen in a leukocyte culture appeared to be normal. Autoradiographic studies in this culture were only partially successful, but the results were not against the interpretation that the extra chromosome was No. 3 and excluded the possibility that the extra element was derived from an X chromosome.

Finally, there were 2 specimens which consistently produced tetraploid cells. Though tetraploidy has never been found in living subjects it has been reported in other studies of aborted specimens.[36, 86] It is well known that tetraploid cells may arise in culture. The 2 tetraploid cultures reported here are believed to represent the true genome of the tissue because tetraploidy was found in the first culture, the culture time was short (7 and 8 days), all cells

were tetraploid, and these cells were seen in cultures derived from several separate tissue explants. In addition, there is conclusive evidence from sex-chromatin studies on the specimen with an XXXX sex-chromosome complex that the uncultured amnion was uniformly tetraploid. Whole mounts of amnion, taken from several areas of the large empty sac, showed duplicated sex chromatin in 95% of the cells. Although occasional cells in amnion are tetraploid and have 2 sex-chromatin masses, the number rarely exceeds 2%.[42] An abortion has been described in which sex-chromatin findings indicated that localized masses of cells were apparently tetraploid, but this was considered to be a cellular abnormality or possibly an early neoplasm.[43] In the tetraploid specimen in this series, the cells containing duplicated sex-chromatin masses were not localized. However, since there were no embryonic remnants, it is not possible to say whether or not the embryo itself was tetraploid. The XXYY tetraploid specimen was chromatin-negative.

It is most likely that the 2 tetraploid specimens in this series arose by suppression of the first cleavage of a fertilized ovum. The sex-chromatin findings are of interest in relation to the time of origin of the tetraploidy. Harnden proposed a formula for calculating the number of sex-chromatin masses in relation to the number of X chromosomes which would apply even in cases of triploidy or tetraploidy. In this formula, $S = X - A/2$, S is the number of sex-chromatin masses, X the number of X chromosomes, and A the number of chromosome sets. The tetraploid states usually cited as examples in support of the Harnden formula presumably arose after the sex-chromatin status of the parent cells was decided upon. The number of sex-chromatin masses in these cells would be expected to be simple multiples of the parent cell. The situation of the 2 tetraploid abortuses in this series is quite different. The tetraploidy probably arose by suppression of the first cleavage, and the sex chromatin in the human is not "decided" until about 16 days after fertilization in the embryo or 12 days after fertilization in the trophoblast.[62] The sex-chromatin findings in these 2 tetraploid specimens support the view of Harnden that the number of contracted X chromosomes, responsible for the formation of the sex-chromatin body, is related in a fundamental way to the number of sets of autosomes in the cell.

It has been known for some time that a high percentage of aborted specimens are structurally abnormal.[17, 34, 40, 56] The incidence of abnormalities reported by different authors varies from 31% to 80%. For the first time, the underlying mechanism for many of these abnormalities has been defined in this and other similar reports. Among the intact sacs without recognizable embryonic remains in this series, 50% produced cells with a chromosome anomaly or, to put it another way, "blighted ova" accounted for 11.5% of the chromosomally normal specimens but over 40% of those with a chromosome abnormality. By selecting for "blighted ova," Thiede and Salm[86] raised the incidence of chromosome anomalies in their series to 65%.

The average length of gestation for the chromosomally abnormal specimens was shorter, to a highly significant degree, than the length of gestation of the chromosomally normal abortions. This finding is supported by the fact that Szulman found 65% chromosome anomalies among 20 abortions from early pregnancies. There have been few reports of chromosome anomalies in later pregnancy. In this study, only 3 chromosomally abnormal specimens were from women who aborted after the sixteenth week of pregnancy, and 2 of these were fetuses with an XO anomaly. Aula and Hjelt[3] found a dicentric chromosome in 20% of cells cultured from a 4-month-old aborted fetus and El-Alfi et al.[24] reported trisomy 17–18 in a fetus aborted at 25 weeks' gestation.

Chromosome anomalies are probably a commoner cause of abortion in man than the 22% incidence indicated by this series. There was an excess of abortions between 16 and 22 weeks' gestation, which formed nearly 30% of the total series. Only 3 of the chromosome anomalies were found in this group. Also, it seems likely that the total zygotic loss in man is at least 30%,[33] as it is in several domestic and laboratory animals which have been studied.[13, 18] Definite proof of chromosome anomalies in early zygotes of man is lacking, but if an analogy may be drawn from animal studies, they may occur commonly.[89] In judging the incidence of chromosome anomalies, the failure to find specimens with one autosome missing (monosomy) should be taken into consideration. Inhorn et al. believe they have 2 specimens with monosomy of an autosome in Group C, but otherwise, autosomal monosomy has not been found in aborted specimens. Kelly et al. claim to have found monosomy for chromosome No. 1 in cells from an abortion, but their published karyotype does not support this claim. Monosomy, other than that involving the X chromosome, must surely occur, though not necessarily as commonly as trisomy.

The excess of males over females at birth varies geographically but is 106:100 in North America.[79] In order to arrive at a figure closer to the primary sex ratio, numerous abortuses have been studied. A report by Tietze in 1948 is the most useful summary of results preceding the discovery of sex chromatin. Using data from 5787 fetuses in the collection of the Department of Embryology of the Carnegie Institution of Washington, Tietze found a sex ratio of 107.9 male to 100 female.

As the sex chromatin appears in the human embryo on the sixteenth day after fertilization,[62] it is valuable for determining sex in abortions. In 1962, Stevenson summarized the published findings on the sex ratio of early abortions computed by the

173

study of sex chromatin.[81] In the 6 series containing more than 100 specimens which Stevenson listed, the sex ratio (male/female), exceeded unity. The ratio varied from 1.34 in Stevenson's own material,[80] to 3.54 in a study reported from Germany. Published series since that time have not altered these figures.[20, 73, 75] There remains only one sizable collection of early abortions in which the sex ratio was less than one. Moore and Hyrniuk studied the sex chromatin in material from 276 abortions between the second and fourth months of pregnancy. There were 145 female and 131 male specimens.

Stevenson[80] stressed the difficulties in sexing abortions from chorionic villi and pointed out that mistakes are likely to be made in the direction of too frequent diagnosis of males. This is the same error made by untrained investigators who sex early fetuses from their external appearance.

The sex ratio in this series shows an excess of females (104:85) though it is not significantly different from a 1:1 ratio on statistical analysis. The XO specimens could not be included in the sex ratio, but their presence is important as a correction factor in interpreting sex-chromatin data. In using this correction factor for sex-chromatin data from aborted specimens, one is faced with an enigma. Should 5.5% be simply deducted from the total of chromatin-negative specimens? Or should this 5.5% be added to the total females, on the basis that, had they come to term, the XO specimens would have been manifest phenotypically as females? The use of this correction could alter the results of the sex ratio, based on sex chromatin analysis of abortions, to a significant degree.

During the period of conception covering the 4 months May to August 1964, only 1 of 25 abortions received was a male, 21 being female and 3 XO. This is most likely a chance event, but the possibility was considered that seasonal factors might affect the sex ratio. For the same 4-month period

in 1963, there were twice as many male abortuses as female. The only other 4-month period with a marked excess of females over males (18:5 and 2 with XO) fell in the winter months October 1962 to January 1963. In order to see whether this variation was found at term, figures were obtained for births by month in Ontario for 1962. There was no variation from the over-all pattern of an excess of male births in any of the 12 months of that year. However, the possibility of some factor other than chance producing the extreme excess of females in certain months should be considered.

346 South St.

London, Ontario, Canada

REFERENCES

1. ASPILLAGA, M. J., SCHLEGEL, R. J., NEU, R., and GARDNER, L. I. Triploid/diploid mosaicism in tissue from an early fetus (abst.). J. Pediat. 65:1098, 1964.
2. ATKIN, N. B., and KLINGER, H. P. The superfemale mole. Lancet 2:727, 1962.
3. AULA, P., and HJELT, L. A structural chromosome anomaly in a human foetus. Ann. paediat. Fenniae 8:297, 1962.
4. AUSTIN, C. R. Anomalies of fertilization leading to triploidy. J. Cell. & Comp. Physiol. 56, Supp. 1:1, 1960.
5. BARR, M. L., and CARR, D. H. Sex chromatin, sex chromosomes and sex anomalies. Canad. M. A. J. 83:979, 1960.
6. BATEMAN, A. J. Maternal v. paternal nondisjunction as the source of sex-chromosome unbalance. Lancet 2:1383, 1962.
7. BEATTY, R. A. Parthenogenesis and Polyploidy in Mammalian Development. Cambridge, London, 1957.
8. BÖÖK, J. A., MASTERSON, J. G., and SANTESSON, B. Malformation syndrome associated with triploidy—further chromosome studies of the patient and his family. Acta Genet. (Basel) 12:193, 1962.
9. CARR, D. H. Chromosome studies in abortuses and stillborn infants. Lancet 2:603, 1963.
10. CARR, D. H. Chromosome abnormalities in spontaneous abortuses. (In press)
11. CARR, D. H., BARR, M. L., and RATHBUN, J. C. A probable isochromosome in a child with multiple congenital anomalies. J. Pediat. 62:696, 1963.
12. CARR, D. H., and WALKER, J. E. Carbol fuchsin as a stain for human chromosomes. Stain Technol. 36:233, 1961.

13. Casida, L. E. "Fertilization Failure and Embryonic Death in Domestic Animals." In *Pregnancy Wastage*, Ed. by Engel, E. Thomas, Springfield, Ill., 1953.

14. Cattanach, B. M. XO mice. *Genet. Res. 3*: 487, 1962.

15. De la Chapelle, A. Cytogenetical and clinical observations in female gonadal dysgenesis. *Acta endocrinol. 65*, Supp., 1962.

16. Clendenin, T. M., and Benirschke, K. Chromosome studies on spontaneous abortions. *Lab. Invest. 12*:1281, 1963.

17. Colvin, E. D., Bartholomew, R. A., Grimes, W. H., and Fish, J. S. Salvage possibilities in threatened abortion. *Am. J. Obst. & Gynec. 59*:1208, 1950.

18. Corner, G. W., and Bartelmez, G. W. "Early Embryos of the Rhesus Monkey." In *Pregnancy Wastage*, Ed. by Engel, E. Thomas, Springfield, Ill., 1953.

19. Court Brown, W. M., Harnden, D. G., Jacobs, P. A., Maclean, N., and Mantle, D. J. *Abnormalities of the Sex Chromosome Complement in Man*. Medical Research Council Special Report. Her Majesty's Stat. Off., London, 1964.

20. Csordas, T., Dömötori, E., Gergely, E., and Rechnitz, K. Über die Geschlechtsproportion der Früchte in den ersten 3 Monaten des intrauterinen Lebens. *Zentralbl. Gynäk. 85*: 1036, 1963.

21. Delhanty, J. D. A., Ellis, J. R., and Rowley, P. T. Triploid cells in a human embryo. *Lancet 1*:1286, 1961.

22. Edwards, J. H., Fraccaro, M., Davies, P., and Young, R. B. Structural heterozygosis in man: analysis of two families. *Ann. Human Genet. 26*:163, 1962.

23. Edwards, J. H., Harnden, D. G., Cameron, A. H., Crosse, V. M., and Wolff, O. H. A new trisomic syndrome. *Lancet 1*:787, 1960.

24. El-Alfi, O. S., Biesele, J. J., and Smith, P. M. Trisomy 18 in a hydrocephalic fetus. *J. Pediat. 65*:67, 1964.

25. El-Alfi, O. S., Powell, H. C., and Biesele, J. J. Possible trisomy in chromosome group 6-12 in a mentally retarded patient. *Lancet 1*:700, 1963.

26. Ellis, J. R., Marshall, R., Normand, I. C. S., and Penrose, L. S. A girl with triploid cells. *Nature, London 198*:411, 1963.

27. Ferrier, P., Ferrier, S., Stolder, G., Bühler, E., Bamatter, F., and Klein, D. Congenital asymmetry associated with diploid-triploid mosaicism and large satellites. *Lancet 1*:80, 1964.

28. Fraccaro, M., Ikkos, D., Lindsten, J., Luft, R., and Kaijser, K. A new type of chromosomal abnormality in gonadal dysgenesis. *Lancet 2*:1144, 1960.

29. Frøland, A., Lykke, A., and Zachan-Christiansen, B. Ovarian dysgenesis (Turner's syndrome) in the newborn. *Acta path. et microbiol. scandinav. 57*:21, 1963.

30. Gropp, A., and Hole, H. Personal communication, 1963.

31. Hall, B., and Kallén, B. Chromosome studies in abortuses and stillborn infants. *Lancet 1*:110, 1964.

32. Harnden, D. G. Nuclear sex in triploid XXY human cells. *Lancet 2*:483, 1961.

33. Hertig, A. T., Rock, J., and Adams, E. C. A description of 34 human ova within the first 17 days of development. *Am. J. Anat. 98*:435, 1956.

34. Hertig, A. T., and Sheldon, W. H. Minimal criteria required to prove prima facie case of traumatic abortion or miscarriage. *Ann. Surg. 117*:596, 1943.

35. Inhorn, S. L., Smith, D. W., Patau, K., and Therman, E. Presumed C-group autosomal trisomy/normal mosaicism in a child with multiple anomalies (abst). In *Second Mammalian Cytogenetics Conference*, pp. 11–12.

36. Inhorn, S. L., Therman, E., and Patau, K. Cytogenetic studies in spontaneous human abortion (abst.) *Am. J. Clin. Path. 42*:528, 1964.

37. International Study Group. A proposed standard system of nomenclature of human mitotic chromosomes. *Acta Genet. (Basel) 10*:322, 1960.

38. Jacobs, P. A. Chromosome studies in the general population. *Sympos. Internat. Soc. Cell Biol. 3*:111, 1964.

39. Jacobs, P. A., Harnden, D. G., Buckton, K. E., Court Brown, W. M., King, M. J., McBride, J. A., MacGregor, T. N., and Maclean, N. Cytogenetic studies in primary amenorrhoea. *Lancet 1*:1183, 1961.

40. Javert, C. T. *Spontaneous and Habitual Abortion*. McGraw-Hill. New York, 1957.

41. Kelly, S., Almy, R., Jakovic, L., and Buckner, L. Autosomal monosomy in a spontaneous abortion. *Lancet 1*:166, 1965.

42. Klinger, H. P. The sex chromatin in fetal and maternal portions of the human placenta. *Acta Anat. (Basel) 30*:371, 1957.

43. Klinger, H. P. "Abnormal Content of Deoxyribonucleic Acid and Sex Chromatin Number in the Nuclei of a Human Amnion." In *Human Chromosome Abnormalities*. Staples, London, 1959.

44. Lafourcade, J., Bocquet, L., Cruveiller, J., Saraux, H., Berger, R., Lejeune, J., Huet de Brachez, Y., and Turpin, R. Aberrations chromosomiques et maladies humaines. Contribution à l'étude anatomique de la trisomie 13. *Soc. méd hôp. (Paris) 114*:383, 1964.

45. Lejeune, J. Les caryotypes de la trisomie 21. *Path. Biol. 11*:1153, 1963.

46. LEJEUNE, J., TURPIN, R., and GAUTIER, M. Études des chromosomes somatiques humains. Technique de culture de fibroblastes in vitro. *Rev. Franc. Étud. Clin. Biol.* 5:406, 1960.

47. LEWIS, A. J. The pathology of 18 trisomy. *J. Pediat.* 65:92, 1964.

48. LEWIS, F. J. W., HYMAN, J. M., MacTAGGART, J., and POULDING, R. H. Trisomy of autosome 16. *Nature, London* 199:404, 1963.

49. LINDSLEY, D. L., EDINGTON, C. W., and VON HALLE, E. S. Sex-linked recessive lethals in Drosophilia whose expression is suppressed by the Y chromosome. *Genetics* 45:1649, 1960.

50. LINDSTEN, J. *The Nature and Origin of X Chromosome Aberrations in Turner's Syndrome.* Almquist, Uppsala, 1963.

51. London Conference on "The normal human karyotype." *Am. J. Hum. Genet.* 16:156, 1964.

52. LYON, M. F. Sex chromatin and gene action in the mammalian X-chromosome. *Am. J. Hum. Genet.* 14:135, 1962.

53. MACLEAN, N., HARNDEN, D. G., COURT BROWN, W. M., BOND, J., and MANTLE, D. J. Sex-chromosome abnormalities in newborn babies. *Lancet* 1:286, 1964.

54. MAKINO, S., KIKUCHI, Y., SASAKI, M. S., SASAKI, M., and YOSHIDA, M. A further study of the chromosomes in the Japanese. *Chromosoma* 13:148, 1962.

55. MAKINO, S., SASAKI, M. S., and FUKUSCHIMA, T. Triploid chromosome constitution in human chorionic lesions. *Lancet* 2:1273, 1964.

56. MALL, F. P., and MEYER, A. W. Studies on abortuses: A survey of pathologic ova in the Carnegie Embryological collection. *Contrib. Embryol. Carnegie. Inst. 12*, No. 56, 1921.

57. MARDEN, P. M., SMITH, D. W., and McDONALD, M. J. Congenital anomalies in the newborn infant, including minor variations. *J. Pediat.* 64:357, 1964.

58. MARIN-PADILLA, M., HOEFNAGEL, D., and BENIRSCHKE, K. Anatomic and histopathologic study of two cases of D₁ (13-15) trisomy. *Cytogenetics (Basel)* 3:258, 1964.

59. McKUSICK, V. A. *On the X Chromosome of Man.* American Institute of Biological Sciences, Washington, 1964.

60. MITTWOCH, U., ATKIN, N. B., and ELLIS, J. R. Barr bodies in triploid cells. *Cytogenetics (Basel)* 2:323, 1963.

61. MOORE, K. L., and HYRNIUK, W. Sex diagnosis of early human abortions by the chromatin method (abst.). *Anat. Rec.* 136:277, 1960.

62. PARK, W. W. The occurrence of sex chromatin in early human and macaque embryos. *J. Anat.* 91:369, 1957.

63. PATAU, K., SMITH, D. W., THERMAN, E., INHORN, S. L., and WAGNER, H. P. Multiple congenital anomalies caused by an extra autosome. *Lancet* 1:790, 1960.

64. PECKHAM, C. H. Abortion: A statistical analysis of 2,287 cases. *Surg. Gynec. & Obst.* 63:109, 1936.

65. PENROSE, L. S. *The Biology of Mental Defect.* Sidwick, London, 1963.

66. PENROSE, L. S., and DELHANTY, J. D. A. Triploid cell cultures from a macerated foetus. *Lancet* 1:1261, 1961.

67. PFEIFFER, R. A., SCHELLONG, G., and KOSENOW, W. Chromosomenanomalien in den Blutzellen eines Kindes mit multiplen Abortungen. *Klin. Wchnschr.* 40:1058, 1962.

68. ROTH, D. B. The frequency of spontaneous abortion. *Internat. J. Fertil.* 8:431, 1963.

69. RUSSELL, L. B. "Chromosome Aberrations in Experimental Mammals." In *Progress in Medical Genetics,* Vol. 2; Ed. by Steinberg, A. G. Grune, New York, 1962.

70. RUSSELL, L. B. Another look at the single-active-X hypothesis. *Trans. New York Acad. Sc.* 26:726, 1964.

71. RUSSELL, L. B., and SAYLORS, C. L. Spontaneous and induced abnormal sex-chromosome number in the mouse (abst.). *Genetics* 46:894, 1961.

72. SCHMIDT, B. J., BECAK, W., BECAK, M. L., SOIBELMAN, I., QUEIROZ, A. DA S., LORGA, A. P., SECAF, F., ANTONIO, C. F., and CARVALHO, A. DE A. Metaphyseal dysostosis. *J. Pediat.* 63:106, 1963.

73. SCHULTZE, K. W. Geschlechtsbestimmungen bei Abortus verschiedener Genese. *Zentralbl. Gynäk.* 83:56, 1961.

74. SERGOVICH, F. R. Personal communication, 1964.

75. SERR, D. M., and ISMAJOVICH, B. Determination of the primary sex ratio for human abortions. *Am. J. Obst. & Gynec.* 87:63, 1963.

76. SHAW, M. W., COHEN, M. M., and HILDEBRANDT, H. M. A familial 4/5 reciprocal translocation resulting in partial trisomy B. *Am. J. Human Genet.* 17:54, 1965.

77. SHAPIRO, S., JONES, E. W., and DENSEN, P. M. A life table of pregnancy terminations and correlates of fetal loss. *Milbank Mem. Fund Quart.* 40:7, 1962.

78. STALDER, G. R., BÜHLER, E. M., and WEBER, J. R. Possible trisomy in chromosome group 6-12. *Lancet.* 1:1379, 1963.

79. STERN, C. *Principles of Human Genetics.* Freeman, San Francisco, 1960.

80. STEVENSON, A. C. Observations on the results of pregnancies in women resident in Belfast. III. Sex ratio with particular reference to nuclear sexing of chorionic villi of abortions. *Ann. Human Genet.* 23:415, 1959.

81. STEVENSON, A. C. Discussion of paper by Heinz, H. A., and Stoll, P. Sex determinations in intra-uterine death by means of sex chromatin. *Acta Cytol. (Phila.)* 6:116, 1962.

82. STEVENSON, A. C., DUDGEON, M. Y., and

McClure, H. I. Observations on the results of pregnancies in women resident in Belfast. II. Abortions, hydatidiform moles and ectopic pregnancies. *Ann. Human Genet. 23*:395, 1959.

83. Stolte, L., Evers, J., and Blankenborg, G. Possible trisomy in chromosome group 6-12 in a normal woman. *Lancet 2*:480, 1964.

84. Szulman, A. E. Chromosomal aberrations in early human abortions (abst.). *Fed. Proc. 23*:499, 1964.

85. Taylor, A. I., and Polani, P. E. Autosomal trisomy syndromes excluding Down's. *Guy's Hosp. Rep. 113*:231, 1964.

86. Thiede, H. A., and Salm, S. B. Chromosome studies of human spontaneous abortions. *Am. J. Obst. & Gynec. 90*:205, 1964.

87. Tietze, C. A note on the sex ratio of abortions. *Human Biol. 20*:156, 1948.

88. Twiesselmann, F., Defrise-Gussenhoven, E. G., and Leguebe, A. Incidence on genetics of mechanism of segregation and disjunction at meiosis in man. *Nature, London 196*:1232, 1962.

89. Vara, P., and Personen, S. Uber Abortiveier. *Acta obst. et gynec. scandinav. 27*:215, 1947.

90. Walker, S., and Harris, R. Familial transmission of a translocation between two chromosomes of the 13-15 group (Denver classification). *Ann. Human Genet. 26*:151, 1962.

91. Warburton, D., and Fraser, F. C. Spontaneous abortion risks in man: Data from reproductive histories collected in a medical genetics unit. *Am. J. Human Genet. 16*:1, 1964.

92. Weber, W. W., Mamunes, P., Day, R., and Miller, P. Trisomy 17-18 (E): Studies in long term survival with respect of two autopsied cases. *Pediatrics 34*:533, 1964.

93. Zellweger, H., Beck, K., and Hawtrey, C. E. Trisomy 18. *Arch. Int. Med. 113*:598, 1964.

13

Reprinted from *Proc. Nat. Acad. Sci.* **58**:1104–1111 (1967)

HUMAN-MOUSE HYBRID CELL LINES CONTAINING PARTIAL COMPLEMENTS OF HUMAN CHROMOSOMES AND FUNCTIONING HUMAN GENES*

By Mary C. Weiss† and Howard Green

DEPARTMENT OF PATHOLOGY, NEW YORK UNIVERSITY SCHOOL OF MEDICINE

Communicated by Boris Ephrussi, June 26, 1967

This paper will describe the isolation and properties of a group of new somatic hybrid cell lines obtained by crossing human diploid fibroblasts with an established mouse fibroblast line. These hybrids represent a combination between species more remote than those previously described (see, however, discussion of virus-induced heterokaryons). They are also the first reported hybrid cell lines containing human components and possess properties which may be useful for certain types of genetic investigations.

Interspecific hybridizations involving rat-mouse,[1] hamster-mouse,[2, 3] and Armenian hamster–Syrian hamster[4] combinations have been shown to yield populations of hybrid cells capable of indefinite serial propagation. Investigations of the karyotype and phenotype of such hybrids have shown that both parental genomes are present[2, 5] and functional.[6] In every case, some loss of chromosomes has been observed; this occurred primarily during the first few months of propagation, usually amounted to approximately 10–20 per cent of the complement present in newly formed hybrid cells, and involved chromosomes of both parents. Recent studies have provided evidence of preferential loss of chromosomes of one parental species in interspecific hybrids.[2, 5]

A more extreme example of this preferential loss has been encountered in the human-mouse hybrid lines to be described, in which at least 75 per cent, and in some cases more than 95 per cent, of the human complement has been lost. It has been possible to relate the human characteristics of the hybrid phenotype to the number of human chromosomes retained. This has been shown for the colonial morphology of the hybrid and for the human antigens in the hybrid cell membrane, which were detected by mixed cell agglutination. Furthermore, information has been obtained with respect to the chromosomal localization of the thymidine kinase gene.

Materials and Methods.—Cell lines: LM (TK⁻) cl 1-D (hereafter referred to as cl 1-D), a subline of mouse L cells, was isolated by Dubbs and Kit[7] and provided by Dr. B. Ephrussi. This clone, deficient in thymidine kinase, is resistant to 30 μg/ml of 5-bromodeoxyuridine (BUDR).

WI-38, a diploid strain of human embryonic lung fibroblasts,[8] was provided by the American Type Culture Collection and was propagated in monolayer culture in this laboratory for 10 to 20 cell generations before hybridization.

Culture method and selection of hybrids: All cultures were maintained in standard growth medium (Dulbecco and Vogt's modification of Eagle's minimal medium containing 10% calf serum), in some cases supplemented with 30 μg/ml of BUDR, or hypoxanthine ($1 \times 10^{-4} M$), aminopterin ($4 \times 10^{-7} M$), thymidine ($1.6 \times 10^{-5} M$) (HAT). The selective system used was originally described by Littlefield[9] and has been modified by Davidson and Ephrussi[10] for use with a biochemically marked cell line combined with a contact-inhibited strain of diploid fibroblasts. The mouse cell line cl 1-D, deficient in thymidine kinase, is unable to grow in medium containing HAT; no spontaneous reversion to HAT resistance has been observed in this cell line, and it seems most likely that a deletion involving this gene has occurred.[11] The human diploid strain, containing thymidine kinase activity, is able to grow in this medium but it forms only a relatively thin cell layer, while hybrid cells are able to pile up against this background and form discrete colonies.

Karyotype of parental cells: Cl 1-D is characterized by the presence of 51 (50–55) chromosomes, 9 (8–10) of which are large metacentrics. Among the latter is the D chromosome[12] which, owing to the presence of a secondary constriction, appears to be dicentric and has no equivalent in the human complement. As can be seen in Figure 2a, most of the chromosomes of cl 1-D are telocentric.

WI-38 is a diploid human strain;[8] the cells contain 46 chromosomes, the pairs of which may be grouped into seven classes (Denver Classification). Of these, only two (A and G) contain chromosomes which are likely to be confused with the mouse chromosomes (Fig. 2b).

Demonstration of cell surface antigens by mixed hemagglutination: These experiments were carried out using a modification of the method described by Kelus, Gurner, and Coombs.[13] Immune sera were obtained from rabbits following injection of WI-38 cells in Freund's adjuvant. Complement was inactivated at 56° and the serum absorbed twice with a total of 6×10^6 cl 1-D cells per ml of serum. In order to test for the presence of human antigens, 2×10^5 trypsinized parental or hybrid cells were washed in buffered salt solution containing 0.1% bovine serum albumin, incubated with 0.1 ml antiserum for 1 hr at room temperature, again washed three times and mixed with 0.1 ml of a 2% suspension of washed human red blood cells. The mixture was centrifuged gently, resuspended, and a drop pipetted onto a glass slide for examination. A cover slip placed over the drop was allowed to settle for 5 min, so that erythrocytes attached to the surface of a test cell were forced to its perimeter and appeared as an encircling ring. In controls for specificity of the agglutination, mouse erythrocytes were employed in place of human. Mouse species-specific antigens were identified by the use of rabbit antimouse ascites tumor antiserum and mouse erythrocytes.

Results.—Production and identification of hybrid cells: Cultures were initiated with mixtures of 2×10^6 cl 1-D cells and 1×10^4 WI-38 cells. After four days of growth in standard medium, the cultures were placed in selective medium (HAT). The cl 1-D cells degenerated within seven days, leaving a single layer of human cells; after 14 to 21 days, hybrid colonies (Fig. 1) could be detected growing on the human cell monolayer. A number of these were isolated and grown to mass culture. In other cases the entire culture was transferred; within a few weeks the hybrid cells over-grew the remaining human fibroblasts and in the course of serial cultivation all human cells disappeared from the population.

Of three independent experiments performed, all yielded hybrid colonies, with a frequency of approximately one per 2×10^4 WI-38 cells. In all hybrids examined 20 generations after their formation, the same karyotypic pattern was found; all (or nearly all) of the expected mouse chromosomes were present, but of the human chromosomes only a minority remained, varying in number from 2 to 15 in the

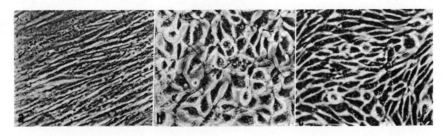

Fig. 1.—Phase contrast photomicrographs of living cells of (a) WI-38; (b) cl 1-D; (c) hybrid clone HM-2 P2. The parallel orientation of the highly elongated WI-38 cells is in contrast to the random orientation of the highly refractile and less fusiform cells of cl 1-D. The cells of the hybrid clone, while highly refractile, have a degree of orientation intermediate between cl 1-D and WI-38.

various populations. The mean numbers of human chromosomes found in six independent hybrid clones were as follows: 6.5, 2.0, 11.6, 10.9, 3.0, and 3.1. An occasional hybrid mitosis contained more than one copy of the same human chromosome, but most contained only one member of a given pair. The karyotype of a hybrid cell of clone HM-2 is shown in Figure 2c.

Loss of human thymidine kinase gene(s) from hybrid cells: Since survival of cells in HAT medium requires the presence of thymidine kinase, all hybrid cells selected in these experiments presumably contain the human gene(s) for this enzyme. This conclusion is supported by the fact that the cells can be killed by growth in the presence of BUDR. During continued propagation in HAT, any variants which may have lost this gene are eliminated. However, such variants occur with high frequency and were obtained selectively in a single step by transfer of the population to a medium containing BUDR, which eliminated most of the population and permitted the growth only of cells without thymidine kinase activity. As expected, these BUDR-resistant variant hybrid cells were not able to grow in HAT medium. They did retain many, but not all, of the human chromosomes which were present before the selection (see below). Cells resistant to BUDR also appeared in cultures propagated in standard medium without HAT and seemed to enjoy some selective advantage, for after several months of serial propagation they amounted to about 50 per cent of the population, as measured by HAT sensitivity. As no revertants to HAT resistance were obtained, BUDR resistance probably occurred by deletion.

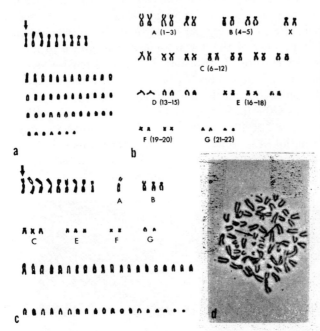

Fig. 2.—Chromosomes of Cl 1-D, WI-38, and hybrids. (*a*) Karyotype of the mouse parental line, cl 1-D. Nine long metacentric chromosomes are present, one of which (*arrow*, the D chromosome) is identifiable in all chromosome preparations. The karyotype also shows 43 telocentric chromosomes, none of which contains distinguishable satellites. (*b*) Karyotype of a cell of the WI-38 strain (see also ref. 8). Comparison with (*a*) shows that only chromosomes of groups A (long metacentrics) and G (small acrocentrics) are likely to be confused with chromosomes of cl 1-D. (*c*) Karyotype of hybrid clone HM-2. All chromosomes have been classified as to their origin, 14 of them being identified as human and 48 as mouse. In this karyotype, representatives from each of the major human chromosome groups except D are present. However, in other mitoses of HM-2, group D chromosomes were present. The A group chromosome is the only large bi-armed chromosome with a nonmedian centromere and is therefore probably a member of the no. 2 pair. (*d*) Mitosis of a hybrid cell of HM-1 after 60 generations of growth in HAT medium. The human chromosomes indicated by arrows are of group 6-12-X and are thought to carry the thymidine kinase gene. Two other human chromosomes are present, one of group D and one of group G.

Since most of the hybrid lines carried few human chromosomes, almost all of which could be distinguished from those of mouse origin, we attempted to correlate thymidine kinase activity with the presence of a specific human chromosome. Table 1 shows that nearly every cell in populations grown in HAT medium contains at least one chromosome of the group 6–12 and X, whereas chromosomes of this group are relatively rare in hybrid cells propagated in BUDR. In some of the hybrid metaphases, a submetacentric chromosome of this group has a peculiar configuration (Fig. 2d), in that there is slight separation of the chromatids in the region of the centromeres.[14] As shown in Table 1, this chromosome was identified in 15 of the 40 mitoses examined from HAT-grown populations and was not detected in any of the 34 BUDR-grown cells. This evidence would suggest that loss of the thymidine kinase gene occurs by loss of the relevant chromosome, and that this chromosome belongs to the 6–12-X group.

Although very soon after their formation human-mouse hybrids already showed extensive loss of human chromosomes, hybrid clones isolated later were found to be relatively stable karyotypically. However, it was expected that their human chromosome content would depend to some extent upon the selective conditions employed; therefore after 30 generations in HAT medium, hybrid cells were propagated in (1) HAT, (2) standard medium, and (3) BUDR. Two clonal hybrid populations were chosen for this study, HM-1 and HM-2, characterized by the presence of 6 and 11 human chromosomes, respectively.

Figure 3 shows the chromosomal changes which occurred during four to six months of serial culture. In both lines there was extensive loss of human chromosomes correlated with number of generations of propagation. The populations grown in HAT medium, checked after 20–60 generations, showed very little change, but after 85 generations there was a decrease by one half in the number of human chromosomes. This change appeared more marked in the absence of HAT or in BUDR medium. Under these conditions, a proportion of the cells (10–25%) of a number of clones lost all identifiable human chromosomes.

Figure 3b also shows the chromosomal composition of three subclones isolated from the HM-2 HAT-grown population after 85 generations. Two of these clones (HM-2 P5 and HM-1 P2), which were selected on the basis of humanlike colonial morphology, contained a larger number (9–12) of human chromosomes than the average of the parent population, while the third clone, which strongly resembled cl 1-D, nevertheless contained approximately six human chromosomes. It is clear that in all hybrid populations very little mouse genetic material was lost;

TABLE 1

Chromosomes of HAT-Resistant and BUDR-Resistant Hybrid Clones

Hybrid line	Total no. of cells counted	No. of cells containing member of 6-12-X group	Average no. of human chromosomes per cell
HM-1 HAT	20	17* (7)	6.0
HM-1 BUDR	20	5 (0)	3.4
HM-2 HAT	20	18* (8)	4.9
HM-2 BUDR	14	2 (0)	1.8

In parentheses are the number of cells containing the chromosome of distinctive appearance described in text.
* Failure to identify 6-12-X chromosome(s) in all cells of these HAT-grown populations could be due to chromosomal rearrangements, such as translocations. A cell which has lost the thymidine kinase-bearing chromosome might still be able to complete one or two cell divisions in HAT medium.

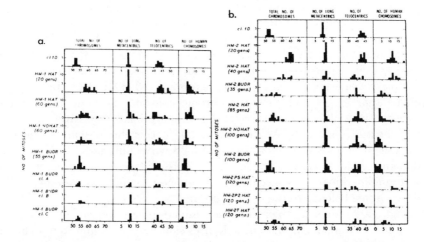

Fig. 3.—Histograms showing evolution of the karyotype of (a) HM-1 and (b) HM-2. The chromosomes of the hybrids are grouped into the following classes: (1) total number of chromosomes; (2) number of long metacentrics (this class is composed almost entirely of chromosomes of mouse origin but might also include one or two group A human chromosomes); (3) mouse telocentrics (in some cases, members of the group G human chromosomes would be placed in this class if the chromatin distal to the centromere were not visible as two distinct masses); and (4) number of human chromosomes. It is likely that these values are low by perhaps one or two chromosomes, due to failure to distinguish group A, and perhaps group G, chromosomes. Numbers on abscissa refer to bars immediately to their right.

on the average there was a decrease in number of mouse telocentrics by three or four and no decrease in number of long metacentrics. Some chromosomal rearrangements may have occurred among the mouse telocentrics in a minority of the population of HM-1, since those cells which contain fewer than expected telocentric chromosomes contain more than the expected number of long metacentrics and it appears likely that these changes were the result of centric fusion of telocentrics. (For a more detailed discussion of this process see ref 2.) However, in most hybrid populations no evidence of chromosomal rearrangement was found.

Cell membrane antigens and hybrid karyotype: The presence of human species-specific antigens in the cell membrane of hybrid cells was determined by mixed cell agglutination. A cell was considered positive if two thirds of its perimeter was covered with erythrocytes. By this criterion 60 per cent of the cells of WI-38 were positive under the standard conditions. Values given by hybrid cells are shown in Figure 4 on an arbitrary scale denoted as relative agglutination index (R.A.I.; WI-38 = 1.0) plotted against the mean number of human chromosomes per cell. Cl 1-D gave a value of 0.005, which can be considered as background due to nonspecific adherence of erythrocytes. All hybrids gave agglutination values in excess of this, and the values increased with increasing number of human chromosomes up to 12, at which point the R.A.I. became equal to that of the WI-38 parent. The R.A.I. rose very slowly with chromosome number up to about five chromosomes per cell, and then more sharply until the maximum was reached at 12 chromosomes. Similar results were obtained when less stringent criteria were used for scoring agglutinations; under these conditions a more nearly linear relation was observed

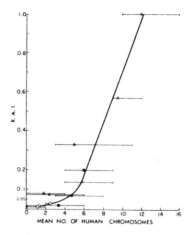

Fig. 4.—Human cell membrane antigens in hybrid lines. Ordinate gives relative agglutination index calculated from the fraction of cells showing positive hemagglutination, taking the values given by WI-38 as equal to 1.0. Abscissa gives mean human chromosome number per cell and the range for each population. Since the subclones have the narrowest range of human chromosome number, their agglutination values are probably the most significant.

● HM-1
○ HM-1 subclones
▲ HM-2
△ HM-2 subclones
× cl 1-D control

between human chromosome number and R.A.I., but the background values given by cl 1-D were much higher.

The agglutination values obtained from the hybrids using antimouse cell serum and mouse erythrocytes were practically identical with those given by cl 1-D under the same conditions. This is consistent with the fact (Fig. 3) that virtually the entire cl 1-D genome is present in all the hybrids.

In control experiments no mixed agglutination was observed when hybrid cells were tested with antimouse serum and human erythrocytes, or with antihuman serum and mouse erythrocytes.

Poliovirus infection of hybrid cells: Somatic cell hybrids may be useful for the study of the process of viral infection. For example, it has been shown in heterokaryons made between cells of two different species that virus unable to multiply in one parental cell type is able to do so in the common cytoplasm.[15] However, in our hybrids containing up to 12 human chromosomes poliovirus infection at high multiplicity did not lead to any detectable cytocidal effect, under conditions in which the parental strain WI-38 was completely destroyed within 24–36 hours. While poliovirus itself does not successfully infect mouse cells, L-cells, from which cl 1-D is derived, are known to support viral multiplication very well following infection with poliovirus RNA.[16] Therefore, it seems most likely that failure of the virus to kill the hybrid cells is due to inability of the virus to penetrate the cell membrane, due either to lack of essential human elements or to interference by mouse elements.

Discussion.—The production and properties of a number of different interspecies hybrid lines, including some made between diploid strains and established cell lines, have been summarized by Ephrussi.[17] It has also been shown by Harris and Watkins[18] that virus-induced heterokaryons between HeLa (human) and Erhlich ascites (mouse) cells underwent at least one or two divisions as some hybrid mitoses could be identified. Though no established lines of hybrid cells developed from these heterokaryons, Yerganian and Nell[4] were able by this method to produce established lines of rodent cell hybrids. It might have seemed doubtful whether a permanent hybrid line could be made between a human fibroblast strain and an established mouse line since, in addition to the remoteness of the two species, human fibroblasts do not spontaneously develop into established lines as do those of rodents, but uniformly die out after 50–100 cell generations.[8, 19] Though it turned out that these factors did not prevent the formation of hybrid lines, they may have been the cause of the very extensive loss of human chromosomes from the hybrid cells.[20] This loss began before 20 generations after fusion and most likely during the first few divisions, as the colonial morphology did not change appreciably from the time colonies were first identified to the time pure cultures were obtained.

After the development of hybrid populations, continued subculture results in continuing though slow elimination of human chromosomes. Since nearly the entire cl 1-D complement is retained, the human genes are probably not essential for viability, with the exception of that providing thymidine kinase activity, which is necessary as long as the cells are cultivated in HAT. Omission of HAT from the medium permits variants lacking thymidine kinase to survive, as indicated by the appearance of HAT-sensitive cells. When a hybrid population is placed in BUDR, cells bearing thymidine kinase are eliminated and the entire population becomes HAT-sensitive.

These losses of the thymidine kinase gene appear to occur by loss of the relevant chromosome(s) since: (1) there is a small decline in mean human chromosome number associated with continued growth in the absence of HAT, and in the BUDR-resistant populations; (2) there is a rather strong correlation between the absence of a definite chromosome of the 6-12-X group, BUDR resistance, and HAT sensitivity; and (3) there are no obvious chromosomal rearrangements by which genetic material of that chromosome might have been retained.

It has been shown in a number of cases that antigens characteristic of the parent cells are expressed in the hybrids. Among the antigens which follow this rule are H-2 antigens of the mouse,[21] and polyoma virus-induced T-antigens and surface antigens.[22] It is therefore not surprising that human species-specific cell membrane antigens could be detected in all of the hybrids described here. Even the presence of a very small number of human chromosomes (<4) produced an R.A.I. of 2–15 times background. The genes for surface antigens appear to be widely distributed among the different human chromosomes, since agglutinability of the hybrid cells increased progressively with number of human chromosomes.

Clones of hybrid cells selected on the basis of humanlike morphology (HM-2P2 and HM-2P5) contained the largest number of human chromosomes (9.5 and 12) and gave the highest agglutination values. On the other hand, a clone selected on the basis of absence of humanlike morphology (HM-2T) had six human chromosomes and a corresponding R.A.I. This suggests that some human chromosomes may have little to do in determining colonial morphology, though they do determine cell membrane composition. Similar hybrids may permit the chromosomal localization of other human genes expressed at the cellular level, such as those for blood group antigens and enzymes whose physical properties are different from those of mouse determination. It should also be possible to isolate clones of hybrid cells which have lost all human chromosomes and which might be useful for the investigation of nonchromosomal genes.

Summary.—Cocultivation of a human diploid cell strain with a thymidine kinase deficient mouse cell line has led to the formation of hybrid cell lines, which were isolated in selective medium. The hybrid lines contained substantially the entire mouse genome and a greatly reduced complement of human chromosomes. The functioning of the human genes was shown by the presence of human antigens on the surface of the hybrid cells. The agglutinability of the cells in mixed hemagglutination tests depended on the number of human chromosomes contained (up to 12), indicating that the genes for surface membrane antigens are widely distributed among the human chromosomes. The human gene for thymidine kinase permits the hybrid cells to grow in medium containing aminopterin, but variants lacking this function may be selected in medium containing 5-bromodeoxyuridine. These variants appear to have lost a human chromosome of the 6-12-X group and it is suggested that this chromosome contains the thymidine kinase gene. Continued growth of hybrid lines results in slow elimination of human chromosomes. Study of clones containing a small number of human chromosomes should permit the localization of other human genes.

It is a pleasure to acknowledge the kind assistance of Mr. C. deSzalay and Dr. L. J. Scaletta and the valuable advice of Drs. G. J. Todaro and Z. Ovary.

* Aided by grant CA 06793, postdoctoral fellowship 1-F2-GM-34,679 (M.W.), and award 4-K6-CA-1181 (H.G.), all from the U.S. Public Health Service.

† Present address: Carnegie Institute of Washington, Department of Embryology, 115 W. University Parkway, Baltimore, Maryland.

[1] Ephrussi, B., and M. C. Weiss, these PROCEEDINGS, **53**, 1040 (1965).

[2] Scaletta, L. J., N. Rushforth, and B. Ephrussi, *Genetics*, in press.

[3] Davidson, R. L., B. Ephrussi, and K. Yamamoto, these PROCEEDINGS, **56**, 1437 (1966).

[4] Yerganian, G., and M. B. Nell, these PROCEEDINGS, **55**, 1066 (1966).

[5] Weiss, M. C., and B. Ephrussi, *Genetics*, **54**, 1095 (1966).

[6] *Ibid.*, p. 1111.

[7] Dubbs, D. R., and S. Kit, *Exptl. Cell Res.*, **33**, 19 (1964).

[8] Hayflick, L., and P. S. Moorhead, *Exptl. Cell Res.*, **25**, 585 (1961).

[9] Littlefield, J., *Science*, **145**, 709 (1964).

[10] Davidson, R., and B. Ephrussi, *Nature*, **205**, 1170 (1965).

[11] Kit, S., D. R. Dubbs, L. J. Piekarski, and T. C. Hsu, *Exptl. Cell Res.*, **31**, 297 (1963).

[12] Hsu, T. C., *J. Natl. Cancer Inst.*, **25**, 1339 (1960).

[13] Kelus, A., B. W. Gurner, and R. R. A. Coombs, *Immunology*, **2**, 262 (1959).

[14] This configuration was not seen in the human parent cell and may have occured as a consequence of the new environment of the chromosome in the hybrid cell. It is also possible that this chromosome arises through translocation and is only part human, but this seems unlikely as it was seen in three hybrid populations of independent origin.

[15] Koprowski, H., F. C. Jensen, and Z. Steplewski, these PROCEEDINGS, **58**, 127 (1967).

[16] Holland, J. J., B. H. Hayer, L. C. McLaren, and J. T. Syverton, *J. Exptl. Med.*, **112**, 821 (1960).

[17] Ephrussi, B., "Phenotypic Expression," *In Vitro* (Baltimore: Williams and Wilkins, 1967) vol. 2, p. 40.

[18] Harris, H., and J. F. Watkins, *Nature*, **205**, 640 (1965).

[19] Todaro, G. J., and H. Green, *Proc. Soc. Exptl. Biol. Med.*, **116**, 688 (1964).

[20] A more general discussion of some of the mechanisms of loss of chromosomes in interspecific somatic hybrids has been given in: Ephrussi, B., and M. C. Weiss, in *Control Mechanisms in Developmental Processes*, ed. M. Locke (New York: Academic Press, in press).

[21] Spencer, R. A., T. S. Hauschka, D. B. Amos, and B. Ephrussi, *J. Natl. Cancer Inst.*, **33**, 893 (1964).

[22] Defendi, V., B. Ephrussi, H. Koprowski, and M. C. Yoshida, these PROCEEDINGS, **57**, 299 (1967).

14

Reprinted from *J. Exp. Med.* **113**:1155–1172 (1961)

STUDIES ON CELL LINES DEVELOPED FROM THE TISSUES OF PATIENTS WITH GALACTOSEMIA*

By ROBERT S. KROOTH, M.D., AND ARNOLD N. WEINBERG, M.D.

(From the National Institute of Neurological Diseases and Blindness, and the Laboratory of Cell Biology, National Institute of Allergy and Infectious Diseases, and the National Institute of Arthritis and Metabolic Diseases, National Institutes of Health, Bethesda)

PLATES 112 AND 113

(Received for publication, January 3, 1961)

The value of studying inherited metabolic abnormalities in cell lines developed from the tissues of affected patients has been emphasized by Böök and Kostman (1956), and more recently by Kalckar (1959a) and Luria (1959), among others. Such cell lines may eventually prove useful for the study of genetic changes such as back mutation, recombination, and transformation. In addition, they should occasionally yield fresh information about the nature of the disease itself.

One consideration, however, sharply limits the usefulness of cell culture in the study of genetically determined metabolic abnormalities. Cells which propagate for long periods in culture usually retain few, if any, of the specialized features of their tissue of origin (see Levintow and Eagle, 1960). Thus, a "liver" or "muscle" culture is not generally a collection of cells which can be shown to resemble liver or muscle. They are merely the cells which grew out when the specialized tissue was cultured.

The reason that cultured cells do not in general resemble their tissue of origin is unknown. It does seem clear, however, that if one wishes at present to study a genetic disease among actively growing cells, the disease must fulfill at least one requirement: the abnormality should be demonstrable in most, if not all, of the tissues of the body. That is, the disease must not be one where the function deranged is peculiar to some single tissue, for that function may well "disappear" in cultures of cells from normal as well as affected donors.

The disease galactosemia appears to fulfill the requirement. The enzyme, galactose-1-phosphate uridyl transferase has been found in all normal tissues where it has been looked for. In the corresponding tissues of galactosemic patients, activity of the enzyme cannot be demonstrated (Kalckar, 1959a). The disease is due to a recessive gene with heterozygotes having enzyme activities about one-half normal (Kirkman and Bynum, 1959).

---

* Some of these investigations have been summarized in a preliminary note published elsewhere (Krooth and Weinberg, 1960).

In the present paper, growth and metabolic studies are described using cell lines developed from the tissues of galactosemics, non-galactosemics, and a presumed heterozygote. Enzymatic studies upon these same cell lines will be reported (Bias *et al.*, 1961).

Materials

Clinical Data.—

A summary on the patients used in these studies is given in Table I. It should be noted that the three patients with congenital anomalies have been karyotyped and found euploid.[1] The presumably heterozygous patient was the mother of patient JDU. Brief clinical descriptions of the two galactosemic patients are as follows:

TABLE I

Clinical Data

Patient (line)	Age of patient	Sex	Tissue biopsied	Diagnosis
RCU	4½ mos.	Female	Skin	Multiple malformations
BY	28 yrs.	Male	"	Galactosemia
BE	21 "	Female	"	Primary amyloidosis
JDU	11 "	"	"	Galactosemia
JDU (R)	11 "	"	Marrow	"
MAD	38 "	"	Skin	Galactosemia heterozygote (Mother of JDU)
SK	3 "	"	"	Multiple malformations
WIM	2½ "	Male	"	"
SE	20 "	"	"	Wilson's disease
MI	27 "	"	Marrow	Psoriasis and psoriatic arthritis

BY: 28 Year Old Male: After an uneventful delivery and neonatal period this patient developed milk intolerence, nausea, vomiting, and failed to gain weight. At 3 months of age he was hospitalized and found to have malnutrition, anemia, hepatosplenomegaly, albuminuria, and a urinary reducing substance which was later identified as galactose. On the institution of a galactose-free diet the melituria and albuminuria cleared, the patient gained weight rapidly, and became symptom-free. Except for the development of cataracts and mild mental retardation he has remained essentially well, though a restriction of galactose was never totally enforced. The patient has a negative family history for galactosemia, and no history of parental consanguinity.

The diagnosis of galactosemia has been confirmed by assays for galactose-1-phosphate uridyl transferase in red blood cells and liver (Anderson *et al.*, 1957).

JDU: 11 Year Old Female: The patient was born into a family (originally reported by Goldbloom and Brickman, 1946) with two older siblings, both of whom were affected with galactosemia. Galactose was omitted from her diet from birth, although it is uncertain how strictly this regimen was enforced. While the patient has never had the clinical symptoms or

---

[1] Karyotyping was carried out in collaboration with Dr. Hin Tjio, and the technique and detailed analysis will be published later.

signs of galactosemia, she has an abnormal galactose tolerance test, frequent urines with reducing substance, and a hemolysate of her red blood cells is unable to oxidize galactose-1-C^{14} to $C^{14}O_2$.

Composition of Media and Solutions.—

    A. Growth medium:

	By volume
Pooled whole human sera[2]	12 per cent
NCTC109 (McQuilkin *et al.*, 1957)	5 " "
Minimum essential medium (Eagle, 1959)	83 " "

Supplement	*Final concentration*
Pyruvate	1 millimolar
"Non-essential"	
Amino acids each at	1 millimolar

    (Alanine, proline, serine, glycine, glutamic acid, asparagine, and aspartic acid)

    B. Experimental Media:

	By volume
Dialyzed pooled human sera	12 per cent
Hexose-free minimum essential medium	88 " "

The supplements are identical with those used for the growth medium, except that hexose is added as a supplement. The kind and final concentration of added hexose will be described subsequently. Sodium pyruvate was kindly prepared for us by Dr. Leon Levintow (Levintow and Price, 1952, Montgomery and Webb, 1954). No commercial pyruvate was used.

Trypsin Solutions.

    Trypsin (1:300, Nutritional Biochemicals, Cleveland) was dissolved at concentrations of 0.05 and 0.25 per cent in a solution described by Puck *et al.* (1958), and is 40 per cent medium N 15 and 60 per cent saline G in their notation. 1 per cent of antibiotic solution (Puck *et al.*) was added.

Conditions of Dialysis.—

    Pooled whole human serum, free of hemoglobin, is dialyzed 1 liter at a time for 21 hour[s] against running tap water and then 3 hours against running distilled water, at about 5°C. The rate of flow is 425 ml./minute. The dialysis bag consists of cellulose casing (The Visking Corporation, Chicago) with an inflated diameter of 30/32 inches. The bag is mounted on a rocking table throughout the whole of the dialysis. Following the dialysis, the serum is free of glucose when tested by the method discussed by Chang (1960) using glucose oxidase on paper sticks (clinistix, Ames Co., Inc., Elkhart, Indiana).

Source of Hexose.—

    D-glucose and/or D-galactose (both Nutritional Biochemicals) were added to the experimental media. The galactose was first recrystallized from 70 per cent ethanol (after the method used by Kalckar *et al.*, 1959b) to remove contaminating glucose. Galactose-1-C^{14} (specific activity 4.72 μc. per mg.) was obtained from the National Bureau of Standards and was chromatographically pure. Glucose-1-C^{14} (specific activity 6.74 μc. per mg.) was obtained from the same source.

---

[2] For cell lines developed from bone marrow, 40 per cent type A_1 B single donor human serum is used; the other constituents of the growth medium are the same.

Methods

Development of the Cell Line From the Biopsy:

A technique for developing permanent euploid cell lines from human biopsies was published by Puck, Ciecura, and Robinson in 1958. Although our methods differ somewhat from theirs, the principles are the same.

A. Skin Biopsies.—The area biopsied in every case was the lateral surface of the shoulder, at about the level of the neck of the humerus. The skin was thoroughly scrubbed with phiso-hex,[3] using a surgical brush, and then prepared first with zephirin[3] and finally with 70 per cent ethanol. As much friction as possible was used in the preparation to ensure absolute sterility. 1 per cent procaine or xylocaine was infused intradermally and then subcutaneously, and a disc of skin 4 mm. in diameter was cut with a skin punch. Care was taken to include dermis, and subcutaneous fat was always visible when the disc was lifted up and its pedicle cut.

The tissue was placed in a dish, cut into 5 to 8 pieces with newly sharpened conjunctival scissors, and the pieces then partially minced. 5 ml. of 0.25 per cent trypsin solution were added, and the suspension of explants was transferred to a 25 ml. Erlenmeyer flask. 5 to 10 additional ml. of trypsin were added. The flask was stoppered and placed in a water bath shaker, adjusted to 37.5°C. Following agitation for 25 minutes, the suspension (now consist-ing of explants and cells) was transferred to a centrifuge tube and spun at 500 to 1500 R.P.M. for 10 minutes. The supernatant trypsin was removed, and the explants and cells were re-suspended in 6 ml. of growth medium. The suspension was divided among 3 Petri dishes (2 inches in diameter) and the dishes were incubated in 5 per cent CO_2 (saturated with water vapor) at 37.5°C. At 48 to 72 hours, the medium was removed (care being taken to leave the explants in the dish) and 0.60 to 0.75 ml. of fresh medium was added to each dish. Thereafter, this volume of medium was used and was changed every 48 hours. Large quantities of medium (of the order of 4 ml.) appeared to prevent growth when working with these minute quan-tities of tissue. In 4 or 5 days, individual "fibroblasts" (fusiform, transparent cells) could be seen attached to the glass, mainly in the immediate vicinity of the explants, from which most of them appeared to migrate. Epithelial outgrowths occurred from a number of explants. The epithelial monolayers eventually disappeared, and no epithelial cells were noted after trypsinization for the first subculture. Fibroblastic and epithelial outgrowths are shown in Fig. 1. This finding is consistent with that reported by Puck *et al.*, 1957.

15 to 25 days after the biopsy, from 10 to 20 per cent of the surface of 2 or 3 primary dishes was covered with one or more colonies of confluent cells; 1 dish was then subcultured with 0.25 per cent trypsin solution, the explants being left in the original dish. After decanting the trypsin suspension, the explants were reincubated with growth medium.

B. Bone Marrow Propagation.—Sternal marrow aspirates were obtained by the usual technique. The aspirate was collected in 5 ml. of a solution of 40 per cent N 15 (Puck *et al.*, 1958) and 60 per cent of their saline G to which 1 per cent of their antibiotic solution and several drops of 100 mg./ml. aqueous heparin had been added. The suspension was centri-fuged for 10 minutes at 500 to 1500 R.P.M. and the fat floating on top of the supernatant and the buffy coat were inoculated into a total of 3 to 6 small tissue culture flasks (T_{15}), each con-taining 3 ml. of marrow growth medium. Thereafter, except for the serum composition of the medium (see Materials) the cultures were handled like those developed from skin. It should be noted that vast numbers of recognizable marrow elements attached to glass (and may even

---

[3] Winthrop-Stearns, Inc., Rensselaer, New York.

have proliferated) in the primary bottles, but these did not come over on subculture. Here, as with skin, experiments were performed with the "fibroblastic" cells which eventually predominated. Marrow fat and the buffy coat appeared to yield these cells about equally.

Subculturing; Propagation; Computation of M-Numbers:

Cells were incubated with trypsin solution (in an air atmosphere) at 37.5°C. for 15 minutes to digest them off the glass. 0.25 per cent trypsin was used for the first subculture and 0.05 per cent trypsin for the subsequent ones. The trypsin suspension was centrifuged, the supernatant removed, and the cells resuspended in growth medium for reinoculation. Following the first subculture, larger volumes of media were employed. At each feeding about 0.2 ml of media were used for every square centimeter of surface area available for cell growth in the containing vessel.

Cells were at all times maintained as monolayers in Petri dishes and bottles, and after the first subculture were fed every 72 hours. When the cells of the layer became confluent, a fraction of them was subcultured into a fresh vessel. The fraction varied from $\frac{1}{2}$ to $\frac{1}{100}$. A continuous record of the fraction by which each culture was split and the size of the surface it covered was kept. From these records the M-number, the minimum number of times the cells had increased as of a given moment, could be computed. All "growth" of cells in the primary dish, the one containing the explant, was assumed to be exfoliation, and growth was not counted until the first subculture. Almost certainly we thereby underestimated cell growth.

After subculturing from the primary dish, the explants were fed until a new monolayer was generated—the new cells apparently coming largely from a fringe of the previous monolayer adhering to the explant. When inadequate growth or other difficulties were encountered with a cell line the line was started again with another wave of cells from the primary dish. In computing M-numbers growth in the primary dish was still ignored. We have thus far found no differences between successive samples of cells obtained in this way. However a discrepancy between the M-number and the age of the culture results from this practice.

Experimental Techniques:

Growth Experiments.—After at least two subcultures, the cells derived from skin were used in growth experiments: a large flask (a T_{60} or a Blake bottle) of cells was subcultured into 20 to 100 smaller flasks (T_{15}'s). 24 hours later, after the cells had attached and spread, they were washed with Eagle's (1959) minimum essential medium from which the hexose had been omitted. A random sample of 4 or 5 bottles was then taken to determine the initial cell protein. The remaining bottles were divided into 5 groups, and each group was overlaid with experimental media containing a different concentration of hexose. The five concentrations of hexose used were:

    100 mg. per cent glucose
    100 mg. per cent galactose
      5 mg. per cent glucose
     95 mg. per cent galactose and 5 mg. per cent glucose (mixed hexose)
    Hexose—free

Cells were not exposed to galactose prior to the experiments. Each kind of medium was changed every 72 hours. Between 9 and 18 days after placing the cells in experimental media, the cell protein was again determined. Measurement of cell protein was by the method of Oyama and Eagle (1956). Cell protein is expressed in terms of the amount of bovine serum albumen giving the same optical density with the reagents used. (In the graphs and tables one unit of protein corresponds to 100 μg. of bovine albumen. All values are based on the mean of 2 to 4 replicate bottles—usually 3).

TABLE II

Growth Experiments with Non-Galactosemic Cells (AA)

Line	BE[11]	RCU[8]	BE[12]	SK[6]	RCU[11]	BE[14]	SK[7,*]	SK[7,*]	WIM[3]	BE[20]	SE[8]	SE[11,*]	SE[11,*]
Duration, days	9	9	9	9	9	9	9	15	18	14	13	9	16
Initial protein	1.0	1.0	1.1	2.1	0.8	0.8	0.8	0.8	0.8	1.0	1.1	0.9	0.9
100 Glu	5.2	3.4	5.5	8.0	2.2	4.5	2.4	7.4	9.4	7.3	7.5	7.0	8.3
100 Gal	6.1	3.2	—	9.4	2.5	5.1	2.4	7.1	9.0	7.1	8.0	6.3	9.7
95/5	5.5	4.4	7.6	9.5	2.7	5.0	3.2	9.6	11.6	8.0	10.5	6.5	9.4
5 Glu	6.9	3.2	6.7	7.0	2.3	5.5	2.6	7.2	—	8.5	9.0	6.6	8.2
Hexose-free	2.5	1.6	1.8	2.2	1.1	1.0	1.0	1.3	2.3	1.9	2.1	1.8	2.1
"Age", days	79	86	86	44	107	100	60	63	61	153	84	100	107
M-No.	1.7×10^5	8×10^3	5×10^5	15	8×10^5	6.4×10^6	290	290	4	4.1×10^{12}	8×10^3	6.4×10^4	6.4×10^4

100 Glu and 100 Gal refer to media containing respectively 100 mg. per cent glucose and 100 mg per cent galactose. 5 Glu and 95/5 refer to media containing respectively 5 mg. per cent glucose and the mixture of 95 mg. per cent galactose and 5 mg. per cent glucose. The superscript on the letters designating the line is 1 plus the number of subcultures. Each unique combination of letters represents a different patient. Age is the time between the date of the biopsy and the date of the experiment.

* Part of a multiple point experiment.

191

TABLE III

Growth Experiments with Galactosemic Cells (aa)

Line	JDU[5]	BY[5]	JDU[7]	JDU[7],*	BY[6]	JDU[7],*	JDU[5]	JDU[10]	JDU,[12]*	JDU,[12]*
Duration, *days*...	9	9	9	9	18	15	18	27	9	16
Initial protein ...	1.3	0.9	0.7	0.6	0.8	0.6	0.9	0.6	0.6	0.6
100 Glu.........	5.2	3.3	2.0	1.5	3.9	2.2	8.5	4.9	4.9	5.7
100 Gal.........	2.0	1.5	1.1	0.9	1.1	1.2	3.0	1.4	1.8	1.8
95/5............	5.3	3.0	1.6	1.2	2.7	1.7	8.7	3.5	3.3	3.9
5 Glu...........	5.9	3.4	2.0	1.7	3.4	2.2	—	4.5	3.8	4.8
Hexose-free......	1.9	1.1	1.0	0.9	1.3	1.0	2.6	1.5	1.7	1.5
"Age" *days*......	55	104	69	71	122	77	77	145	147	154
M-Nos...........	60	12	640	120	80	120	80	1.1×10^4	4.8×10^5	4.8×10^5

See Table II for notation.

* Part of a multiple point experiment.

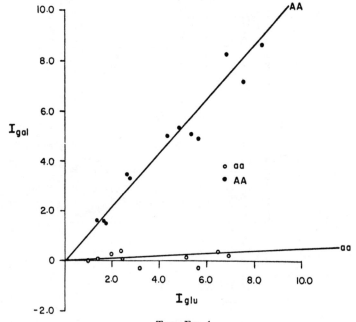

TEXT-FIG. 1

Two types of growth experiment were performed: in the "2-point" experiments only the initial cell protein and the cell protein at the end of the experiment was determined. In the "multiple point" experiments, cell proteins were determined on bottles removed every 72 hours. In most of the two point experiments and all the multiple point experiments, galactosemic cells were run concurrently with non-galactosemic cells, the two kinds of cells being fed with the same media.

Isotope Experiments.—Equal aliquots of cells were incubated with glucose-*1*-C^{14} or galactose-*1*-C^{14}, and the activity of the $C^{14}O_2$ produced was determined. The techniques were those of Weinberg and Segal (1960).

Growth Experiments:

A. The Homozygous Lines.—

Relative growth in glucose and galactose: Tables II and III contain the data from our two point experiments. Included also in the Tables are the data from the 9th and final day of those multiple point experiments in which the cells were grown in all 5 experimental media. Note that among the non-galactosemic cells (Table II) the growth in galactose is about equal to the growth in glucose. The galactosemic cells, however, grow better in glucose. The growth in galactose is about equal to the growth in medium which is hexose-free.

In Text-fig. 1, using the data from Tables II and III, growth ascribable to glucose is plotted against growth ascribable to galactose for galactosemic (aa) and non-galactosemic cells (AA). The index I_{glu} is defined:

$$I_{glu} = \frac{\text{Growth (final cell protein) in medium containing 100 mg. per cent glucose } \textit{minus} \text{ growth in hexose-free medium}}{\text{Inoculum (initial cell protein)}}$$

I_{gal} is the corresponding function for galactose.

The two types of cells appear to differ sharply. The aa line may not have a zero slope, perhaps due to the presence of minute quantities of glucose in the galactose; glucose is known to persist as a contaminant of galactose even after two crystallizations in 70 per cent ethanol (Kalckar, 1960).

In Tables II and III, note the marked variation from experiment to experiment in the absolute growth rate. Some of this variation is probably spurious. A few of the experiments (*e.g.*, one JDU[5] and one JDU[10] in Table III) were deliberately allowed to run for a long period of time after the cells had built up to see if the galactosemic cells in galactose would ever tend to catch up with the ones in glucose. In addition, we have evidence that there is variation from experiment to experiment in the population density at which the cells plateau, a phenomenon which is easily confused with a slow growth rate in two point experiments. Beyond this, however, there is a true variation in the absolute parameter of growth from experiment to experiment. It is of interest that the differences in growth rate and in the density at which the cells plateau usually characterize the *experiment* rather than the line. Cells from an individual donor have not in most cases tended consistently to grow well or poorly. It is our impression that the variation we observe may be due chiefly to the character of the serum in the medium (Puck *et al.*, 1957), possibly involving immunological reactions between serum and cell, rather than to some intrinsic property of the cell. Superior and more uniform growth appears to result when pretested serum is employed in our hands as well as Puck's (1957). However, adequate screening has not always been possible.

In Text-fig. 2, data from a multiple point experiment are given. Both the

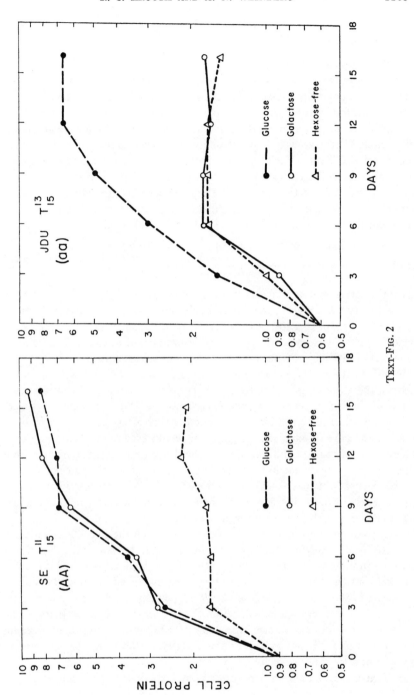

TEXT-FIG. 2

galactosemic and non-galactosemic cells grew well in this experiment. Note that the galactose curve in the case of the non-galactosemic cells follows the glucose curve, whereas in the galactosemic cells it winds about the hexose-free curve.

In Fig. 2, photographs are shown of galactosemic and non-galactosemic cells growing in the two sugars. The photographs are of replicate bottles and were taken at 9 days.

Galactose sensitivity: Growth in media containing 5 mg. per cent glucose was compared with growth in media containing the mixture of 95 mg. per cent galactose and 5 mg. per cent glucose. We wished to see if galactose sensitivity of the sort observed in the Gal-1-P uridyl[4] transferase mutants of *Escherichia coli* (Kurahashi and Wahba, 1957 and 1958, Kalckar *et al.*, 1957) occurred among human galactosemic cells.

Our multiple point experiments when pooled yielded 14 observations in which mean growth was simultaneously recorded in the medium containing the mixture of 95 mg. per cent galactose and 5 mg. per cent glucose (95/5 medium) and the medium containing 5 mg. per cent glucose (0/5 media). The data[5] are as follows:

Presumed genotype	Growth in 95/5 greater than growth in 0/5	Growth in 0/5 greater than growth in 95/5	Growth equal	Total
AA	8	3	3	14
aa	1	11	2	14

It would appear from the data on the non-galactosemic cells that growth in the mixed media is, if anything, better than growth in 5 mg. per cent glucose alone. Among the galactosemic cells, however, growth in the mixed medium is usually poorer than growth in glucose alone. Perhaps the superior growth of the *AA* cells in 95/5 is due to the greater concentration of hexose which seems important at high population densities or perhaps to some synergistic combination of the two hexoses—we are not sure which. In either case one would not expect better growth in 95/5 among the *aa* cells since they appear to be unable to utilize galactose. However, inability to utilize galactose does not explain the fact that the *aa* cells do *less well* in the medium containing galactose. Therefore, these data at least suggest that the growth of galactosemic cells is inhibited by galactose.

The significance of these data can be tested by chi square, letting the assumption of equal growth in 95/5 and 0/5 generate the expected numbers. On this

---

[4] Galactose-1-phosphate uridyl transferase.

[5] Growth was considered greater in one medium than in another if the means of 2 to 4 replicates in the two differed by an amount of protein equivalent to at least 10 μg. of bovine albumen.

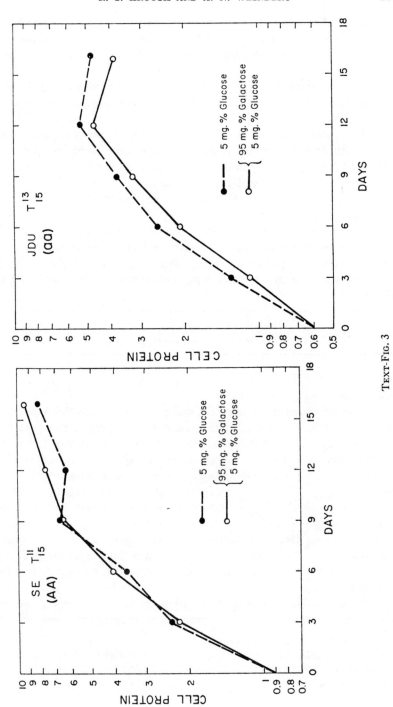

TEXT-FIG. 3

assumption,[6] there should be as many cases where growth is better in 95/5 as cases where growth is poorer, that is, 6 and 6, rather than 1 and 11. Chi square (using Yates' correction) is 6.8 for one degree freedom, which is significant.

In Text-fig. 3, a multiple point experiment is graphed. The effect is small. Note that the action of galactose appears to delay growth for the first 72 hours or less. Thereafter, the curves rise with equal slope. This type of an effect is similar to that reported by Kurahashi and Wahba (1957) in certain of their experiments where transferase mutants of *E. coli* were grown in a mixture of glucose and galactose. Our effect, considering the 19:1 ratio of galactose to glucose we use, is much smaller than the one they noted. However, they also noted that the galactose sensitivity could be abolished by the addition to the medium of 500 mg. per cent whole yeast extract. It is possible that the dialyzed serum in our experimental media is partially supplying the same unknown principle.

The two point experiments seem (to us) to be less suited for detecting a small difference in growth rate. The data are as follows:

Presumed genotype	Growth in 95/5 greater than growth in 0/5	Growth in 0/5 greater than growth in 95/5	Equal growth	Total
AA	8	3	1	12
aa	0	5	0	5

The data therefore resemble the findings in the multiple point experiments. However, the 5 observations do not differ significantly from the expected values of 2.5 in each class on the null hypothesis. If one takes all the data in Table III (which includes 4 entries from multiple point experiments), there are 9 observations, with growth in 0/5 being greater than growth in 95/5 in every case. Summing the terms of a binomial expansion, a two-tailed probability of getting such data or a poorer fit on the null hypothesis is 1/256. The data on galactose sensitivity in Tables II and III are graphed in Text-fig. 4.

If, in Table III, one compares growth in media containing 100 mg. per cent glucose with growth in 95/5, we find better growth in the former case in 8 instances, and in the latter in 2, which is non-significant. Table III also reveals that the *aa* cells generally yield slightly less protein in hexose-free medium than in medium containing 100 mg. per cent galactose. This *prima facie* is

---

[6] A non-parametric test rather than an analysis of variance of the actual protein values is used. Occasionally, 3 or 4 replicate bottles in the same medium will contain one value markedly different from the others. Sometimes the reason is discernible (contamination, inadequate washing of the bottle before use, etc.). Other times it is not. In both cases the extreme value is omitted. We are confident this practice gives us better estimates of the mean than would the inclusion of extreme observations. It deprives us, however, of the right to estimate the variance among replicate bottles.

inconsistent with a galactose sensitivity effect. We suspect that minute quantities of glucose contaminating our galactose may make the comparison unfair.

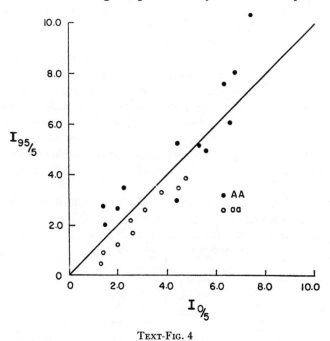

TEXT-FIG. 4

TABLE IV

Growth Experiments with the Heterozygous Line (Aa)

Line	MAD[5]	MAD[6]	MAD[7]	MAD[10]	MAD[4]	MAD[8]	MAD[14]
Duration, *days*	9	18	13	35	32	28	9
Initial protein	0.7	0.9	1.1	0.6	0.9	0.7	1.1
100 Glu	2.1	3.4	4.0	3.1	3.0	2.6	3.9
100 Gal	1.5	2.3	3.8	1.4	3.0	2.3	3.9
95/5	2.2	3.2	4.0	2.4	2.7	2.5	—
5 Glu	—	—	3.9	1.9	1.6	2.9	4.6
Hexose-free	1.2	1.7	1.8	1.4	1.5	1.4	1.6
"Age" (*days*)	76	93	139	153	155	151	195
M-Nos	128	64	36	512	40	108	2.3×10^4

See Table II for notation.

Certainly the inclusion of pyruvate and the non-essential amino acids (including alanine) in the medium renders the cells exquisitely sensitive to glucose (Chang and Geyer, 1957).

The Heterozygous Line.—Data on the growth of the heterozygous line are given in Table IV. For some time marked technical problems were encountered with the handling of this line, and the absolute growth in most of the experiments has been intractably slow. Only two point experiments have thus far been done. However, all were terminated before the cell density reached the usual plateau levels. It is clear from Table IV that the heterozygous line was more variable than the homozygous ones in its relative growth in glucose and

TABLE V

Isotope Experiments

	Line	"Age"	M-No.	Tissue of origin	Millions of cells counted	(1) $C^{14}O_2$ from C^{14} gal: counts per 10^6 cells per minute	(2) $C^{14}O_2$ from C^{14}glu counts per 10^6 cells per minute	Ratio (1):(2)
		days						
	(AA) MI[4]	27	16	Marrow	7.8	220	380	0.6
	(aa) JDU[4]	27	10	"	3.9	0	870	0
	(AA) Be[7]	65	8 × 10³	Skin	4.3	352	807	0.4
	(AA) RCU[4]	79	32	"	7.7	87	298	0.3
AA	(AA) BE[11]	72	2 × 10⁵	"	8.3	157	529	0.3
	(AA) BE[20]	160	3.8 × 10¹²	"	1 Blake bottle	1235	3544	0.3
	(AA) SE[8]	71	9.6 × 10³	"	" "	266	944	0.3
	(aa) By[4]	56	10	Skin	6.3	0	—	(0)
aa	(aa) JDU[6]	55	240	"	5.3	7	590	0.01
	(aa) JDU[9]	119	9.6 × 10⁴	"	1 Blake bottle	0	2644	0
	(aa) JDU[17]	192	2.3 × 10¹⁰	"	1 " "	1	4541	0
Aa	(Aa) MAD[4]	68	40	Skin	1.5	67	720	0.09
	(Aa) MAD[(5+6)]	76	128	"	2.8	66	735	0.09

AA denotes non-galactosemic patients, *Aa* the presumably heterozygous mother of galactosemic patient JDU, and *aa* denotes galactosemic patients. The superscript on the letters designating the line is 1, plus the number of subcultures. Age is the time between the date of biopsy and the date of the experiment. Each unique combination of letters represents a different patient. Each flask contained of the order of 10^6 cells, and 0.472 μc. of either galactose-*l*-C^{14} or glucose-*l*-C^{14} (approximately 250,000 c.p.m.).

* Counts entered here are c.p.m. per 1/3.2 of a Blake bottle.

galactose. However, in three experiments, all terminated before plateau levels, growth in glucose and galactose were so close that a true difference appears unlikely. However, there must be factors influencing the growth of this cell line over which we still exercise imperfect control.

Isotope Experiments.—The results of the isotope experiments are given in Table V. Note that the ratio of counts as $C^{14}O_2$, from galactose to counts from glucose appears to be fairly constant and to reflect genotype. The effect seems to persist over many months, and after the cells have increased many billion-fold. The variation within genotype of counts per million cells is probably due largely to our crude method of enumerating cells, which is accurate to but a factor of 2.

GENERAL REMARKS

In addition to galactosemia, a number of other diseases are probably susceptible to study in this way, *e.g.*, orotic aciduria (Huguley *et al.*, 1959, Smith and Huguley, 1960), acatalasemia (Takahara 1952, Takahara *et al.*, 1960, Wyngaarden, 1960), cystathioninuria (Harris *et al.*, 1959) and perhaps some of the other hexose and pentosurias (see Stanbury *et al.*, 1960, Rapaport, 1959, and Hsia, 1959). Indeed the number of possible diseases appears to be increasing sharply with time.

Cells of this kind are, as noted earlier, of potential use in the demonstration of genetic exchange. The possibility of transforming human cells has been made particularly attractive by Gartler's (1960) recent work suggesting that Earle's "L" cells will incorporate polymerized DNA *as such* from the medium into their nuclei. "Primary" human cells offer certain advantages. They are at present the sole source of mutant mammalian cells, where both the genetics and the identity of the abnormal protein (a particular enzyme in the present case) are known.

SUMMARY

Cell lines were developed from biopsies on galactosemic and non-galactosemic patients. It was shown that one can discriminate between lines from the two types of donors by their relative growth in glucose and galactose and by their ability to oxidize galactose-1-C^{14}. The latter method was successful in distinguishing a heterozygous cell line from the normal ones. Sensitivity of galactosemic cells to galactose was suggested by some of the experiments. The kinetics of growth were in some ways reminiscent of a similar phenomenon in the transferase mutants of *E. coli*, though in the human cells the effect was much less marked.

The authors are grateful to Dr. Harry S. Eagle and Dr. L. T. Kurland for their interest and encouragement. The two patients JDU and MAD were referred to us through Dr. Julius Metrakos, to whom we are most grateful. Dr. Helen Brickman, who has followed the family for some time, was most generous in her interest, and encouraged the patients to come to the National Institutes. We are also deeply indebted to Dr. Ntinos Myrianthopoulos for assistance in locating this family. Dr. Stanton S. Segal kindly permitted us to perform the biopsy on the patient BY. Finally, we should like to acknowledge with gratitude the technical assistance of Miss Mary Jane Madden.

BIBLIOGRAPHY

Anderson, E. P., Kalckar, H. M., Kurahashi, K., and Isselbacher, K. J., A specific enzymatic assay for the diagnosis of congenital galactosemia, *J. Lab. and Clin. Med.*, 1957, **50,** 469.

Bias, W., Troedson, H., and Kalckar, H., Studies of enzymes in galactose metabolism in sonicates of human fibroblasts, 1961, in preparation.

200

Böök, T. A., and Kostman, R., Prospects of biochemical genetics in medicine, *Ann. Human Genet.*, 1956, **20,** 251.

Chang, R. S., Genetic study of human cells *in vitro*. Carbohydrate variants from cultures of Hela and conjunctival cells, *J. Exp. Med.*, 1960, **111,** 235.

Chang, R. S., and Geyer, R. P., Propagation of conjunctival and Hela cells in various carbohydrate media, *Proc. Soc. Exp. Biol. and Med.*, 1957, **96,** 336.

Eagle, H., Amino acid metabolism in mammalian cell cultures, *Science*, 1959, **130,** 432.

Gartler, S., Demonstration of cellular uptake of polymerized DNA in mammalian cell cultures, *Biochem. and Biophysic. Research Commun.*, 1960, **3,** 127.

Goldbloom, A., and Brickman, H. F., Galactemia, *J. Pediat.*, 1946, **28,** 674.

Harris, H., Penrose, L. S., and D. H. H. Thomas, Cystathioninuria, *Ann. Human Genet.*, 1959, **23,** 442.

Hsia, D. Y., Inborn Errors of Metabolism, Chicago, Yearbook Publishers Inc., 1959, 358.

Huguley, C. M., Bain, J. A., Rivers, S. L., and Scoggins, R. B., Refractory megaloblastic anemia associated with excretion of orotic acid, *Blood*, 1959, **14,** 615.

Kalckar, H. M., Biochemical genetics as illustrated by hereditary galactosemia, *in* Wolstenholme, G. E. W., and O'Connor, C. M., Biochemistry of Human Genetics (Ciba Foundation Symposium), Boston, Little, Brown and Co., 1959a, 347.

Kalckar, H. M., 1960, personal communication.

Kalckar, H. M., Kurahashi, K., and Jordan, E., Hereditary defects in galactose metabolism in *Escherichia coli* mutants. I. Determination of enzyme activities, *Proc. Nat. Acad. Sc.*, 1959b, **45,** 1776.

Kalckar, H. M., Szulmajster, H. de R., and Kurahaski, K., Galactose metabolism in mutants of man and micro-organisms, *Proc. Internat. Symposium in Enzyme Chem.*, Tokyo and Kyoto, 1957, 52.

Kirkman, H. N., and Bynum, E., Enzymic evidence of a galactosemic trait in parents of galactosemic children, *Ann. Human Genet.*, 1959, **23,** 117.

Krooth, R. S., and Weinberg, A., Properties of galactosemic cells in culture, *Biochem. and Biophysic. Research Commun.*, 1960, **3,** 518.

Kurahashi, K., and Wahba, A. J., Inhibition of growth of *E. coli* mutants by galactose, *Fed. Proc.*, 1957, **16,** 207 (abstr. No. 889).

Kurahashi, K., and Wahba, A. J., Interference with growth of certain *E. coli* mutants by galactose, *Biochim. Biophysica. Acta*, 1958, **30,** 298.

Levintow, L., and Price, V. E., Sodium pyruvate, *in* Biochemical Preparations, (G. E. Ball, editor), 1952, **2,** 22.

Levintow, L., and Eagle, H., Biochemistry of cultured mammalian cells, *Ann. Rev. Biochem.*, 1960, in press.

Luria, S. E., in Panel discussion of the approach to tissue culture *in* Wolstenholme, G. E. W., and O'Connor, C. M., Biochemistry of Human Genetics (Ciba Foundation Symposium), Boston, Little, Brown and Co., 1959, 374.

McQuilkin, W. T., Evans, V. L., and Earle, W. R., The adaptation of additional lines of NCTC Clone 929 (Strain L) cells to chemically defined protein-free medium NCTC 109, *J. Nat. Cancer Inst.*, 1957, **19,** 885.

Montgomery, C. M., and Webb, J. L., Detection of a new inhibitor of the tricarboxylic acid cycle, *Science* 1954, **120,** 843.

Oyama, V. G., and Eagle, H., Measurement of cell growth in tissue culture with a phenol reagent (Folin-Ciocalteau), *Proc. Soc. Exp. Biol. Med.*, 1956, **91**, 305.

Puck, T. T., Ciecura, S. J., and Fisher, H. W., Clonal growth in vitro of human cells with fibroblastic morphology, *J. Exp. Med.*, 1957, **106**, 145.

Puck, T. T., Ciecura, S. L., and Robinson, A., Genetics of somatic mammalian cells, *J. Exp. Med.*, 1958, **108**, 945.

Rapaport, M., Inborn errors of metabolism *in Textbook of Pediatrics*, (W. E. Nelson, editor), Philadelphia, W. B. Saunders Co., 1959.

Smith, L. H., and Huguley, C. M., The enzymatic defect of orotic aciduria, *J. Clin. Invest.*, 1960, **39**, 1029.

Stanbury, J. B., Wyngaarden, J. B., Frederickson, D. S., *The Metabolic Basis of Inherited Disease*, New York, McGraw-Hill Book Company, Inc., 1960, 1477.

Takahara, S., Progressive oral gangrene probably due to lack of catalase in the blood (acatalasemia), *Lancet*, 1952, **2**, 1101.

Takahara, S., Hamilton, H. B., Neel, J. V., Kobara, T. Y., Ogura, Y., and Nishimura, E. T., Hypocatalasemia, a new genetic carrier state, *J. Clin. Invest.*, 1960, **39**, 610.

Weinberg, A., and Segal, S., Effect of galactose-1-phosphate on glucose oxidation by normal and galactosemic leucocytes, *Science*, 1960, **132**, 1015.

Wyngaarden, J. B., and Howell, R. R., Acatalasia, *in* The Metabolic Basis of Inherited Disease, (J. B. Stanbury, J. B. Wyngaarden, and D. S. Frederickson, editors), New York, McGraw-Hill Book Company, Inc., 1960.

202

EXPLANATION OF PLATES

PLATE 112

FIG. 1. Exfoliation from biopsy. SK line. × 150.

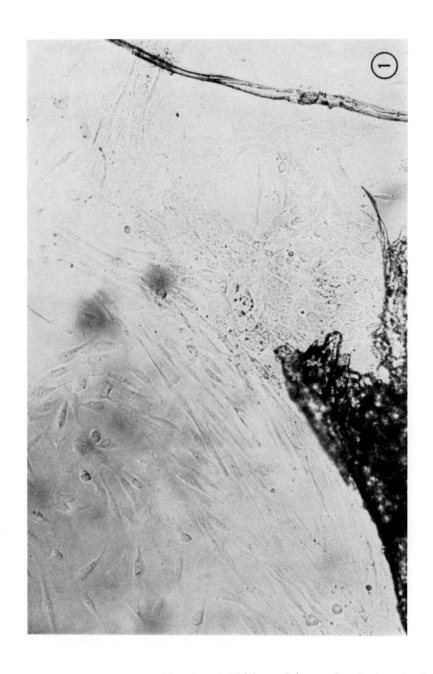

(Krooth and Weinberg: Galactosemic cells in culture)

PLATE 113

FIG. 2: Growth in 100 mg. per cent glucose or galactose. Growth in galactose is shown in the two left hand panels and growth in glucose in the right hand ones. *AA* refers to the normal BE line and *aa* to the galactosemic JDU line. $\times$ 75.

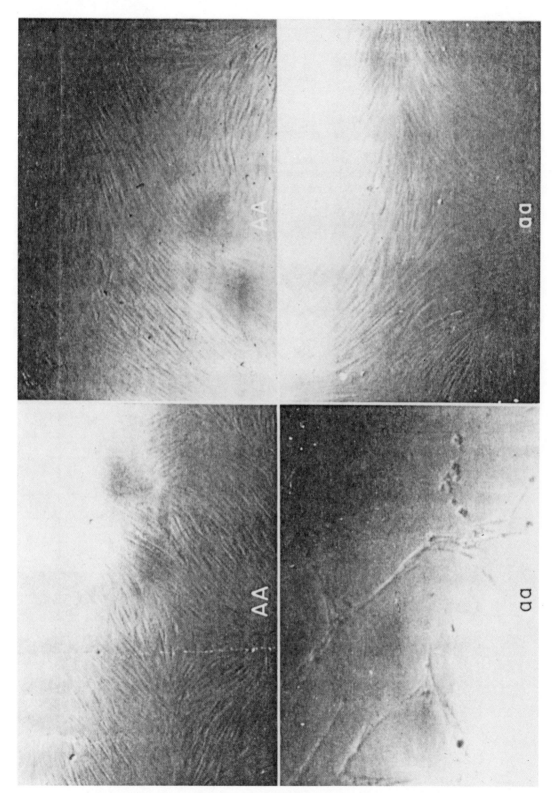

Fig. 2

(Krooth and Weinberg: Galactosemic cells in culture)

15

Reprinted from *Science* 155:1682–1684 (1967)

Enzyme Defect Associated with a Sex-Linked Human Neurological Disorder and Excessive Purine Synthesis

Abstract. *A sex-linked familial neurological disease consisting of cerebral palsy, mental retardation, choreoathetosis, and compulsive aggressive behavior is associated with a loss of an enzyme that participates in purine metabolism, namely, hypoxanthine-guanine phosphoribosyltransferase. The production of excessive uric acid in this disorder implies that the enzyme is involved in the normal regulation of purine biosynthesis. This is the first example of a relation between a specific enzyme defect and abnormal compulsive behavior. It is also the first enzyme defect in purine metabolism demonstrated in a neurological disease.*

A familial neurological disorder consisting of choreoathetosis, spasticity, mental retardation, aggressive behavior, and a compulsive biting resulting in mutilation of the lips and fingers was first described by Lesch and Nyhan (*1*). These findings on two young brothers were associated with hyperuricemia and excessive uric acid synthesis. Subsequently other cases have been described (*2, 3*). The original patients excreted in their daily urine three to six times the amount of uric acid found in urine from control subjects of similar body size and age. They incorporated between 100 and 200 times as much uniformly labeled glycine-¹⁴C into urinary uric acid as nonaffected control subjects did. This degree of overproduction of uric acid is far greater than that encountered in adults with clinical gouty arthritis, yet development of clinical gouty arthritis is a relatively late manifestation of their disease. Gouty nephropathy, however, has been a contributing factor to the death of some of these patients in early puberty (*4*). The apparent familial distribution and limitation to males is compatible with X-linked inheritance (*3*).

The drug azathioprine [6-(1′-methyl-4′-nitro-5′-imidazolyl)thiopurine] (*5*), a purine analog used clinically as an immunosuppressive agent, has been reported to inhibit the excessive purine synthesis found in some gouty adults (*6*). We compared the effectiveness of this drug in inhibiting uric acid production in two children affected with the neurological disease and in two gouty men who also produced abnormally large quantities of uric acid. Both of the children were 14 years of age, unrelated, and had the complete clinical syndrome. Purine synthesis was assessed by the daily excretion of uric acid (while the four patients were on a purine-free diet) and the incorporation of 5 μc of orally administered glycine-1-¹⁴C into urinary uric acid (*7*). As shown in Table 1, azathioprine at a dose of 4 mg per kilogram of body weight per day markedly suppressed purine synthesis in the two gout patients but did not diminish it in the children. Sorensen (*8*) found similar results.

Since some of the pharmacological actions of azathioprine may result from its degradation to 6-mercaptopurine, the effect of the latter drug on purine synthesis by fibroblasts grown in vitro from skin biopsies obtained from a child with the neurological disorder was compared with its effect on cells from a normal subject. Cells were grown in Eagle's minimal medium supplemented with 10 percent fetal calf serum, nonessential amino acids, and neomycin (50 μg/ml), for bacteriostasis. Purine synthesis by fibroblasts growing in monolayer was blocked at the stage of formylglycinamide ribonucleotide (FGAR) by use of azaserine (*9*), and the activity of the first three enzymes of purine biosynthesis was assessed by determining the incorporation of sodium formate-¹⁴C into FGAR (*10*). As shown in Table 2, 10⁻⁵M 6-mercaptopurine greatly inhibited purine biosynthesis in normal cells but had no effect on cells derived from a child with the neurological disorder.

Resistance to the action of 6-mercaptopurine, ascribed to a deficiency of the enzyme hypoxanthine-guanine phos-

phoribosyltransferase (E.C. 2.4.2.8), has been described in mutant mammalian tumor cells selected in vitro (*11*) and in leukemic leukocytes selected in vivo during the course of treatment of leukemia with 6-mercaptopurine (*12*). These considerations led us to examine the activity of this enzyme in fibroblasts grown in vitro as well as in dialyzed hemolyzates of washed erythrocytes obtained from children affected with this neurological disease and from spastic quadriplegic, gouty, and normal subjects. This enzyme converts the free bases hypoxanthine or guanine to their respective ribonucleotides by reaction with 5-phosphoribosyl-1-pyrophosphate (PRPP) (Fig. 1).

The incorporation of hypoxanthine-8-¹⁴C into inosinic acid, of guanine-8-¹⁴C into guanylic acid, and of adenine-8-¹⁴C into adenylic acid was determined after separation of these substances in the incubated reaction mixture by high-voltage electrophoresis. Both fibroblasts and erythrocytes (Table 3) from unrelated children manifesting the complete syndrome were unable to synthesize inosinic acid or guanylic acid from the respective bases but were able to synthesize adenylic acid from adenine. We conclude that the activity of the enzyme in erythrocytes from these chil-

Table 1. Effects of azathioprine on purine metabolism in patients manifesting overproduction of uric acid, showing the cumulative incorporation of isotopic glycine into urinary uric acid in 7 days (expressed as percentage of administered dose).

Subject	Drug	Serum urate (mg/100 ml)	Excretion, uric acid (mg/day)	Glycine incorporation (%)
Adult overproduction—Gout				
B. P.	None	9.7	1241	1.24
B. P.	Azathioprine	6.5	740	0.22
T. B.	None	10.5	1529	1.11
T. B.	Azathioprine	6.8	922	0.28
Childhood overproduction with neurological disease				
D. F.	None	12.4	659	2.49
D. F.	Azathioprine	11.8	719	2.70
F. H.	None	10.0	828	3.16
F. H.	Azathioprine	10.9	926	3.15

Table 2. Effects of 6-mercaptopurine on incorporation of formate-¹⁴C into formylglycinamide ribonucleotide in fibroblasts in tissue culture in the presence of azaserine. Growth medium (6 ml) deficient in glutamine and containing the appropriate concentration of 6-mercaptopurine and azaserine (10 μg/ml) was overlayered on cells in stationary monolayer culture (the cells were not confluent). After 30 minutes at 37°C, glutamine (0.5 mg/ml) and sodium formate-¹⁴C (50 μc/flask, specific activity 52 mc/mmole) were added. Eighteen hours later the reaction was stopped by washing away the labeled substrate with 25 volumes of cold isotonic saline. The cells were treated with trypsin and harvested; the nucleotide was extracted by the addition of five volumes of ethanol and separated from other radioactive compounds by high voltage electrophoresis on 3-MM Whatman paper in 0.05M borate buffer, *p*H 9.0, at 2000 volts for 45 minutes. The portion of the paper containing radioactive FGAR was identified by radioautography cut out, and counted in a liquid scintillation counting system at 52 percent efficiency of counting.

6-Mercaptopurine (mole/liter)	FGAR-¹⁴C accumulation in the soluble ethanol fraction*	
	Normal	Mutant
0	100†	100‡
10⁻⁶	101	109
10⁻⁵	19	99
10⁻⁴	18	86

* Expressed as percentage of control without 6-mercaptopurine. † 7,500 count/min. ‡ 12,550 count/min.

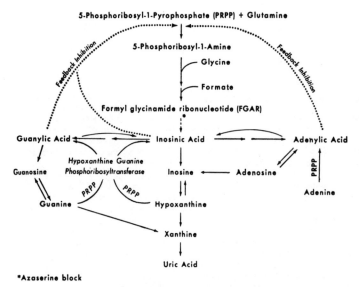

*Azaserine block

Fig. 1. Purine biosynthesis and interconversions.

dren is less than 0.05 percent of normal activity. Hemolyzate from affected children added to the hemolyzate of a normal subject did not alter normal enzyme activity. From this we conclude that the presence of an inhibitor does not explain the absence of enzyme activity.

The biochemical function of the enzyme hypoxanthine-guanine phosphoribosyltransferase has been assumed until now to be solely that of allowing utilization of preformed purines. The precise mechanism by which the absence of this enzyme gives rise to overproduction of purines is not yet clear. The association of an excessive formation of a metabolic end product (uric acid) with an enzyme deficit rather than with a primary increase in activity or amount of an enzyme is of considerable interest. To our knowledge this is a unique finding among human mutants. These findings suggest that the deficient enzyme may be concerned with some aspect of the normal regulatory mechanism of purine synthesis (Fig. 1).

The absence of hypoxanthine-guanine phosphoribosyltransferase in these patients readily explains their failure to respond in the normal manner to azathioprine and to 6-mercaptopurine, since the latter compound must be converted to its ribonucleotide by this enzyme before it can function as a pseudo feedback inhibitor of purine synthesis (11).

The sex-linked pattern of inheritance in this disorder suggests that the genetic locus for this enzyme may be on the X chromosome, and suspected female carriers should therefore show mosaicism for this enzyme defect similar to that shown for glucose-6-phosphate dehydrogenase deficiency (13).

The association of a specific enzyme defect with a neurological disease, mental retardation, and a characteristic compulsive aggressive behavior may serve to reorient our fundamental approach to other behavioral disorders.

J. Edwin Seegmiller
Frederick M. Rosenbloom
William N. Kelley
Section on Human Biochemical Genetics, National Institute of Arthritis and Metabolic Diseases, Bethesda, Maryland 20014

Table 3. Specific activities of hypoxanthine, guanine, and adenine phosphoribosyltransferase from erythrocyte hemolyzates and fibroblast extracts. The erythrocyte hemolyzate assay system was made up as follows. Incubation mixtures contained 55 mM tris buffer, pH 7.4; 5 mM MgCl$_2$; 1 mM 5-phosphoribosyl-1-pyrophosphate (PRPP); 0.6 mM hypoxanthine-8-^{14}C (4.1 mc/mmole); 0.6 mM adenine-8-^{14}C (3.7 mc/mmole), or 0.14mM guanine-8-^{14}C (12.6 mc/mmole); and 0.6 to 1.2 mg of protein from dialyzed erythrocyte hemolyzate in a final volume of 200 μl. Each sample was incubated for 20 minutes at 38°C. The reactions were terminated by the addition of 4 μmole of neutralized ethylenediaminetetraacetate (EDTA) and chilling on ice. The fibroblast extract assay system was made as follows. The reaction mixture (total volume 120 μl) contained 330 μg of crystalline bovine albumin; 0.27 mM adenine-8-^{14}C (3.7 mc/mmole), or 0.27 mM hypoxanthine-8-^{14}C (6.6 mc/mmole); 1.67 mM PRPP; 9 mM MgCl$_2$; 6 mM tris buffer, pH 7.4; and between 150 to 225 μg of protein of cell extract consisting of the supernatant after centrifugation at 19,000g of disrupted washed fibroblasts. Incubation was at 25°C for 1 hour.

Source	Sex	Age	Phosphoribosyltransferase activity (count/min) per milligram of protein per hour as:		
			Inosinic acid	Guanylic acid	Adenylic acid
Erythrocyte					
Normal adults					
J.E.S.	M	46	639	1188	141
J.S.,Sr.	M	28	511	1881	146
N.F.	F	51	574	1560	158
J.A.	F	32	467	1548	107
Spastic quadriplegia					
S.S.	M	6	653	1656	200
C.L.	F	4	526	1515	164
B.R.P.	M	2	443	1638	132
Adult overproduction—gout					
B.P.	M	27	592	1104	105
A.P.	M	48	477	1194	136
Childhood overproduction with neurological disease					
J.S.*	M	3	< 0.2	< 0.2	322
M.W.†	M	8	< 0.2	< 0.2	295
F.H.‡ (NIH 069353)	M	15	< 0.2	< 0.2	426
Fibroblast					
Normal adult					
F.R.	M	30	13		31
Childhood overproduction with neurological disease					
D.F.§ (NIH 067235)	M	15	< 0.2		8

* Patient has been reported by Shapiro *et al.* (3). † This patient is the propositus of the original case description of Lesch and Nyhan (1). ‡ This patient was one of two affected siblings described in family A by Hoefnagel, *et al.* (3). § Case is being reported (14).

References and Notes

1. M. Lesch and W. L. Nyhan, *Amer. J. Med.* 36, 561 (1964).
2. D. Hoefnagel, *J. Ment. Defic. Res.* 9, 69 (1965); W. L. Nyhan, W. J. Oliver, M. J. Lesch, *J. Pediat.* 67, 257 (1965); W. B. Reed and C. H. Fish, *Arch. Dermatol.* 94, 194 (1966).
3. D. Hoefnagel, E. D. Andrew, N. G. Mireault, W. O. Berndt, *New Engl. J. Med.* 273, 130 (1965); S. L. Shapiro, G. L. Sheppard, Jr.,

F. E. Dreifuss, D. S. Newcombe, *Proc. Soc. Exp. Biol. Med.* **122**, 609 (1966).

4. J. K. Sass, H. H. Itabashi, R. A. Dexter, *Arch. Neurol.* **13**, 639 (1965).

5. Imuran, Burroughs Wellcome & Co., Inc., New York City.

6. L. B. Sorensen, *Proc. Nat. Acad. Sci. U.S.* **55**, 571 (f966).

7. J. E. Seegmiller, A. I. Grayzel, L. Laster, L. Liddle, *J. Clin. Invest.* **40**, 1304 (1961).

8. L. B. Sorensen and P. J. Benke, *Nature*, in press.

9. L. L. Bennett, Jr., F. M. Schabel, Jr., H. E. Skipper, *Arch. Biochem. Biophys.* **64**, 423 (1956); B. Levenberg, I. Melnick, J. M. Buchanan, *J. Biol. Chem.* **225**, 163 (1957).

10. G. A. LePage and M. Jones, *Cancer Res.* **21**, 642 (1961); J. F. Henderson, *J. Biol. Chem.* **237**, 2631 (1962); R. W. Brockman, *Cancer Res.* **23**, 1191 (1963).

11. R. W. Brockman, *Cancer Res.* **25**, 1596 (1965).

12. J. D. Davidson and T. S. Winter, *ibid.* **24**, 261 (1964).

13. R. G. Davidson, H. M. Nitowsky, B. Childs, *Proc. Nat. Acad. Sci. U.S.* **50**, 481 (1963).

14. M. G. Smith, J. H. Bland, W. N. Kelley, J. E. Seegmiller, in preparation.

15. We thank Drs. J. Bland, E. D. Andrew, D.

Chrisman, S. Bessman, Z. H. B. Koppanyi, D. Greene, and J. Boyd for making patients available for study. Dr. J. F. Henderson kindly reviewed the manuscript, Dr. W. B. Uhlendorf and M. F. Damuth gave generous assistance and advice in details of tissue culture, and J. M. Miller assumed responsibility for the chemical determinations in the clinical studies. We thank Drs. E. Anderson and R. W. Brockman for providing an authentic sample of FGAR-^{14}C as a reference compound.

16 January 1967

Reprinted from *Lancet* II:1–7, 73–79, 142–148, 214–220 (1908)

𝔗𝔥𝔢 ℭ𝔯𝔬𝔬𝔫𝔦𝔞𝔫 𝔏𝔢𝔠𝔱𝔲𝔯𝔢𝔰

ON

INBORN ERRORS OF METABOLISM.

Delivered before the Royal College of Physicians of London on June 18th, 23rd, 25th, and 30th, 1908,

BY ARCHIBALD E. GARROD, M.A., M.D. OXON., F.R.C.P. LOND.,

ASSISTANT PHYSICIAN TO, AND LECTURER ON CHEMICAL PATHOLOGY AT, ST. BARTHOLOMEW'S HOSPITAL; SENIOR PHYSICIAN, HOSPITAL FOR SICK CHILDREN, GREAT ORMOND STREET.

LECTURE I.

Delivered on June 18th, 1908.

GENERAL AND INTRODUCTORY.

MR. PRESIDENT AND FELLOWS,—It is my first agreeable duty to offer my sincere thanks for the honour conferred upon me in the invitation to deliver the Croonian lectures of the current year before this College. I trust that the subject which I have selected will be found to conform closely to the instructions to the lecturer, for it is one which lies upon the very border-line of physiology and pathology and pertains to both sciences alike ; nor is it without bearing upon the control and cure of disease, in so far as no study which helps to throw light upon the complex chemical processes which are carried out in the human organism can fail in the long run to strengthen our hands in the combat with the pathogenic influences which make for its destruction.

The differences of structure and form which serve to distinguish the various genera and species of animals and plants are among the most obvious facts of nature. For their detection no scientific training is needed, seeing that they cannot escape the notice of even the least cultivated intelligence. Yet with the growth of knowledge we have learned to recognise the uniformity which underlies this so apparent diversity and the genetic relationship of form to form. With regard to the chemical composition of the tissues of living organisms and the metabolic processes by which those tissues are built up and broken down, the advance of knowledge has been in the opposite direction, and the progress of chemical physiology is teaching us that behind a superficial uniformity there exists a diversity which is no less real than that of structure, although it is far less obvious. The differences of ultimate composition and crystalline form which distinguish the hæmoglobins of animals of distinct genera have long been known. That the fats of animals are not alike in composition is well recognised, as also are the differences of their bile acids, to quote only a few of the most conspicuous examples. As instances of distinctive end-products of metabolism may be mentioned kynurenic acid, which is present in the urine of animals of the canine tribe and which bears witness to a generic peculiarity in the manner of dealing with the tryptophane fraction of proteins, and the excretion by birds and reptiles of the bulk of their nitrogenous waste in the form of uric acid, whereas in the urine of mammals urea is the chief nitrogenous constituent.

A more extended study even by strictly chemical methods will doubtless serve to reveal innumerable minor differences, such as are foreshadowed by Przibram's[1] work on muscle proteins. The delicate ultra-chemical methods which the researches of recent years have brought to light, such as the precipitin test, reveal differences still more subtle, and teach the lesson that the members of each individual species are built up of their own specific proteins which resemble each other the more closely the more nearly the species are allied. Obviously it is among the highly complex proteins that such specific differences are to be looked for rather than in the simple end-products of their disintegration. The many amino-acids which enter into the structure of the protein molecules are capable of almost innumerable groupings and proportional representations, and each fresh grouping will produce a distinct protein ; but all alike in their breaking down will yield the same simple end-products, urea, carbon dioxide, and others.

Nor can it be supposed that the diversity of chemical structure and process stops at the boundary of the species,

and that within that boundary, which has no real finality, rigid uniformity reigns. Such a conception is at variance with any evolutionary conception of the nature and origin of species. The existence of chemical individuality follows of necessity from that of chemical specificity, but we should expect the differences between individuals to be still more subtle and difficult of detection. Indications of their existence are seen, even in man, in the various tints of skin, hair, and eyes, and in the quantitative differences in those portions of the end-products of metabolism which are endogenous and are not affected by diet, such as recent researches have revealed in increasing numbers. Even those idiosyncrasies with regard to drugs and articles of food which are summed up in the proverbial saying that what is one man's meat is another man's poison presumably have a chemical basis.

Upon chemical as upon structural variations the factors which make for evolution have worked and are working. Evidences of this are to be detected in many directions, as, for example, in the delicate selective power of the kidneys, in virtue of which they are enabled to hold back in the circulation the essential proteins of the blood but at the same time allow free passage to other proteins which are foreign to the plasma, such as hæmoglobin, egg albumin, and the Bence-Jones protein, when these are present in any but quite small amounts. The working of these factors is also seen in the various protective mechanisms against chemical poisons, such as that which averts the depletion of the fixed alkalies of the organism by the neutralisation of abnormal supplies of acids by ammonia. This mechanism is well developed in the carnivora and in man, but in vegetivorous animals which from the nature of their diet are little exposed to acidosis it appears to be wanting.

Even in the normal metabolic processes the working of such influences may be traced, as in the power which the organism possesses of destroying the benzene ring of those aromatic amino-acids which enter into the composition of proteins and cannot therefore be regarded as substances foreign to the body ; whereas the benzene ring of foreign aromatic compounds, with very few exceptions, are left intact, and such compounds require to be rendered innocuous by being combined with sulphuric acid to form aromatic sulphates, or with glycocoll to form the acids of the hippuric group and so combined are excreted in the urine and got rid of. The few exceptions referred to are compounds which so closely resemble the protein fractions in their structure that they fall victims with these to the normal destructive processes.

The great strides which recent years have witnessed in the sciences of chemical physiology and pathology, the newly-acquired knowledge of the constitution of proteins and of the part played by enzymes in connexion with the chemical changes brought about within the organism, have profoundly modified our conceptions of the nature of the metabolic processes and have made it easier to understand how these changes may differ in the various genera and species. It was formerly widely held that many derangements of metabolism which result from disease were due to a general slackening of the process of oxidation in the tissues. The whole series of catabolic changes was looked upon as a simple combustion and according as the metabolic fires burnt brightly or burnt low the destruction of the products of the breaking down of food and tissues was supposed to be complete or imperfect. A very clear setting forth of such views will be found in the lectures of Bence Jones[2] on Diseases of Suboxidation, delivered and published in 1855, but the thesis in question is chiefly associated with the name of Bouchard,[3] who expounded it in his well-known lectures on Maladies par Ralentissement de la Nutrition, published in 1882. The so frequent clinical association of such maladies as gout, obesity, and diabetes was involved in its support, nor was it regarded as a serious obstacle to the acceptance of such views that there is but scanty evidence to show that failure to burn any particular metabolic product, such as glucose, is associated with inability to deal with others.

Nowadays, very different ideas are in the ascendant. The conception of metabolism in block is giving place to that of metabolism in compartments. The view is daily gaining ground that each successive step in the building up and

---

[1] Hofmeister's Beiträge, 1902, Band ii., p. 143.

[2] Medical Times and Gazette, 1855, vol. ii., pp. 29–83.
[3] Maladies par Ralentissement de la Nutrition, Paris, 1882.

breaking down, not merely of proteins, carbohydrates, and fats in general, but even of individual fractions of proteins and of individual sugars, is the work of special enzymes set apart for each particular purpose. Thus the notion of general suboxidation is reduced to very narrow limits, to the recognition of a controlling influence exercised by certain glandular organs, such as the thyroid, upon metabolism as a whole. For example, it is known that lævulose is not dealt with in the human organism in the same way as dextrose is but follows its own path. A patient whose power of burning dextrose is seriously impaired may yet utilise lævulose in the normal manner. Again, there is evidence to show that the several fractions of proteins, tyrosin, cystin, tryptophane, and the rest, do not merely serve as fuel for a common furnace but are dealt with each in a special manner and in successive stages.

It may well be that the intermediate products formed at the several stages have only momentary existence as such, being subjected to further change almost as soon as they are formed ; and that the course of metabolism along any particular path should be pictured as in continuous movement rather than as series of distinct steps. If any one step in the process fail the intermediate product in being at the point of arrest will escape further change, just as when the film of a biograph is brought to a standstill the moving figures are left foot in air. All that is known of the course of catabolism tends to show that in such circumstances the intermediate product in being is wont to be excreted as such, rather than that it is further dealt with along abnormal lines. Indeed, it is an arguable question whether, under abnormal conditions, the metabolic processes are ever thrown out of their ordinary lines into entirely fresh ones, with the result that products are formed which have no place in the normal body chemistry. It is commonly assumed that this happens, but if the conception of metabolism in compartments, under the influence of enzymes, be a correct one, it is unlikely, *a priori*, that alternative paths are provided which may be followed when for any reason the normal paths are blocked. It is far easier to suppose that in such circumstances normal intermediate products are excreted without further change and that processes which in health play but small parts in metabolism are called into unwonted activity.

This conception of the permanency of the metabolic paths is no new one, for it may be read between the lines in the writings of some physiologists of the last century, and especially in those of Claude Bernard,[4] from which the following passage is translated :—

It used to be supposed that in diabetes wholly new conditions were developed in the economy, under the influence of the morbid state, and that there resulted therefrom a special morbid product—namely, glucose. But it is admitted nowadays that the observed phenomena are to be explained by an augmentation, pure and simple, of a normal function in virtue of which glucose is formed in all subjects, even in health. It is clear that the malady is nothing more than a physiological phenomenon perturbed and exaggerated.

Still more striking is the following passage[5] :—

Et maintenant oserait on soutenir qu'il faut distinguer les lois de la vie à l'état pathologique des lois de la vie à l'état normal ? Ce serait vouloir distinguer les lois de la mécanique dans une maison qui tombe, des lois de la mécanique dans une maison qui tient debout.

To prove the truth of the contention put forward it would be necessary to show that every abnormal product found in the tissues or in the excreta, und r morbid conditions, can be ascribed to other causes than the deflexion of the metabolic processes into new and unwonted paths. It must be admitted that this cannot yet be asserted of all such products. For example, we are still ignorant of the parent substance and mode of origin of the remarkable Bence Jones protein which is excreted in the urine of patients with multiple myeloma, nor is there any evidence as yet forthcoming that it is a normal intermediate product of protein metabolism. Nevertheless, when an endeavour is made to classify the unusual constituents which are occasionally present in that most important animal excretion, the urine, it is found that there are few of them which cannot be accounted for as intermediate products incompletely burnt, or as exaggeration of traces normally present ; if we exclude such as are merely foreign substances absorbed from the alimentary canal or derivatives of these, or are products of bacterial life and action in the intestines or in the tissues. A number of unusual constituents of urine, and of normal

constituents also, are derived from the alimentary canal. Thus foreign substances administered in food or as drugs may be excreted unchanged or may undergo oxidation or reduction in the intestine or after absorption, or again may appear in the urine in combination with products of metabolism. These last compounds, which result from the working of the chemical protective mechanisms, cannot be regarded as abnormal excreta. Even in health some tenth part of the total sulphuric acid of the urine is in combination with aromatic substances as ethereal sulphates. Traces of compound glycuronates are also normally present, and the aromatic constituents of vegetable foods are in part excreted in combination with glycocoll as hippuric acid and its allies. When other harmful substances, with which these protective mechanisms are competent to deal, are introduced in abnormal quantities the protective processes are stimulated to unwonted activity.

It does not necessarily follow that the substances which are utilised for such combinations are themselves intermediate products of normal metabolism, for it may well happen that combination has preceded oxidation. Thus the glycuronic acid which is excreted in compound glycuronates may not represent an intermediate stage in the breaking down of glucose as it has been supposed to do, for, as Emil Fischer has pointed out, the oxidation of the alcohol grouping of dextrose, whilst the less stable aldehyde grouping remains intact, is more easily explained on the supposition that the primary combination of the foreign substance is with glucose itself, and that, the aldehyde group being thus protected from change, oxidation to glycuronic acid occurs as a subsequent event.

Some excreta are due to the action of bacteria in the alimentary canal upon the proteins of food or upon constituents of the bile. Thus urobilin is formed by the intestinal bacteria from bilirubin and is abundantly present in the fæces so long as bile enters the intestine. From the intestine some urobilin is absorbed and is excreted in part in the bile and in part in the urine, whilst some of it is probably destroyed in the tissues. Again, from the tryptophane of ingested proteins the intestinal bacteria form indol, which after absorption is oxidised to indoxyl and is excreted in the urine, mainly as indoxyl sulphate, but in part in combination with glycuronic acid. Disease of the actual organs of excretion has conspicuous effects upon the excreta. Thus diseased kidneys may hold back in part some constituents of the urine or, on the other hand, may allow passage to the normal proteins of the blood serum which it is their function to retain in the circulation. Again, by the blockage of a duct, as in jaundice, the products of glandular activity may be thrown back into the blood and appear in the urine, but the presence of such abnormal constituents is in no sense due to errors of metabolism.

Actual derangements of the metabolic processes follow almost any deviations from the normal health, but our interpretation of the urinary changes which result is in many instances greatly hampered by the scantiness of our knowledge of the intermediate steps of the paths of metabolism. Such knowledge as we have of these steps is derived from casual glimpses afforded when, as the outcome of one of Nature's experiments, some particular line is interfered with, and intermediate products are excreted incompletely burnt. Many of the substances which rank as abnormal constituents are present in traces in normal urine as by-products of the metabolic processes, and it may safely be assumed that we are not cognisant of all the traces which so occur. Exceptional methods will reveal traces previously unsuspected. Thus Dombrowski, working with enormous volumes of normal urine (100 litres), was able to demonstrate the presence of minute quantities of cadaverin ; and that very delicate instrument the spectroscope reveals traces of hæmatoporphyrin in normal urine which would escape detection by rougher means. Only recently it has been shown that certain sulphur-containing acids, previously unknown, occur in no inconsiderable quantities in normal urine, and even now we do not know with any certainty all the constituents which go to make up the so-called neutral sulphur and residual nitrogen.

The effects of disease of the great laboratory glands, of which the liver is the chief, upon the chemical processes of which they are the seats, are less conspicuous than might be expected. This is perhaps due to the power of a small intact residue of an organ to carry on the functions of the whole,

---

[4] Pathologie Expérimentale, second edition, 1880, p. 15.
[5] Ibid., p. 568.

nor must it be forgotten that any very grave interference with the metabolic activities of the liver is incompatible with life. On the other hand, the phenomena of exophthalmic goitre and of myxœdema bear witness to the profound effects of atrophy or disease of a gland which exerts a controlling influence over the metabolic processes as a whole. Some abnormal constituents of urine are believed to be products of undue breaking down of tissues, of autolysis *intra vitam.* Such an origin is now usually ascribed to the tyrosin and leucin excreted in acute yellow atrophy of the liver, and to the albumoses met with in urine.

There is a group of maladies in which metabolic disturbances are by far the most conspicuous features, whereas the structural changes behind them are scanty or even inappreciable. Of such "diseases of metabolism," diabetes, gout, and obesity are the most important. It is still uncertain how far the accumulation of uric acid in the blood and the deposition of sodium biurate in the tissues which are the characteristic features of gout, are actually due to derangement of metabolism, as distinct from a mere excretory defect. In diabetes mellitus, under which name we probably include more than one morbid condition attended by persistent glycosuria, the metabolic derangements, primary and secondary, dominate the clinical picture. At the outset sugar may be excreted in very small quantity and only after a meal rich in carbohydrates; at a later stage the glycosuria tends to become continuous and the percentage of glucose in the urine rises until, in grave cases, the excretion almost ceases to be controlled by diet and the tissue proteins are called upon to supply sugar. By the destruction of fats and proteins the acetone bodies and especially β-oxybutyric acid are formed in increasing amounts, and to them the fatal ending is commonly attributed at the present day, although the accumulation of unburnt glucose in the blood is itself productive of serious evils. The liability to develop diabetes or gout is often inherited but the diseases themselves are not inherited, for they are never congenital. Developing at any period of life, the mischief, once begun, tends to become aggravated as time goes on, but the rate of aggravation differs widely in individual cases and is often conspicuously controlled by appropriate treatment.

Quite unlike that of the above metabolic diseases is the course of the anomalies of which I propose to treat in these lectures and which may be classed together as inborn errors of metabolism. Some of them are certainly, and all of them are probably, present from birth. The chemical error pursues an even course and shows no tendency to become aggravated as time goes on. With one exception they bring in their train no serious morbid effects, do not call for treatment, and are little likely to be influenced by any therapeutic measures at our disposal. Yet they are characterised by wide departures from the normal of the species far more conspicuous than any ordinary individual variations, and one is tempted to regard them as metabolic sports, the chemical analogues of structural malformations. It is interesting to note that as far back as the earlier years of the nineteenth century, one of them, albinism, was classed by Mansfeld [6] and by Meckel [7] as a "Hemmungsmissbildung" or malformation by arrest.

It may be pointed out that the epithets inborn and congenital are by no means synonymous. Structural abnormalities may be present at birth which owe their origin to intra-uterine disease or intra-uterine injury and are in no sense developmental errors. Again, an infective disease may be congenital but cannot be inborn. It has merely been acquired in utero. Even true developmental errors are of several distinct kinds. In some there is malposition or transposition of organs, partial or complete; in others doubling of parts or inclusion of twin structures. Some structural anomalies are malformations by excess such as polydactyly, and some are malformations by defect, such as absence of the middle phalanx of each digit. In one large class, the so-called malformations by arrest, the process of development meets with a check and some portion of the body is left unfinished. To this group belong such abnormalities as hare lip, cleft palate, and spina bifida. Speaking of such arrests Professor J. A. Thomson writes:[8] "These abnormalities occasionally recur repeatedly in a family tree, but it seems probable

that what is really inherited is a deficiency in 'developmental vigour' accentuated by nutritive defects on the parts of the mothers during the period of gestation." No extraneous causes, such as intra-uterine injury or disease, can be assigned to the metabolic errors which are under discussion. As far as our present knowledge of them enables us to judge they apparently result from failure of some step or other in the series of chemical changes which constitute metabolism, and are in this respect most nearly analogous to what are known as malformations by defect. It is not only in the field of metabolism that inborn derangements of function are met with, and Daltonism and night-blindness may be quoted as examples of such anomalies having no obvious chemical basis.

At first sight there appears to be little in common between inborn derangements of function and structural defects, but on further consideration the difference is seen to be rather apparent than real. Almost any structural defect will entail some disorder of function; sometimes this is almost inappreciable, but, on the other hand, the resulting functional disorder may be so conspicuous that it completely overshadows the defect to which it is due. Very slight structural changes may lead to profound functional derangements, as witness the effects of atrophy of the thyroid gland, whether congenital or acquired in later life, and the stormy metabolic disorders which may ensue upon comparatively insignificant morbid changes in the pancreas. By selective breeding there has been produced a race of waltzing mice, but their bizarre dance is merely the functional manifestation of an inborn and hereditary malformation of the semicircular canals. In the same way beneath each chemical sport may well exist some abnormality of structure so slight that it has hitherto escaped detection. Among the complex metabolic processes of which the human body is the seat there is room for an almost countless variety of such sports, but the examples which can be adduced are very few in number. Indeed, up to now the only known anomalies which can, with any good show of reason, be assigned to this class are albinism, alkaptonuria, cystinuria, and pentosuria, and even as regards these the reasons for their inclusion in the group are not of equal cogency.

We should naturally expect that among such abnormalities those would earliest attract attention which advertise their presence in some conspicuous way, either by some strikingly unusual appearance of surface tissues or of excreta, by the excretion of some substance which responds to a test habitually applied in the routine of clinical work, or by giving rise to obvious morbid symptoms. Each of the known inborn errors of metabolism manifests itself in one or other of these ways, and this suggests that others, equally rare, which do not so advertise their presence, may well have escaped notice until now. One man in 20,000 whose urine of 24 hours habitually contained a few grammes of aspartic acid might well be overlooked. Theoretically any anomaly which claims a place in the group should be present from birth and should persist throughout life, but it must be confessed at the outset that this cannot as yet be definitely asserted of all the four metabolic errors which I have mentioned. Some of them produce no obvious effects which compel attention, and may only be accidentally detected in adult life, and neither the evidence of the patient himself nor that of his parents can be of any help in endeavouring to trace back the peculiarity to infancy.

That albinism is congenital and persists through life is self-evident and admits of no dispute, for the condition is as obvious as any structural malformation and much more so than many such. Its rarity in man is also evident, although by artificial selection it may be reproduced indefinitely in lower animals.

As regards alkaptonuria its lifelong persistence is equally well established, although it may be that this error, which is in the great majority of instances inborn, may occasionally occur as a temporary phenomenon in disease. The remarkable staining property of alkapton urine allows of its recognition at the very beginning of life, and I have been able to obtain reliable evidence with regard to two cases that the staining of clothing was noticed on the second day of life, and in one of these instances had an opportunity of examining some urine passed during the first ten days after birth.[9] On the other hand an individual who exhibits the anomaly may

[6] Archiv für Anatomie und Physiologie, 1826, p. 96.
[7] Handbuch der Pathologischen Anatomie, 1816, Band ii., 2, p. 3.
[8] Heredity, London, 1908, p. 289.
[9] THE LANCET, Nov. 30th, 1901, p. 1484; Transactions of the Royal Medical and Chirurgical Society, 1902, vol. lxxxv., p. 69.

reach adult life without being aware of any special peculiarity of his urine and the condition may only attract attention when he is an applicant for life insurance or comes under treatment on account of some illness. Winternitz [10] has recorded the interesting fact that the mother of seven children, three of whom were alkaptonuric, was convinced that whereas two of her children had passed alkapton urine from the earliest days of life, this had not been the case with the youngest child, in whom she had only noticed the peculiarity from the age of five years. Such information supplied by a mother familiar with the symptoms of alkaptonuria carries as much weight as any hearsay evidence can carry, but nothing short of actual chemical examination of the urine would suffice to prove a point of so much importance.

Turning now to cystinuria, the evidence of its congenital occurrence is much more difficult to obtain, for this anomaly is little likely to attract attention in a young infant. Seeing that cystinuria is not infrequently transmitted from parent to child, examination of the urine of the infant children of those who manifest the peculiarity offers the most likely means of settling the point. That it may be present in early childhood there is abundant evidence to show. The first cystin calculus ever examined [11] was taken from the bladder of a child of five years. Abderhalden [12] has placed on record the detection of cystinuria in two children of the same family at the ages of 21 months and 14 months respectively, and Ultzmann [13] the case of a child of two with a cystin calculus who had exhibited symptoms of stone from the age of 12 months. Not a few cases in older children have been described. As affording evidence of the long persistence of cystinuria a case which came under the observation of the late Sir Henry Thompson [14] may be quoted that of an old man, aged 81 years, for whom a cystin calculus was crushed, and who had passed a stone of the same nature 39 years previously. There is reason to believe that cystinuria, like alkaptonuria, is occasionally temporary or intermittent.

No direct evidence of the congenital occurrence of pentosuria, the latest discovered and least known member of the group, is yet forthcoming, and its inclusion is to be justified on other grounds. That it may persist unchanged for years is certain, but the youngest pentosuric yet observed was a boy aged 15 years.[15]

It is probable that the rarity of albinism in man, of which each of us is able to judge from personal observation, is no greater than that of the other inborn errors. My belief is that cystinuria is decidedly the least rare of the four, but C. E. Simon [16] met with only one cystin sediment among some 15,000 urines examined, and Primavera [17] one in 20,000 urines. Another feature which all four anomalies share in common is their more frequent occurrence in males than in females. Of 38 cases of alkaptonuria, presumably congenital, 31 were males and only seven in females. Of 93 cystinurics collected from the records by Simon 63 were males and 30 females. Of 26 pentosurics, 19 were males and seven females. That males preponderate among albinos has often been stated, but among the cases collected by Ascoleo [18] in Sicily the disparity was comparatively slight, the numbers being 34 males and 28 females. It does not appear that there is any such unequal liability of the sexes to structural malformations, as far as can be judged from figures collected from the records of two hospitals, and in the remarkable family described by Farabee [19] in which many members exhibited a curious malformation by defect—namely, absence of the middle phalanges—the majority of the affected members were females. In a similar family recorded by Drinkwater [20] male cases slightly preponderated.

To be harmless is no essential attribute of an inborn

abnormality, but it stands to reason that an error of metabolism which persists from birth into adult and even into advanced life must needs be relatively innocuous. Albinism, although inconvenient, is certainly harmless, and the most serious direct result of alkaptonuria is the peculiar pigmentation, a variety of ochronosis, which some of its subjects develop in later life. Evidence is accumulating of the harmlessness of pentosuria, and that the excretion of a sugar with five carbon atoms in its molecule has none of the sinister significance of glycosuria. Only cystinuria can be classed as actually injurious, but even its ill-effects, serious as they often are, are not due to the deranged metabolism as such but are secondary, and result from the unsuitability of so sparingly soluble a substance as cystin for excretion by way of a urinary apparatus constructed upon the mammalian plan. Even in the single recorded instances in which death may possibly have resulted from cystinuria, apart from urinary complications, that of an infant aged 21 months, which Abderhalden describes, the deposition of cystin in the tissues was the conspicuous lesion found at the necropsy.

There remains to be mentioned yet another feature which the abnormalities under discussion possess in common—namely, the liability for each of them to occur in several members of a family, most often in collaterals of the same generation. In this respect, of course, they do not stand alone. In connexion with many actual diseases, as well as with structural malformations, hereditary influences come into play, and among diseases with none more strikingly than with such metabolic disorders as diabetes and gout. Only when taken in conjunction with all the other common features which have already been discussed can their mode of incidence be adduced in support of the view here taken of their nature.

To the students of heredity the inborn errors of metabolism offer a promising field of investigation but their adequate study from this point of view is beset with many difficulties. Save in the case of albinism one is driven to rely upon the casual mating of human beings and the conclusions based thereon cannot be checked by experimental breeding of animals. It is true that cystinuria is known to occur in dogs. A calculus obtained from a dog was described by Lassaigne [21] in 1823, and other examples have since been recorded,[22] but hitherto the diagnosis has only been made after the death of the animal and no opportunity has presented itself of utilising this fact for the advancement of the study of the anomaly. Whether alkaptonuria and pentosuria occur among lower animals is unknown. If they do they are little likely to be recognised.

Again it is naturally far more difficult to collect information as to the occurrence of chemical than of structural anomalies in past generations of a family, save in the case of albinism and possibly of alkaptonuria. Even as regards the relative numbers of normal and abnormal members, a knowledge of which is so important in connexion with questions of heredity, the information available is scanty and unreliable unless based upon personal examination. However, one point which stands out clearly is the remarkable similarity of the modes of incidence of alkaptonuria and albinism, which suggests that the manifestation of both is governed by the same laws. Both are apt to occur in several brothers and sisters of a family whose parents do not exhibit the anomaly and direct transmission of either from parent to child is very rare. It has been repeatedly stated that a considerable proportion of human albinos are the offspring of consanguineous marriages. Thus Ascoleo found that of 24 families which included 60 albino members five were the offspring of the mating of first cousins. In only two instances was albinism directly transmitted from parent to child. Of the cases of alkaptonuria, concerning which the necessary information is forthcoming, a very large proportion have been in children of first cousin marriages. In a paper published in 1902 [23] I called attention to this point and more recent cases, although they have somewhat lowered the proportion, have borne out the fact, as will be seen from the following table, in which the more recent cases have been incorporated.

[10] Münchener Medicinische Wochenschrift, 1899, Band xlvi., p. 749.
[11] Philosophical Transactions of the Royal Society, 1810, vol. c., p. 223.
[12] Zeitschrift für physiologische Chemie, 1903, Band xxxviii., p. 557.
[13] Wiener Medicinische Wochenschrift, 1871, Band xxi., pp. 286, 307.
[14] Transactions of the Pathological Society, 1870, vol. xxi., p. 272.
[15] Chobola: Centralblatt für Innere Medicin (abstract), 1907. Band xxviii., p. 864.
[16] American Journal of the Medical Sciences, 1900. vol. cxix., p. 39.
[17] Quoted by Piccini e Conti, Lo Sperimentale, 1891, vol. xlv., p. 353.
[18] Archivio per l'Anthropologia, 1871, vol. i., p. 367.
[19] Papers of the Peabody Museum of American Archæology and Ethnology, Harvard, 1905.
[20] Proceedings of the Royal Society of Edinburgh, 1908, vol. xxviii., p. 35.

[21] Annales de Chimie, 1823, 2e s., tome xxiii., p. 328.
[22] Gross, S.W.: North American Medico-Chirurgical Review, 1861, vol. v., p. 311.
[23] THE LANCET, Dec. 13th, 1902, p. 1616.

Families the offspring of first cousins.		Families of parents who were not blood relations.	
Names of observers.	Number of alkaptonuric members.	Names of observers.	Number of alkaptonuric members.
1. Pavy	4	1. W. Smith and Garrod	2
2. R. Kirk	3	2. Ewald Stier	1
3. Garrod	2	3. Noccioli and Domenici	1
4. Erich Meyer	1	4. Marshall and Futcher	3
5. Ogden	1	5. Langstein and E. Meyer	1
6. Hammarsten	2	6. Garrod and T. W. Clarke	1
7. Grutterink and van der Bergh	1	7. Grutterink and van der Bergh	2
8. Cronvall	1	8. ,, ,,	4
	—	9. Schumm	1
Number of families, 8.	15		—
		Number of families, 9.	16*

* In some instances private information has supplemented the published records. For some of the more recent cases the required information is not forthcoming. The new references will be found in the bibliography of Lecture II.

Thus of 17 families eight were the offspring of first cousins and nine were the children of parents who were not related, and of the total of 31 cases 15 fall into the first group. There appears to be a close connexion between the occurrence of an anomaly in several children of normal parents and consanguinity of the parents, a connexion which has been emphasised by Feer [24] in a recent paper. No one would suggest nowadays that the mere fact that the parents are of one blood would cause an anomaly to appear *de novo* in their children but it is obvious that the reappearance of a latent character which both parents tend to transmit is likely to be favoured by the mating of members of certain families.

The statistics as to the proportion of first-cousin marriages in this country are very scanty. Sir George Darwin has calculated [25] that less than 3 per cent. of all marriages are of this class, and Professor Karl Pearson [26] has recently collected some figures which give a percentage of 4·9 first-cousin marriages among the professional classes, a figure which for reasons which he states he regards as probably somewhat too high, and 0·86 among the classes from which patients in London hospitals are drawn. The totals of consanguineous marriages of all degrees in the two classes were 7·76 per cent. and 1·3 per cent. respectively. Hence it is obvious that the proportion of such marriages among the parents of alkaptonuric families is altogether abnormal. On the other hand, the proportion of alkaptonurics among children of such marriages must be very small indeed. Only some 50 to 60 cases of alkaptonuria have been recorded in Europe and America, whereas in London alone there are probably many thousands of children of first cousins.

It was pointed out by Bateson,[27] and has recently been emphasised by Punnett,[28] that the mode of incidence of alkaptonuria finds a ready explanation if the anomaly in question be regarded as a rare recessive character in the Mendelian sense. Mendel's law asserts that as regards two mutually exclusive characters, one of which tends to be dominant and the other recessive, cross-bred individuals will tend to manifest the dominant character, but when they interbreed the offspring of the hybrids will exhibit one or other of the original characters and will consist of dominants and recessives in definite proportions. Mendel's theory explains this by the supposition that the germinal cells or gametes of each generation are pure as regards the qualities in question and accounts for the numerical results observed by the production of dominant and recessive gametes in equal numbers. Of the offspring of two hybrids, one quarter will result from the union of two dominant gametes and will produce such gametes only; another quarter will result from the union of recessive gametes and will produce only recessive gametes. The remaining half will themselves manifest the dominant character, but will be hybrids like their parents and will produce gametes of both varieties. Only when two recessive gametes meet in fertilisation will the resulting individual show the recessive character.

If the recessive character be a rare one many generations

may elapse before the union of two such gametes occurs, for the families in which they are produced will be few in number and the chance that in any given marriage both parents will contribute such gametes will be very small. When, however, intermarriage occurs between two members of such a family the chance will be much greater, and of the offspring of such a marriage several are likely to exhibit the peculiarity. The rarer the anomaly the more conspicuous should be the influence of consanguinity. When a recessive individual mates with an apparent dominant who produces gametes of both kinds a larger proportion of the offspring will be recessives, and we should expect that recessive children of a recessive parent, but whose other parent is apparently normal, will occasionally be met with. Of such direct transmission of alkaptonuria from parent to child, the other parent not being alkaptonuric, two examples are known. One of these was observed by Osler.[29] An alkaptonuric father whose brother also showed the anomaly had an alkaptonuric son. The second case, which was recorded by Ossi,[30] was that of a mother and her son and daughter, all alkaptonuric. Lastly, when recessive mates with recessive all the offspring should manifest the recessive character, but no such marriage of alkaptonurics is known to have occurred. Whereas in animals, such as mice, which produce a numerous progeny, the proportions of dominants and recessives among their offspring can be readily observed, the results of the chance matings of human beings, who are so far less prolific, are far less demonstrative.

It must be confessed, indeed, that as regards human characteristics the relative numbers of dominant and recessive offspring have often departed widely from these required by Mendel's law, but a number of sources of error will tend to vitiate such results. Experience has shown that the information supplied as to the normality or otherwise of other members of a patient's family can seldom be relied upon, and this is especially the case with chemical anomalies. An individual in middle life seldom knows much about his brothers and sisters who died in infancy. Miscarriages must be taken into account, and again the figures supplied may relate to incomplete families and may be profoundly modified by subsequent births. For the above reason little importance is to be attached to the figures available with regard to alkaptonuria, but those contained in the following table, which relate to 17 families, are given for what they are worth:—

—	Observers.	Total number in family.	Normal members.	Alkaptonuric members.
Family No. 1	F. W. Pavy.	14	10	4
,, 2	Noccioli e Domenici.	10	9	1
,, 3	Ogden.	8	7	1
,, 4	Zimper.	8	6	2
,, 5	Winternitz.	7	4	3
,, 6	Langstein and E. Meyer.	6	5	1
,, 7	Schumm.	6	4	2
,, 8	Garrod.	5	3	2
,, 9	R. Kirk.	4	1	3
,, 10	Bandel.	4	2	2
,, 11	Hammarsten.	4	2	2
,, 12	W. Smith and Garrod.	3	1	2
,, 13	Baumann and Embden.	2	0	2
,, 14	Ewald Stier.	2	1	1
,, 15	Erich Meyer.	1	0	1
,, 16	Garrod and Clarke.	1	0	1
,, 17	Cronvall.	1	0	1
Totals		86	55	31

Although the totals show that the normal members largely preponderate they do not approach the requirements of Mendel's law, according to which a recessive character should

[24] Jahrbuch für Kinderheilkunde, 1907, vol. lxvi., p. 188.
[25] G. Darwin: Journal of the Statistical Society, 1875, vol. xxxviii., p. 153.
[26] K. Pearson: Brit. Med. Jour., 1908, vol. i., p. 1395.
[27] Report of the Evolution Committee of the Royal Society, 1902, No. 1, p. 133, note.
[28] Proceedings of the Royal Society of Medicine, 1908, vol. i., p. 148.
[29] See Garrod, THE LANCET, Dec. 13th, 1902, p. 1617.
[30] Gazzetta Medica Lombarda, 1889, vol. xlviii., p. 115.

appear in one quarter of the offspring; the alkaptonurics should number 21 or 22 instead of 31. It is clear that the figures as they stand are not fair to Mendel, for in the last three families the affected members were only children, and in families 15 and 16 might be classed as first children. One would expect to obtain a more accurate estimate by confining our attention to the larger families, and it is noteworthy that if one takes families 1 to 8, each containing five or more children, the totals work out in strict conformity to Mendel's law, in the proportions 4 : 3 : 1. However, to draw the line at families of five and upwards is a purely arbitrary proceeding.

The patients of Baumann and Embden were a brother and sister born out of wedlock, but it is not quite clear from the record whether they were the only children of the union. Both parents subsequently married away and had children, none of whom were alkaptonuric. Two recorded alkaptonurics were twins and in one case the second twin was certainly and in the other was probably normal.[31] In each instance the normal twin died and the alkaptonuric survived. The twins were of the same sex in both cases.

It appears to me that the strongest argument which can be adduced in favour of this view that alkaptonuria is a Mendelian recessive character is afforded by the fact that albinism, which so closely resembles it in its mode of incidence in man, behaves as a recessive character in the experimental breeding of animals.[32] Nor do the figures quoted by Bateson[33] relating to the proportion of albino members in human families show any more close conformity to the requirements of Mendel's law than do those above quoted for alkaptonuric families.

Evidence is accumulating of the occurrence of pentosuria in brothers and sisters,[34] and no instance of its transmission from parent to child has yet been recorded. Of consanguinity of parents of the subjects of this abnormality nothing is yet known. There is some evidence of a special liability of the Jewish race.

The available evidence regarding the inheritance of cystinuria is much more scanty than that relating to alkaptonuria. This is largely due to the less obvious character of the anomaly, for it is only by careful examination of the urine of each member of a family that any certainty can be reached as to the numbers of cystinuric and normal members. A cystinuric does not necessarily form calculi, and at any given time his urine may deposit no crystals, nor are the statements of patients as to other members of their families of any value in this connexion save that a history of several cases of stone may be suggestive. Such information as is forthcoming points to a greater frequency of direct transmission from parent to child than is met in connexion with the other metabolic errors, and cystinuria has been traced with certainty in three successive generations. Thus in the family investigated by Abderhalden the paternal grandfather and father were cystinuric, whereas the mother was normal. Of five children one had died with symptoms of inanition at 21 months and deposits of cystin were found in its tissues; two others had died with like symptoms at nine and 17 months respectively, but were not known to excrete cystin. The surviving children, aged five and a half years and 14 months respectively, were both cystinuric. Such large proportion of cystinuric members as was here met with has been observed in other families. In one which Cohn[35] described a cystinuric mother and a normal father had 12 children. The urine of two of them could not be obtained, but of the remaining ten no less than seven excreted cystin. Again, Pfeiffer[36] records four children of normal parents who were, as he has informed me, first cousins, all of whom were cystinuric. The two children of one of the affected daughters were normal. The frequency of direct inheritance and the large proportion of the offspring affected suggest that if cystinuria be transmitted on Mendelian lines it is probably a dominant rather than a recessive characteristic.

[31] Noccioli e Domenici: Gazzetta degli Ospedali, 1898, vol. xix., p. 303. D. Gerhardt: Münchener Medicinische Wochenschrift. 1904, Band li., p. 176.
[32] Castle and Allen: Proceedings of the American Academy of Arts and Sciences, 1903, vol. xxxviii., p. 603.
[33] Brit. Med. Jour, 1908, vol. i., p. 1252.
[34] Brat: Zeitschrift für Klinische Medicin, 1902, Band xlvii., p. 499. Bial: Berliner Klinische Wochenschrift, 1904, Band xli., p. 556, and others.
[35] Berliner Klinische Wochenschrift, 1899, Band xxxvi., p. 503.
[36] Centralblatt für Krankheiten der Harn- und Sexual-Organe, 1897, Band viii., p. 173.

Hitherto we have been considering these inborn errors of metabolism collectively, the points which they have in common, and the grounds for regarding them as constituting a distinct group of anomalies. But each one of them presents peculiar features of much interest which amply repay detailed consideration, if only on account of the light which their study throws upon the chemical processes at work in the normal human organism.

ALBINISM.

Of albinism I propose to speak very briefly, for its study from the chemical side has scarcely begun. In the fairly extensive literature of the subject much space is devoted to the ocular troubles which are connected with it; some authors have treated of its hereditary aspects; the question whether albinism carries with it any impairment of bodily or mental powers has been widely discussed but the actual nature of the anomaly has hardly been touched upon. Clearly it is an abnormality of a different kind from the others here under discussion which manifest themselves by errors of excretion. In albinism there is a defect of substances which are normal constituents of certain specialised tissues and which serve purposes of much utility to the organism. It is not necessary to enter upon any discussion of the question whether it should be classed as an error of anabolism or of catabolism. The work of F. G. Hopkins[37] upon the utilisation of excretory products in ornament, as exemplified in the employment of uric acid in the pigmentation of white butterflies of the genus Pieris, is of interest in this connexion, but it is probable that in insects, as in birds and reptiles, the excreted uric acid is largely formed by synthesis, as the breaking-down of nuclein does not provide an adequate source for the yield.

The essential phenomenon of albinism is the absence of the pigments of the melanin group which play the chief part in the colouration of man and lower animals and which serve the important function of rendering the eye a dark chamber. There are various kinds of melanins, but all alike are wanting in albinos, as witness the white hair, pink eyes, and unpigmented skin which characterise such individuals. Pigments of other kinds are not wanting, such as the lipochromes which impart their yellow tints to fats and blood serum, and hæmoglobin and its derivatives. In the urine of albinos I have found the same pigments as are ordinarily present. The fact that albinos of certain birds, such as the peacock, which are normally characterised by the brilliancy of the colouring of their plumage, are quite white, does not invalidate the truth of this statement. Their brilliant tints are interference colours, due to physical structure and not to pigmentation, and the absence of such colours is merely due to the lack of a dark background for their display, and such a background the melanins with their sombre hues provide. In the feathers of the albino peacock the iridescent tints are faintly visible in certain lights.

Our knowledge of the chemistry of melanins is still very imperfect. Their sulphur content varies widely; some contain iron, others do not. The presence of iron has been held to indicate an origin from hæmoglobin, but the trend of opinion at the present day is towards the view that they are derived from proteins in general, perhaps by the action of a tyrosinase upon tyrosin, and that the natural melanins are allied to the melanoid substances which are formed during the hydrolysis of proteins. If, as has been suggested, white hair contains a white chromogen and is pigmented and not merely devoid of colour; if also this is true of albino hair as well as that of white animals not albinos, we may suppose that the chromogen merely fails to be converted into the dark pigments in the ordinary way.

Three possible explanations of the phenomenon of albinism suggest themselves. We might suppose that the cells which usually contain pigment fail to take up melanins formed elsewhere; or that the albino has an unusual power of destroying these pigments; or again that he fails to form them. Some experiments of Kobert,[38] who injected solutions of melanin into rabbits, seem to negative the notion of an unusual destructive power. The animals excreted melanin, or rather its chromogen melanogen, in their urine. Kobert mentions that some of the animals injected were albinos, but

[37] Philosophical Transactions of the Royal Society, 1895, clxxxvi., B. p. 661.
[38] Ueber Melanine, Wiener Klinik, 1901, Band xxvii., p. 99.

does not say that they behaved any differently in this respect from the pigmented rabbits. The injection caused no pigmentation of the hair or eyes of albino rabbits, nor is it to be expected that it would do so whichever of the above hypotheses be correct. It is very unlikely that the melanin is conveyed to the pigmented cells and there deposited, for all the evidence available indicates that the pigment is formed *in situ*, probably by the action of intracellular enzymes. If the melanin were formed in albinos and merely not utilised we should expect it to be excreted by them in the urine, which is not the case. Only certain specialised cells appear to have the power of forming melanin, and in cases of melanotic sarcoma these are enormously multiplied, and there is a correspondingly increased production of pigment which may find its way into the blood and be excreted in the urine. Such an increase is only to be explained on the supposition that the pigment is actually formed in the tumour cells. In favour of this view is also the fact that melanotic tumours apparently originate only in structures in which melanins are normally present, such as the eye and skin, and possibly in the pineal gland, which as a vestigial remnant of the pineal eye of certainly lowly animal forms may retain some pigment forming power.

The hypothesis of local formation also supplies the easiest explanation of the phenomena of partial or local albinism, such as is seen in the Himalayan rabbit, which has the pink eyes of an albino but the hair of which is always pigmented at certain points. The partial albinism seen in man, in which the hair retains a certain amount of colour, and especially the rare form in which red hair is associated with pink eyes, a variety which has been met with in one albino of several in a family, the others conforming to the ordinary type,[39] is a phenomenon calling for careful investigation.

The ordinary physiological causes of pigmentation are not operative in albinos. In them exposure to the sun does not cause the usual tanning of the skin which is seen in normal individuals; a mere hyperæmia results. It has also been observed that in female subjects the pigmentation of the areolæ of the nipples and of other parts, which usually accompanies pregnancy, is not developed.[40] I know of no observations on the occurrence in albinos of tumours originating in the eye and running the course of melanotic sarcomata. One would expect that such growths, if they occur, would fail to be pigmented. Nor does there appear to be any record of Addison's disease occurring in such subjects which might throw important light upon the question whether the Addisonian pigment is a true melanin. Taking all the known facts into consideration, the theory that what the albino lacks is the power of forming melanin which is normally possessed by certain specialised cells is that which has most in its favour and is probably the true one. If so, an intracellular enzyme is probably wanting in the subjects of this anomaly, an explanation which, as we shall see later, brings albinism into line with some other inborn metabolic errors, of which a similar explanation is at least a possible one.

I must not omit to mention that there are indications that the differences between albinos and normal individuals are not confined to the absence or presence of melanins in the tissues. In animals differences have been observed in the matter of liability to certain infections, and clinical observations of the incidence of infective diseases upon human albinos would be of considerable interest. Halliburton, Brodie, and Pickering[41] found that intra-vascular injections of nucleo-proteins failed to produce in albino animals such clotting as they cause in pigmented ones. G. R. Mudge,[42] who has carried out a number of such experiments, found that all albinos do not behave alike in this respect. Differences were also observed between the results of the injection of nucleo-proteins derived from albino and pigmented animals respectively. He arrived at the conclusion that an albino animal requires a larger dose of nucleo-protein per kilogramme of body weight to cause death by intra-vascular clotting. Pickering's[43] remarkable observation that the Norway hare when in its winter coat behaves as an albino when injected with nucleo-protein, but in its summer coat as a pigmented animal, opens up the most interesting question of the relation, if any, of the change of coat in Arctic animals to albinism.

It will be clear from the above fragmentary sketch that even the lines along which the systematic study of albinism may profitably be directed are only beginning to indicate themselves. The carrying out of such a research remains as a task for future workers.

---

[39] Folker: THE LANCET, May 31st, 1879, p. 795. Nettleship: Transactions of the Ophthalmological Society, 1906, vol. xxvi., p. 244.
[40] B. W. Richardson: Dublin Hospital Gazette, 1856, vol. iii., p. 73.
[41] Journal of Physiology, 1894, vol. xvii., p. 148.
[42] Proceedings of the Royal Society, 1907, Series B., vol. lxxix., p. 103.
[43] Journal of Physiology, 1896, vol. xx., p. 310.

The Croonian Lectures

ON

INBORN ERRORS OF METABOLISM.

Delivered before the Royal College of Physicians of London on June 18th, 23rd, 25th, and 30th, 1908,

By ARCHIBALD E. GARROD, M.A., M.D.
OXON., F.R.C.P. LOND.,

ASSISTANT PHYSICIAN TO, AND LECTURER ON CHEMICAL PATHOLOGY AT, ST. BARTHOLOMEW'S HOSPITAL; SENIOR PHYSICIAN, HOSPITAL FOR SICK CHILDREN, GREAT ORMOND STREET.

LECTURE II.[1]

Delivered on June 23rd.

ALKAPTONURIA.

MR. PRESIDENT AND FELLOWS,—Of inborn errors of metabolism, alkaptonuria is that of which we know most, and from the study of which most has been learnt. In itself it is a trifling matter, inconvenient rather than harmful, which only attracts attention because an infant stains its clothing, or because an adult fails to effect an insurance of his life. The medical man merely needs to be aware of its existence and to be acquainted with the methods for its recognition in order that he may not mistake it for troubles of graver kinds; but for the chemical physiologist and pathologist it is one of the most interesting of metabolic abnormalities. Not only has the study of alkaptonuria thrown much light upon the fate of the aromatic fractions of the proteins of the food and tissues but has also helped materially to reveal a fact of far wider significance—namely, that for each protein fraction, or group of allied fractions, a special catabolic path has been evolved.

When freshly passed the urine of an alkaptonuric seldom exhibits any abnormality of tint, but it quickly begins to darken in contact with the air. The darkening, which is associated with absorption of oxygen, commences at the free surface of the liquid and passes through various shades of brown to actual blackness. Alkalinity of reaction greatly hastens the change. Linen and woollen fabrics moistened with the urine become stained as by a photographic developer. On heating the urine with Fehling's solution a deep brown colour develops and copious reduction occurs, but the browning of the liquid in which the orange precipitate is suspended gives to the test a quite peculiar appearance, which should be recognised by anyone who has once seen it. An ammoniacal solution of silver nitrate is rapidly reduced, even in the cold. On heating the urine with Nylander's solution, a darkening is produced by the alkaline reagent, but no reduction of bismuth is brought about. With Millon's reagent a yellow precipitate is formed. The most striking reaction is observed when a dilute solution of ferric chloride is allowed to fall into the urine drop by drop. The fall of each drop is followed by the appearance of a deep blue colour, which lasts but for a moment, and the phenomenon is repeated until oxidation is completed. With yeast no fermentation occurs and the polarised ray is not rotated either to the right or left.

Our knowledge of alkaptonuria is dated from the year 1858, in which year Bödeker[2] detected, in the urine of a patient with glycosuria, a second reducing substance, not a sugar, to which, on account of its behaviour towards alkalies, he assigned the name of alkapton, a bilingual word derived from alkali and κάπτειν. However, indications of the anomaly may be detected in much earlier medical writings. Thus there can be no doubt that the case of an infant who passed black urine, described by Alexander Marcet[3] in 1823, was of this nature. It is true that Marcet knew nothing of the reducing properties of the urine, but he describes accurately its darkening in colour on standing, the staining of napkins, and the effect of the addition of an alkali; and mentions that the condition was present from the earliest days of the child's life. Until

the early years of the nineteenth century no distinction was drawn in medical writings between urines which were black when passed and such as darkened on exposure to air, but it is difficult to suggest any other diagnosis than that of alkaptonuria for some cases referred to in works of the sixteenth and seventeenth centuries, such as that mentioned by G. A. Scribonius[4] (in 1584) of a schoolboy who, although he enjoyed good health, continuously excreted black urine, and that cited by Schenck[5] (in 1609) of a monk who exhibited a similar peculiarity and stated that he had done so all his life. The most interesting record of this kind is to be found in the work of Zacutus Lusitanus,[6] published in 1649. The patient was a boy who passed black urine and who, at the age of 14 years, was submitted to a drastic course of treatment which had for its aim the subduing of the fiery heat of his viscera, which was supposed to bring about the condition in question by charring and blackening his bile. Among the measures prescribed were bleedings, purgation, baths, a cold and watery diet, and drugs galore. None of these had any obvious effect, and eventually the patient, who tired of the futile and superfluous therapy, resolved to let things take their natural course. None of the predicted evils ensued, he married, begat a large family, and lived a long and healthy life, always passing urine black as ink.

That alkaptonuria is a very rare anomaly admits of no question, and many medical men of large experience have never met with it. Of its occurrence in several members of a family and of its mode of incidence I have already spoken at sufficient length in my first lecture. In the great majority of instances it is present from birth and persists throughout life, but has been said to have been developed as a temporary morbid sign in a very few exceptional cases. It gives rise to no symptoms save occasional dysuria with undue frequency of micturition, but some few alkaptonurics have developed in later life that peculiar tissue staining which Virchow[7] first described and to which he gave the name of ochronosis.

The essential feature of ochronosis is the staining of the cartilaginous structures of an inky blackness, but surface or sub-surface pigmentation is sometimes present and has rendered possible the recognition of the condition in living subjects. The tint of the blackened aural cartilages may show through the skin and give a slaty-blue colour to the hollows of the ears, black flecks may appear upon the conjunctivæ or even upon the skin of the face, and the knuckles may have a bluish tint. Less than 20 cases of ochronosis are on record, but it is becoming evident that rare as the condition is it yet has several distinct causes. Albrecht[8] first suggested that alkaptonuria was one of these, and Osler[9] described the development of surface pigmentation of the characteristic distribution and character in three elderly men who were well-authenticated alkaptonurics. Clemens[10] and A. Wagner[11] have recently supplied the conclusive demonstration of the presence of ochronotic pigmentation of the cartilages in the body of a man who was a subject of the urinary anomaly under discussion. In four of the recorded cases of ochronosis the pigmentation was apparently due to the application of carbolic acid to ulcers of the legs during long series of years.[12] In some of the remaining cases the urine has shown no peculiarity of tint, or if present it is not mentioned in the records, and in others dark urine was passed which does not seem to have been either of the alkaptonuric or carboluric variety.[13]

The evidence available leaves no doubt in the mind that alkaptonuria is one of the causes of ochronosis and that prolonged absorption of carbolic acid is another, but it would appear that not all alkaptonurics develop ochronosis in later life, and in the records of post-mortem examinations of

[1] Lecture I. was published in THE LANCET of July 4th, 1908, p. 1.
[2] Zeitschrift für Rationale Medicin, 1859, vol. vii., p. 130. Annalen der Chemie und Pharmacie, 1861, vol. cxvii., p. 98.
[3] Transactions of the Royal Medical and Chirurgical Society, 1823, vol. xii., p. 37.

[4] De Inspectione Urinarum, 1584, p. 50.
[5] Observationes Medicæ, 1609, Lib. iii., p. 558.
[6] Praxis Medica Admiranda, 1649, Lib. iii., cap. cxxxiv.
[7] Virchow's Archiv, 1866, vol. xxxvii., p. 212.
[8] Zeitschrift für Heilkunde, 1902, vol. xxiii., p. 366.
[9] THE LANCET, Jan. 2nd, 1904, p. 10.
[10] Verhandlungen des Congresses für innere Medicin, 1907, vol. xxiv., p. 249.
[11] Zeitschrift für innere Medicin, 1908, vol. lxv., p. 119.
[12] Pope: THE LANCET, Jan. 6th, 1906, p. 24. Pick: Berliner Klinische Wochenschrift, 1906, vol. xliii., p. 478. Graeffner: Ibid., 1907, vol. xliv., p. 1015. Reid: Quarterly Journal of Medicine, 1908, vol. i., p. 199.
[13] Hansemann: Berliner Klinische Wochenschrift, 1892, vol. xxix., p. 660. Hecker und Wolf: Festschrift des Stadtkrankenhauses zu Dresden-Friedrichstadt, 1899, p. 325.

alkaptonurics by Fürbringer [14] and Moraczewski [15] no mention is made of staining of the cartilages which, if present, could hardly have been overlooked. It is not yet possible to explain why in some cases oxidation and blackening should occur within the tissues during life.

The substance which Bödeker isolated from the urine of his patient, and which he called alkapton, contained nitrogen and was obviously an impure material. In some cases afterwards recorded the abnormal constituent was thought to be pyrocatechin [16] and in others protocatechuic acid.[17] Marshall [18] obtained from the urine of his patient a substance which he named glycosuric acid, and R. Kirk,[19] investigating a group of cases in a single family, isolated an acid which he called uroleucic acid, which yielded on analysis percentages of carbon and hydrogen conforming closely to the requirements of the formula $C_9H_{10}O_5$. These two investigators, Marshall and Kirk, approached very nearly to the recognition of the actual nature and composition of the abnormal constituent. There are no sufficient grounds for supposing that the reducing substances present in these earlier cases were different from that found in all the more recent ones, and where re-examination of the urine or of material extracted therefrom has been possible the presence of homogentisic acid has since been demonstrated.

Homogentisic acid, the excretion of which is the essential feature of alkaptonuria, was isolated, analysed, and fully investigated by Wolkow and Baumann,[20] as is set forth in their classical paper, published in 1891, some five years later than the investigations of Marshall and Kirk. It was shown to have the empirical formula $C_8H_8O_4$; and the work of its discoverers, which has been confirmed by Huppert [21] and by syntheses effected in three different manners, by Baumann and Fränkel,[22] Osborne,[23] and Otto Neubauer and Flatow [24] respectively, has definitely proved that its constitution is that of paradioxybenzene acetic acid (hydroquinone-acetic acid).

The acid was originally isolated from the urine as a lead salt, which may be obtained by an even far simpler method [25] than that employed by Wolkow and Baumann. The urine having been heated nearly to boiling, five grammes of solid neutral lead acetate are added for each 100 cubic centimetres of urine taken. A dense precipitate forms which is filtered off whilst the liquid is still hot, and the clear yellow filtrate is allowed to stand in a cool place. After a time lead homogentisate begins to separate out in crystalline form, and after 24 hours the crystals are filtered off, washed, and dried. The free acid may be obtained by passing sulphuretted hydrogen through ether in which the powdered lead salt is suspended. When the solvent, freed from lead sulphide by filtration, is allowed to evaporate colourless crystals of homogentisic acid are left and these melt at 146° to 147° C.

As alternative methods of extraction that of Wolkow and Baumann, in which the concentrated urine, acidified with sulphuric acid, is repeatedly extracted with ether, and the lead salt is thrown down from an aqueous solution of the residue from the ethereal extracts; or that of Erich Meyer,[26] which yields ethyl homogentisate, may be employed.

14 Berliner Klinische Wochenschrift, 1875, vol. xii., pp. 313, 390.
15 Centralblatt für innere Medicin, 1896, vol. xvii., p. 177.
16 Ebstein und Müller: Virchow's Archiv, 1875, vol. lxii., p. 554.
17 Fleischer: Berliner Klinische Wochenschrift, 1875, vol. l., pp. 529, 547. W. Smith: Dublin Journal of the Medical Sciences, 1882, vol. lxxiii., p. 465.
18 Medical News, Philadelphia, 1887, vol. l., p. 35.
19 Journal of Anatomy and Physiology, 1889, vol. xxiii., p. 69. Brit. Med. Jour., 1888, vol. ii., p. 232.
20 Zeitschrift für physiologische Chemie, 1891, vol. xv., p. 228.
21 Deutsches Archiv für klinische Medicin (Festschrift), 1899, vol. lxiv., p. 129.
22 Zeitschrift für physiologische Chemie, 1895, vol. xx., p. 219.
23 Journal of Physiology, 1903, xxix.; Proc. Physiol. Soc., xiii
24 Zeitschrift für physiologische Chemie, 1907, vol. lii., p. 375.
25 Garrod: Journal of Physiology, 1899, vol. xxiii., p. 512.
26 Deutsches Archiv für klinische Medicin, 1901, vol. lxx., p. 433.

For the purpose of quantitative estimation of homogentisic acid in urine the volumetric method of Baumann [27] is made use of. This somewhat tedious method is based upon the reduction of a decinormal solution of silver nitrate in the presence of ammonia, but it is necessary to employ a stronger solution (8 per cent.) of ammonia than that prescribed by Baumann (3 per cent.) in order to obtain complete reduction in the allotted period of five minutes.[28]

Solutions of homogentisic acid yield all the characteristic reactions of alkapton urines: darken on exposure to air and more quickly when an alkali is added, reduce Fehling's solution on boiling and ammoniacal silver nitrate in the cold, and yield a transient blue colour with ferric chloride. Abderhalden and Falta [29] were able to detect the presence of the acid in the blood of an alkaptonuric man and to obtain the lead salt from the serum. From the fæces it is certainly absent, and I have failed to find it in the sweat. Ewald Stier [30] and Bandel [31] describe an unusual blackness of the aural wax of their patients, and from it the former obtained a substance which yielded the alkapton reactions.

In all the accounts of alkaptonuria written in recent years the statement will be found that in some cases there has been present in the urine, in addition to homogentisic acid, a second acid possessed of similar properties—viz., uroleucic acid, and that this substance is probably hydroquinone α-lactic acid.

This statement I believe to be grounded upon a misapprehension and to be incorrect, and the grounds for this belief have been fully set out in a recent paper written in conjunction with Dr. Hurtley.[32]

It will be remembered that the name of uroleucic acid was assigned by Kirk to the material which he isolated from the urine of his patients at a time before homogentisic acid was known. The late Dr. Kirk never claimed that this was a second distinct alkapton acid, and, indeed, in a letter to me he expressed his opinion that his uroleucic acid was merely impure homogentisic acid. Kirk's substance, to which as the result of his analyses he assigned the formula $C_9H_{10}O_5$, melted at about 133·3° C., whereas, as has already been mentioned, the melting point of homogentisic acid is 146° to 147° C. The analytical figures agreed very closely with the requirements of the above formula, and titration of a solution with alkali under a layer of petroleum ether gave a result which indicated a molecular weight corresponding to that of a monobasic acid of the above constitution. However, there is no room for doubt that Kirk's later surmise was correct and that in spite of these coincidences the substance which he analysed was impure homogentisic acid. Thus we found that when Kirk's method of extraction was applied to an alkapton urine in which a second acid had been sought for without success the product obtained agreed with his description of the lead salt which he obtained, and the free acid isolated from the lead salt melted at 134° to 136° C., but examination showed that it consisted mainly of homogentisic acid. It is known that the urines which Kirk examined contained this acid. Huppert obtained it from some of Kirk's original material, and we also found it in a further specimen of that material, blackened with age, but labelled "uroleucic acid." Moreover, when in 1902 I was enabled, by the kindness of Dr. Kirk, to examine fresh specimens of the urine of his patients much homogentisic acid was obtained from them but there was no indication of the presence of a second alkapton acid.

The view that the uroleucic acid of Kirk was a distinct

27 Zeitschrift für physiologische Chemie, 1892, vol. xvi., p. 268.
28 Garrod and Hurtley: Journal of Physiology, 1905, vol. xxxiii., p. 206.
29 Zeitschrift für physiologische Chemie, 1903, vol. xxxix., p. 143.
30 Berliner klinische Wochenschrift, 1898, vol. xxxv., p. 185.
31 Deutsche medicinische Wochenschrift, 1906, vol. xxxii., p. 287.
32 Journal of Physiology, 1907, vol. xxxvi., p. 136.

substance had its origin in some investigations of the late Professor Huppert,[33] carried out upon some of the original material sent to him in 1897. After separation of as much as possible of the homogentisic acid which it contained a residue remained which melted at 130·5°C., and this residue Huppert regarded as uroleucic acid. Further investigations led him to the conclusion that it was a derivative of hydroquinone and was probably hydroquinone-α-lactic acid, which acid has the formula $C_9H_{10}O_6$ assigned by Kirk to uroleucic acid.

The accuracy of Professor Huppert's results does not admit of question, but it is evident that the scanty material at his disposal did not allow of a fresh analysis of the fraction of lower melting point, and if, as I believe, after careful comparison of Kirk's account of his investigations with Huppert's description of the material sent to him, and after personal examination of a further portion of the material, the substance examined was not a crude one, as Huppert supposed, but the actual uroleucic acid analysed by Kirk, it is obvious that the results of analyses of the material as a whole could not be applied to the fraction of lower melting point. Otto Neubauer and Flatow,[34] who have succeeded in effecting the synthesis of hydroquinone α-lactic acid, have shown that it differs from the supposed uroleucic acid in melting point and in other respects; and it is a significant fact that with the exception of a minute residue, which melted at 133°C., but which only sufficed for the determination of the melting point, obtained by Langstein and E. Meyer[35] from the urine of their patient, no indication of the presence of uroleucic acid has been found in any of the alkapton urines since described. Therefore the conclusion appears to be justified that no sufficient evidence is forthcoming of the occurrence in some alkapton urines of a second abnormal acid (uroleuic acid), a conclusion which has the advantage of making for simplicity. Beyond the presence in it of homogentisic acid the urine of alkaptonurics shows no obvious deviation from the normal. Some earlier investigators described a conspicuous diminution of the uric acid output, but more recent work has failed to confirm this. In a number of cases the excretion of uric acid has been found not to be below the average, and crystals of uric acid stained by the brown pigment are not infrequently deposited from such urines. We may therefore confine our attention to the consideration of the parent substances and mode of origin of homogentisic acid in the human organism.

Seeing that there is no evidence that synthesis of the benzene ring ever occurs in the animal economy Wolkow and Baumann looked to the proteins of the food and tissues as the most likely sources of the alkapton acid and to the aromatic fractions which proteins contain—viz., tyrosin and phenyl-alanin—as its special precursors. This conjecture was shown to be correct by the result of the administration of tyrosin by the mouth to their alkaptonuric subject. Such administration caused a very conspicuous increase of the output of homogentisic acid. Since then this observation has frequently been repeated by other investigators, and the result has been shown to hold good for alkaptonurics in general. A corresponding increase follows an augmented intake of protein food and especially of such proteins as are unusually rich in the aromatic fractions.

Wolkow and Baumann were not able to test the effect of the administration of the phenyl-alanin, but at a later period Langstein and Meyer[36] arrived at the conclusion that the tyrosin of the proteins broken down in the metabolism of their alkaptonuric did not suffice to account for the quantity of alkapton acid which he excreted. This conclusion was based upon a comparison of the estimated daily output of homogentisic acid with that of nitrogen, which latter affords a measure of the protein destruction going on in the body. It was therefore probable that phenyl-alanin must also serve as a parent substance, and that it does so was afterwards demonstrated by Falta and Langstein[37] who found that phenyl-alanin given by the mouth increases the homogentisic output just as tyrosin does.

Falta[38] afterwards showed by a long series of feeding experiments with different proteins added to a constant diet, that the excretion of alkapton acid varies directly with the richness of the proteins given in phenyl-alanin and tyrosin, and arrived at the conclusion that on any given diet the output corresponds closely with that to be expected if the aromatic fractions of the catabolised proteins are wholly converted into homogentisic acid and excreted as such.

The yield of alkapton acid after feeding with tyrosin or phenyl-alanin varies with the mode of administration, and when small doses are given at short intervals, instead of a single large dose, the output is practically quantitative. This observation we owe to Mittelbach.[39]

Yet one other of the known protein fractions has a benzene ring in its molecule, but in tryptophane the ring forms part of the more complex, heterocyclic, indol grouping. It appears unlikely a priori that the catabolism of tryptophane will follow the same path as tyrosin, and that it will become converted into homogentisic acid, but in order to test the point I recently administered one gramme of tryptophane, for which I was indebted to the kindness of Dr. F. G. Hopkins, to an alkaptonuric boy.

$$C_6H_4 \diagdown \begin{array}{c} C-CH_2-CH(NH_2)-CO.OH \\ \\ CH \\ \\ NH \end{array}$$

Tryptophane.

The dose was a small one, and any effect might be expected to be slight. Therefore an equivalent dose of tyrosin (0·9 gramme) was given on two occasions as a control, the patient being kept upon a constant diet. The effect of each dose of tyrosin upon the $\dfrac{\text{homogentisic acid}}{\text{nitrogen}}$ ratio was clearly marked, but no disturbance of the ratio was observed on the day on which the tryptophane was given. Hence I feel justified in concluding that tryptophane administered by the mouth does not increase the output of homogentisic acid. However, it would be desirable that this should be confirmed by the giving of larger doses. I may add that Dr. Hopkins, who was good enough to examine the urine of the tryptophane day, failed to find in it any abnormal constituent the presence of which could be ascribed to an error of tryptophane catabolism. To sum up, it would appear that the tyrosin and phenyl-alanin of proteins are the only parent substances of the alkapton acid.

A glance at the respective formulæ of tyrosin and homogentisic acid suffices to show that the change from the one to the other is a complex one, for it involves the removal of the hydroxyl group from the para-position and substitution of two others in the 2 : 5 position in relation to the side chain, or possibly a shifting of the side chain itself.

$$\begin{array}{ccc} \text{OH} & & \\ & \text{CH}_2 & \text{CH}_2 \\ \text{CH.NH}_2 & \text{CH.NH}_2 & \text{CH}_2 \\ \text{COOH.} & \text{COOH.} & \text{CO.OH.} \\ \text{Tyrosin.} & \text{Phenyl-alanin.} & \text{Homogentisic acid.} \end{array}$$

Such a change presented greater chemical difficulties when Baumann wrote than it does now and could not be paralleled in animal metabolism. As, however, such successive reduction and oxidation were known to be brought about by bacterial action, Wolkow and Baumann suggested that it might have its seat in the intestine of alkaptonurics, being there brought about under the influence of a rare specific micro-organism. Nowadays this infective theory, which was at one time widely accepted, has been completely abandoned, for it has been abundantly disproved. Intestinal disinfection has no influence upon the excretion

[33] Zeitschrift für physiologische Chemie, 1897, vol. xxiii., p. 412.
[34] Loc. cit., sub. 24.
[35] Deutsches Archiv für Klinische Medicin, 1903, vol. lxxviii., p. 161.
[36] Deutsches Archiv für Klinische Medicin, 1903, vol. lxxviii., p. 161.
[37] Zeitschrift für physiologische Chemie, 1903, vol. xxxvii., p. 574; see also Falta: Deutsches Archiv für klinische Medicin, 1904, vol. lxxxi., p. 250.

[38] Deutsches Archiv für klinische Medicin, 1904, vol. lxxxi., p. 231.
[39] Ibid., 1901, vol. lxxi., p. 50.

of homogentisic acid by alkaptonurics, nor can any organism be grown from their fæces which is able to effect such a conversion of tyrosin. Moreover, it has been shown by Mittelbach,[40] and afterwards by Langstein and Meyer and by Falta,[41] that abstinence from protein food, and indeed from all food, does not arrest the excretion of the acid, although it naturally greatly reduces its amount. Hence we must conclude that not the food proteins alone but also those of the tissues contribute their quota to the total output of homogentisic acid, which would not be the case if the conversion were effected in the alimentary canal and the fact, which we owe to Abderhalden, Bloch, and Rona,[42] that the soluble dipeptide glycyl-l-tyrosin when injected subcutaneously into an alkaptonuric produces the same effect as tyrosin introduced into the alimentary canal affords equally conclusive evidence that the change is not merely due to the action of intestinal bacteria.

Lastly, if the aromatic fractions of the proteins were thus changed in the intestine before undergoing absorption therefrom, seeing that they are not synthesised in the animal organism, the tissue proteins of alkaptonurics should exhibit a shortage of tyrosin and phenyl-alanin, if, indeed, such a deviation from normality be compatible with the maintenance of life. Abderhalden and Falta,[43] who have investigated the blood proteins of such subjects as well as their hair and nails, have found no evidence of deficiency of the fractions in question.

The change from phenyl-alanin to homogentisic acid presents less difficulty than that from tyrosin, for that compound has no hydroxyl group upon the benzene ring, and it is not improbable that with tyrosin the removal of the para-hydroxyl group and the oxidation in the 2 and 5 positions occur at quite distinct stages of the catabolic process, so that no shifting of the hydroxyl or of the side chain is involved. Such a view gains support from the fact, ascertained by L. Blum,[44] that metatyrosin, which he has prepared synthetically, and which has its hydroxyl group in the same position relatively to the side chain as one of the hydroxyls of homogentisic acid, is actually not converted

Metatyrosin. Homogentisic acid.

into homogentisic acid when administered by the mouth, for it does not increase the output of that acid by an alkaptonuric. This behaviour of metatyrosin affords a further proof of a fact which had previously been demonstrated by Falta, namely, that substitutions upon the benzene ring are capable of preventing the formation of homogentisic acid from tyrosin. Falta[45] showed that a similar arrest occurs as the result of substitution of halogens, for neither dibrom-tyrosin nor the halogen proteins increase the homogentisic excretion of an alkaptonuric. On the other hand, as we shall see presently, the factor which determines such conversion is the structure of the side chain and the presence of a grouping which is readily attacked in connexion with the a carbon atom.

It will be obvious, from all that has gone before, that the error of metabolism which is at the back of alkaptonuria is a failure to deal with the aromatic fractions of proteins in the ordinary way, and that both the proteins of the food and those of the tissues are implicated in the error. Before proceeding further it is, therefore, necessary to consider what is known as to the ordinary way of dealing with these fractions.

It is an unquestionable fact that the great majority of aromatic compounds, when introduced into the human organism, escape with their benzene ring intact and are excreted in the urine in combination with sulphuric acid, as aromatic sulphates, or with glycocoll, as the acids of the hippuric group. Not so tyrosin and phenyl-alanin, which are in no sense foreign substances but important constituents of proteins, for these suffer disintegration of the aromatic nucleus and are completely destroyed. It is true that they do not stand quite alone in this respect, nor is it to be expected that they should do so, for any aromatic substance which is an intermediate product of their catabolism will obviously be in like manner broken up, as also will compounds so closely allied to them in their molecular structure that when exposed to the same destructive influences they share their fate. Some recent investigations have supplied a clue to the seat of the destruction of such substances and the nature of the changes which they undergo. Thus, G. Embden, Saloman, and F. Schmidt[46] found that when certain protein substances are perfused through the liver acetone is formed. Glycocoll, alanin, glutaminic acid, and asparagin led to no such formation of acetone, but with leucin, tyrosin, and phenyl-alanin, as well as with phenyl a-lactic and homogentisic acids, a conspicuous yield of acetone was obtained. The aromatic acids on this list are all broken up in the normal organism, whereas phenyl β-lactic, phenyl-propionic, and phenyl-acetic acids, which are not so broken up, do not yield acetone in the perfused liver. Furthermore, J. Barr and L. Blum[47] have found that substances which figure in the above list of acetone-yielding compounds cause an increased excretion of β-oxybutyric acid when administered to diabetics. Hence they regard it as probable that the formation of acetone from them is a secondary event and that β-oxybutyric acid is first formed.

The question which next calls for consideration is whether in alkaptonuria the failure to deal with tyrosin and phenyl-alanin is or is not complete. In diabetes we are confronted with a failure to burn glucose, which may be of any degree, from such as leads to a small excretion of glucose only after a meal rich in carbohydrates, to a failure so nearly complete that the proteins are called upon to contribute to the urinary output of glucose. In congenital cases of alkaptonuria, at any rate, and concerning temporary cases no information on this point is forthcoming, the excretion of homogentisic acid, although it varies considerably, as it obviously must do, according to the nature of the diet taken, is fairly uniform in different cases, provided always that the food of the subjects is similar in kind and in proportions.[48] There is no evidence that the alkapton acid ever occurs in traces in the urine; an output of several grammes in the day is observed or no homogentisic acid is found. Four or five grammes a day is the average on an ordinary mixed diet. The evidence afforded by the ratio of homogentisic acid to nitrogen is of far greater value in this connexion than that derived from the measure of the daily output of homogentisic acid alone. A mere increased supply of a given protein in the food does not disturb this ratio, for with the increased destruction nitrogen and alkapton acid in the urine increase *pari passu*. But though unaffected by quantitative changes in the diet the ratio is profoundly disturbed by changes in the quality of the proteins taken, as must needs be the case, seeing that some proteins are far richer in tyrosin and phenyl-alanin than others, whereas their yield of nitrogen is comparatively constant. Thus the substitution of casein, which is rich in the aromatic fractions, for egg albumin, in which they are much more scantily present, will greatly increase the output of homogentisic acid relatively to that of nitrogen, as Langstein and Meyer and Falta have demonstrated.

If in alkaptonuria the error be complete and maximal, all subjects of that anomaly when fed upon a given standard diet should excrete equal quantities of homogentisic acid, and changes in the quantities consumed, although they will

[40] Loc cit., sub. 39.
[41] Loc. cit., sub. 36 and 38.
[42] Zeitschrift für Physiologische Chemie, 1907, vol. lii., p. 435.
[43] Ibid., 1903, vol. xxxix., p. 143, and 1907, vol. lii., p. 445.
[44] Verhandlungen des Congresses für innere Medicin, 1907, vol. xxiv., p. 240.
[45] Loc. cit., sub. 39, p. 254.

[46] Hofmeister's Beiträge, 1906, vol. viii., pp. 121, 129.
[47] Archiv für Experimentelle Pathologie und Pharmakologie, 1906, vol. lv., p. 89.
[48] Most unusually high figures may be ascribed to the diet taken; but it should be mentioned that Zimper records figures for his case which are out of all proportion to those obtained in any other. The homogentisic acid output amounted to as much as 18 grammes per diem. The tyrosin and phenyl-alanin of proteins broken down could not account for such an excretion.

conspicuously affect the total output of the acid, should leave the H : N ratio $\left(\dfrac{homogentisic\ acid}{nitrogen}\right)$ undisturbed. In order to obtain conclusive evidence upon this point it would be necessary to place a number of alkaptonurics upon a standard diet, in which not only are the proteins the same but in which also the several proteins are represented in uniform proportions. These conditions would be best fulfilled by a diet of milk alone. Unfortunately, observations so carried out are not available, but such observations as are forthcoming strongly suggest that the error is in all cases of one degree, and that the failure to deal with tyrosin and phenyl-alanin is complete. The best available standard, although far from a uniform one, is a mixed diet, and the following figures are drawn from the observations of Langstein and Meyer and Falta upon one alkaptonuric, of Schumm upon another, and of Hele and myself upon yet three others. All the patients were, at the times to which the figures refer, upon a mixed diet, and the results serve to show that under roughly uniform conditions of diet the ratios show at least a striking similarity.[49]

a given quantity of a given protein all subjects of the anomaly tend to form and to excrete the same amount of homogentisic acid, and it is probable that that quantity is a maximal one, seeing that Falta has shown that the output of homogentisic acid corresponds roughly to the estimated amounts of the aromatic fractions in the proteins broken down.

Two explanations are possible of the fact that alkaptonurics excrete homogentisic acid whereas normal persons do not. Either the alkapton acid is a strictly abnormal product formed by a perverted metabolism of tyrosin and phenyl-alanin, in virtue of which these protein fractions experience a wholly different fate from that which overtakes them in the normal organism, or it is an intermediate product of normal metabolism which is usually completely destroyed and does not come to excretion, but which in alkaptonuria escapes destruction.

It may be premised that the behaviour of homogentisic acid in the organism is rather that of a normal product than that of an interloper. None of the chemical protective mechanisms are called into play to cope with it in alkap-

Table showing the Daily H : N Ratios in Five Cases of Alkaptonuria, during Periods in which Mixed Diets were Taken.

Anton M., in Basle, aged 50 years.				Male, 20 years. O. Schumm.	Thomas P., aged eight years.		Albert P., aged four years.	Minnie L., aged three years.
Langstein and E. Meyer.		Falta.						
46 : 100	36 : 100	48·1 : 100	40·0 : 100	44·5 : 100	43·8 : 100	44·1 : 100	48·4 : 100	42·7 : 100
44 ,,	42 ,,	44·9 ,,	38·0 ,,	42·2 ,,	39·9 ,,	48·1 ,,	47·6 ,,	27·0 ,,
42 ,,	38 ,,	40·3 ,,	40·5 ,,	47·6 ,,	35·9 ,,	44·7 ,,	44·9 ,,	48·3 ,,
42 ,,	34 ,,	44·3 ,,	40·8 ,,	42·6 ,,	40·0 ,,	49·0 ,,	52·7 ,,	39·6 ,,
43 ,,	34 ,,	50·4 ,,	41·1 ,,	49·5 ,,	37·7 ,,	—	48·8 ,,	39·1 ,,
46 ,,	40 ,,	44·5 ,,	—	42·9 ,,	44·4 ,,	—	50·7 ,,	34·5 ,,
46 ,,	53 ,,	40·9 ,,	—	45·8 ,,	35·9 ,,	—	45·4 ,,	44·8 ,,
44 ,,	40 ,,	40·8 ,,	—	41·3 ,,	38·7 ,,	—	54·3 ,,	—
—	45 ,,	41·8 ,,	—	—	43·9 ,,	—	53·5 ,,	—
—	54 ,,	—	—	—	—	—	—	—
Averages :—								
44 : 100	41·6 : 100	44·0 : 100	40·1 : 100	44·0 : 100	40·0 : 100	46·5 : 100	49·6 : 100	39·4 : 100

Note.—The above ratios are all some 5·8 per cent. too low, owing to the use of 3 per cent. ammonia in estimating homogentisic acid. The error so caused is fairly constant. The figures for Minnie L. are obtained by substracting 5·8 per cent. from each ratio, 8 per cent. ammonia having been used in this case. The first four columns relate to the same individual.

Some other observations bearing upon this point have been published in the last year or two. Cronvall[50] obtained ratios varying between 45 : 100 and 61 : 100 in his case, the mean being 55 : 100. As he used 8 per cent. ammonia in estimating the homogentisic acid the percentages of that acid must be reduced by 5·8 per cent. to render them comparable with those given in the table, and he himself suggested that the ratios were rendered higher by a considerable amount of milk in his patient's diet.

Gross and Allard[51] in a recent investigation obtained very high H : N ratios, 60-70 : 100, which they regarded as showing a want of uniformity in different cases, but their patient was taking large quantities of milk, on one day as much as three litres, and changes in the protein content of the diet appear to have been chiefly brought about by the addition or withdrawal of milk. Their ratios are comparable with those worked out by Langstein and E. Meyer during a period of milk and plasmon diet (average 62·6 : 100), and that the above explanation is correct is shown by the fact that even the addition of plasmon (sodium casein) to the diet scarcely affected the ratio, whereas when superposed upon an ordinary mixed diet it greatly increases the homogentisic output relatively to that of nitrogen. Abderhalden, Bloch, and Rona[52] also give some figures relating to yet another case, that of a boy, and the ratios obtained from these figures are somewhat low, the average for five days being 35·7 : 100. The nature of the diet is not mentioned.

So far as the available evidence goes it tends to show that there is only one degree of alkaptonuria, and that from

tonuria save that which is called out by any acid which is not destroyed in metabolism. It is excreted in simple combination with bases and not as an aromatic sulphate or in combination with glycocoll. Its next homologue, gentisic acid, on the other hand, although it is for the most part destroyed in passage through the body, is in part excreted as aromatic sulphate, as Likhatscheff[53] showed and as Neubauer and Falta[54] also found.

As an acid, homogentisic acid is in part combined with ammonia, for the protection against acid is in no wise selective. Erich Meyer[55] obtained evidence of an increased excretion of ammonia by an alkaptonuric child and Schumm[56] made a like observation in the case of an adult. In neither instance was the increase of urinary ammonia comparable with that observed in some pathological conditions, such as grave diabetes, for, as we have seen, the output of the acid itself is never large as compared with that of the acids of the acetone group in some morbid states.

It stands to reason that an intermediate product of catabolism which is normally absent from the excreta must needs be completely destroyed under ordinary conditions, and that homogentisic acid conforms to this requirement there is abundant evidence to show. Thus, although H. Embden[57] succeeded in producing a transitory alkaptonuria in himself by swallowing eight grammes of the acid, he found that smaller doses had no such effect. Falta,[58] too, failed to induce an excretion of the acid in himself and two others by

[49] Journal of Physiology, 1905, vol. xxxiii., p. 198, and 1906, vol. xxxv. (Proceedings of the Physiological Society).
[50] Upsala Läkareförenings Förhandlingar, 1907, vol. xii., p. 402.
[51] Zeitschrift für klinische Medicin, 1907, vol. lxiv., p. 359.
[52] Zeitschrift für physiologische Chemie, 1907, vol. lii., p. 444.

[53] Zeitschrift für physiologische Chemie, 1895, vol. xxi., p. 422.
[54] Ibid., 1904, vol. xlii., p. 92.
[55] Deutsches Archiv für klinische Medicin, 1901, vol. lxx., p. 463.
[56] Münchener medicinische Wochenschrift, 1904, vol. p. 1599.
[57] Zeitschrift für physiologische Chemie, 1893, vol. xvii., p. 182, and vol. xviii., p. 304.
[58] Loc. cit., sub. 38, p. 264.

taking quantities of from four to six grammes in repeated half-gramme doses at short intervals. Hence it is evident that homogentisic acid is a member of that small group of aromatic compounds the benzene ring of which is broken down in their passage through the body, and, as has been mentioned, further evidence of this is afforded by the fact that it yields acetone when perfused through the liver. In these respects it behaves as a normal intermediate product might be expected to do. That this destructive power may be overtaxed is shown by Embden's experiment upon himself and by those of Wolkow and Baumann upon dogs.[59] In normal persons such overtaxing can never occur, since the maximum daily output taken at a single dose will hardly give rise to experimental alkaptonuria; but in disease the power of destroying homogentisic acid is sometimes conspicuously lowered—for example, in grave cases of diabetes, as Langstein,[60] Falta,[61] and others have demonstrated.

Garnier and Voirin,[62] who were the first to suggest that homogentisic acid is a product of normal metabolism, were inclined to ascribe its presence in the urine to excessive production, the power of destroying it being overtaxed. Whether the power of destroying it were merely overtaxed or wholly lost it might be expected that homogentisic acid given by the mouth to an alkaptonuric will be excreted nearly quantitatively, being added to the ordinary output, and H. Embden showed that this is the case. If, however, it were merely a question of overtaxing, the output of the acid should be considerably less than the amount which corresponds to the whole of the aromatic fraction of the proteins broken down, but, as we have seen, there is reason to believe that the failure to deal with tyrosin and phenyl-alanin in the usual way is complete. On such grounds the more recent adherents of the intermediate product theory have held that the alkaptonuric differs from the normal individual in that he does not possess the power of destroying homogentisic acid.

This theory has lately been challenged by Grutterink and Hijmans van der Bergh,[63] and upon the following grounds. To patients who were found to have much impaired powers of destroying homogentisic acid taken by the mouth, being sufferers from diabetes or hepatic disease, and in one of whom as small a dose as two grammes caused a temporary alkaptonuria, these observers administered tyrosin in doses of 10 to 15 grammes. They argue that after such large doses of tyrosin sufficient homogentisic acid should have been formed to overtax the limited destructive power, supposing that it were a normal intermediate product, and that some should have been excreted in the patient's urine, whereas in no instance did such a result follow. This evidence cannot certainly be lightly set aside. However, we cannot be sure that at any moment sufficient alkapton acid would be in existence to overtax the destructive power, which although diminished was certainly not abolished. Nor is it certain that when such impairment results from disease the formation of homogentisic acid, as well as its destruction, is not interfered with. Grutterink and van der Bergh satisfied themselves that the tyrosin given was well absorbed, but they do not state whether tyrosin itself, or derivatives thereof, other than homogentisic acid, were sought for in the urine. Knoop[64] had previously raised a similar objection. He fed dogs with phenyl α-lactic acid, a compound which, like tyrosin, increases the homogentisic output of alkaptonurics. He, too, argued that, as the power of dogs to destroy homogentisic acid is known to be limited, if it were a normal product it should have appeared in the urine. However, the fact that some of the phenyl α-lactic acid was recovered as such from the urine suggests that a block occurred at an earlier stage in the catabolic series.

It appears to me that at present the evidence in favour of the theory of an intermediate product far outweighs that which can be brought against it. Perhaps the most serious objection which can be raised to the view that homogentisic acid is an abnormal product peculiar to alkaptonurics is that such a view involves the assumption that the alkaptonuric, who alone has the power of forming homogentisic acid, is

also peculiar in having no power of destroying it when formed.

The impaired destruction of the alkapton acid which results from certain morbid conditions has also interesting bearings upon the question of temporary or intermittent alkaptonuria. The records of such cases are very few, only four or five in all, and further work upon such cases is much to be desired. In some of them the evidence that the condition was temporary is not conclusive and in others the evidence of the nature of the abnormal excretory product is not sufficiently complete for the establishment of so important a point. In none of them save in that of Zimnicki,[65] whose paper appeared in a Russian journal which I have not been able to obtain, have quantitative estimations been carried out. His patient, whose alkaptonuria was intermittent, suffered from hypertrophic cirrhosis of the liver. In Geyger's[66] case, also intermittent, the patient was a diabetic. Of strictly temporary cases that described by Carl Hirsch[67] is the most remarkable. A girl, aged 17 years, with a febrile gastro-enteric catarrh, passed for three days only a urine which darkened on standing, contained indican, and also yielded the alkapton reactions. From it Professor Siegfried extracted an acid which formed a sparingly soluble lead salt, but neither the melting point of the acid nor any analytical figures are given.

A temporary or intermittent excretion of homogentisic acid seems more compatible with the theory that it is a normal metabolic product than with the opposite theory. It is noteworthy that the diseases from which Geyger and Zimnicki's patients suffered were such as have been found to be sometimes attended by impaired destructive power of the acid in question, and is conceivable that in rare instances the impairment may proceed further and become practically complete. I realise, however, that what has been suggested above as to diminution of power of forming as well as of destroying the alkapton acid in such diseases may be equally urged against such an interpretation of the intermittent cases. In support of the intermediate product theory certain observations of Gonnermann,[68] Bartels,[69] and others have been adduced as pointing to the formation of homogentisic acid in plants by the action of a tyrosinase upon tyrosin. The evidence brought forward has never appeared to me to be at all conclusive, and the investigations of Schultze,[70] who has failed to obtain any evidence of the formation of homogentisic acid under such conditions, have deprived this argument of all cogency.

Assuming it to be a fact that homogentisic acid is a product of normal metabolism, the results of the administration of various aromatic acids to alkaptonurics may reasonably be expected to throw light upon the higher intermediate steps between the parent protein fractions and that substance. Any compound which represents a link in the chain should, on the one hand, be destroyed, as tyrosin and homogentisic acid are in the normal organism, and on the other hand should increase the output of the latter by alkaptonurics. Any substance which does not behave in the manner indicated cannot form such an intermediate link.

Many aromatic acids have been administered to alkaptonurics at various times, but very few have been found to fulfil the above conditions. The most complete set of observations of the kind have been carried out by Otto Neubauer and Falta,[71] and their findings have recently been fully confirmed by Grutterink and van der Bergh.[72] It was found that compounds such as phenyl-acetic and phenyl-propionic acids, which have simple side chains, have no effect upon the output. Those which increase homogentisic excretion resemble tyrosin and phenyl-alanin in having an easily attacked grouping in the α position upon the side chain, whereas when the substituted group occupies the β position no such result follows. Thus phenyl α-lactic and phenyl-pyruvic acids are excreted as homogentisic acid, whereas phenyl β-lactic acid is not. Even the presence of a

[59] Loc. cit., sub .20, p. 282.
[60] Deutsche medicinische Wochenschrift, 1905, vol. xxxi., p. 457.
[61] Loc. cit., sub. 38, p. 265.
[62] Archives de Physiologie, 1902, 5e s, vol. iv.,p. 225.
[63] Nederlandsh Tijdschrift voor Geneeskunde, 1907, vol. ii., p. 1117.
[64] Hofmeister's Beiträge, 1905, vol. vi., 150.

[65] Jeshenedelnik, 1899, No. 4; abstract Centralblatt für Stoffwechsel und Verdauungskrankheiten, 1900, vol. i., p. 348.
[66] Pharmakeutische Zeitung, 1892, p. 488.
[67] Berliner klinische Wochenschrift, 1897, vol. xxxiv., p. 866.
[68] Archiv für gesammte Physiologie (Pflüger), vol. lxxxii., p. 289.
[69] Berichte der Deutschen Botanischen Gesellschaft, 1902, vol. xx., p. 454.
[70] Zeitschrift für physiologische Chemie, 1906, vol. xlviii., p. 396; 1907, vol. i., p. 508.
[71] Zeitschrift für physiologische Chemie, 1904, vol. xlii., p. 81.
[72] Loc. cit., sub. 63.

second substituted group in β position, as in phenyl-glyceric acid, suffices to prevent the transformation.

CH₂	CH₂	CH₂	CH.OH	CH.OH
CH.NH₂	CH.OH	C=O	C.H₂	CH.OH
COOH.	CO.OH.	CO.OH.	CO.OH.	CO.OH.
Phenyl-alanin.	Phenyl-α-lactic acid.	Phenyl-pyruvic acid.	Phenyl-β-lactic acid.	Phenyl-glyceric acid.

Again, when the substituted grouping in the α position is rendered more stable by benzoylation the formation of homogentisic acid is prevented, for, as L. Blum [73] has found, benzoyl-phenyl-alanin does not increase the alkapton output.

There is good reason to believe that desamination—that is to say, the removal of the amino group—is a very early stage in the catabolism of the amino-acids of which the molecules of proteins are built up, and of the aromatic fractions amongst others. After a meal rich in proteins the resulting excretion of homogentisic acid was found by Falta [74] to commence more promptly and to come to an end sooner than the corresponding increase of the output of nitrogen, and this has been confirmed by Langstein and Meyer. Mittelbach [75] placed the maximum excretion of homogentisic acid within the two or three hours following the protein meal, but in some observations which I made [76] upon specimens of urine passed at short intervals throughout the day, although an obvious increase occurred shortly after a meal rich in protein, the excretion was still larger during the second period of four hours than during the four hours immediately following the meal. It is highly probable that desamination is effected by substitution of hydroxyl for the amino group, and that the formation of α-lactic acid is the first step in the breaking down of phenyl-alanin. Not only does phenyl α-lactic acid fulfil the conditions laid down above, but it is one of the compounds which have been found to yield acetone when perfused through the liver.

Between phenyl α-lactic acid and homogentisic acid at least one other stage must intervene, for the introduction of the two hydroxyl groups in the 2 and 5 positions upon the benzene ring must precede the oxidation of the side chain. Were this not the case phenyl-acetic acid would be formed and the series of changes would be arrested, for phenyl-acetic acid has not the properties requisite for an intermediate product, being neither burnt by a normal individual nor converted into homogentisic acid by the alkaptonuric. The most likely intermediate product is hydroquinone α-lactic acid, which was formerly thought to be the uroleucic acid of Kirk.

CH₂	CH₂	CH₂	CH₂
CH.NH₂	CH.OH	CH.OH	CO.OH.
CO.OH.	CO.OH.	CO.OH.	
Phenyl-alanin.	Phenyl α-lactic acid.	Hydroquinone α-lactic acid.	Homogentisic acid.

With certain reserves as to substitutions upon the ring, the structure of the side chain clearly plays the chief part in determining the formation of homogentisic acid from tyrosin and phenyl-alanin, but apparently that which determines the disruption of the ring, and the destruction of homogentisic acid when it has been formed, is the presence of the two hydroxyl groups in the 2 and 5 positions upon the benzene ring.

Thus, Neubauer and Falta found that of the three isomeric dioxybenzoic acids gentisic acid alone was to a large extent burnt in the human organism, although some 15 per cent. of the dose given appeared in the urine as aromatic sulphate. In gentisic acid, as in its homologue the alkapton acid, the hydroxyl groups occupy the 2 and 5 positions.

COOH.	COOH.	COOH.
Gentisic acid.	2 : 4 dioxybenzoic acid.	Protocatechuic acid.

When the alkaptonuric took gentisic acid by the mouth Neubauer and Falta found that the reducing power of his urine was conspicuously increased not owing to a greater output of homogentisic acid which could hardly have resulted but to the excretion of gentisic acid as such.

This indicates that in alkaptonuria the failure to break up the benzene ring extends to acids with hydroxyl groups in the 2 : 5 position other than homogentisic acid, and that the essential error resolves itself into an inability to destroy the ring of acids so constituted. Homogentisic acid is apparently the only compound formed in normal metabolism which offers itself for such disruption, and accordingly the alkaptonuric excretes it. It has been pointed out above that a higher intermediate compound of such structure must be formed, but this would appear to be changed to homogentisic acid before it comes into the grip of the destructive process.

This conception of the anomaly locates the error in the ante-penultimate stage of the catabolism of the aromatic protein fractions which is in accord with the fact that all the tyrosin and phenyl-alanin, both exogenous and endogenous, is swept into the net and goes to contribute to the excreted homogentisic acid in alkaptonuria.

We may further conceive that the splitting of the benzene ring in normal metabolism is the work of a special enzyme, that in congenital alkaptonuria this enzyme is wanting, whilst in disease its working may be partially or even completely inhibited.

The experiments of G. Embden and others upon perfusion of the liver suggest that organ as the most probable seat of the change. If, however, the acetone-forming enzyme is that which is at fault, it is worthy of note that not only the aromatic fractions of proteins but leucin also contributes to the formation of acetone and might be expected to be likewise involved. Dr. Hurtley was good enough to examine a large quantity of alkapton urine for me by extraction with ether after acidification, and distillation in steam of the residue of the ethereal extracts, but was unable to detect the presence of any abnormal acid of the fatty series, and there is at present no evidence forthcoming of any disturbance of the path of leucin catabolism in alkaptonuria.

[73] Loc. cit., sub. 44.
[74] Verhandlungen der Naturforschenden Gesellschaft in Basel, 1903, vol. xv.
[75] Loc. cit., sub. 39.
[76] Transactions of the Royal Medical and Chirurgical Society, 1903, vol. lxxxv.

The Croonian Lectures

ON

INBORN ERRORS OF METABOLISM.

*Delivered before the Royal College of Physicians of London
on June 18th, 23rd, 25th, and 30th, 1908,*

BY ARCHIBALD E. GARROD, M.A., M.D.
OXON., F.R.C.P. LOND.,

ASSISTANT PHYSICIAN TO, AND LECTURER ON CHEMICAL PATHOLOGY
AT, ST. BARTHOLOMEW'S HOSPITAL; SENIOR PHYSICIAN, HOSPITAL
FOR SICK CHILDREN, GREAT ORMOND STREET.

LECTURE III.*

Delivered on June 25th, 1908.

CYSTINURIA.

MR. PRESIDENT AND FELLOWS,—Our knowledge of cystin and of cystinuria alike dates from the year 1810, in which year Wollaston [1] described a previously unknown variety of urinary calculus, of which two specimens had come into his hands. The first of these had been removed from the bladder of a young child. Seeing that both stones had been found in the urinary bladder Wollaston assigned to the new compound of which they were composed the name of cystic oxide.

It was Berzelius [2] who suggested the change of name from cystic oxide to cystin, and of this new name, which has since been universally adopted, Civiale wrote,[3] in 1838, that although it corrected an error of chemistry it perpetuated an error of physiology, for cystin is excreted by the kidneys and does not have its origin in the bladder.

The sediments of hexagonal crystals found in urine, upon which the diagnosis of cystinuria is so often based, were first observed by Stromeyer [4] in 1824 and by Prout [5] in 1825.

Prout made the earliest analysis of cystin, but being unaware of the presence of sulphur in it he reckoned the contained sulphur as oxygen. Baudrimont and Malaguti [6] announced that cystin contains sulphur in 1837, no less than 27 years after Wollaston's discovery, and in the year following Thaulow [7] published a complete analysis which was in accord with that of Prout, save that half the oxygen of his reckoning was replaced by sulphur.[8]

To Goldmann and Baumann [9] we owe the recognition of the fact that cystin is an amino-derivative of thio-lactic acid, in which two molecules are linked together by their sulphur atoms, whereas cystein, which is easily obtained by reduction of cystin, contains an SH group in its single molecule. In Baumann's formula the sulphur atom and the amino-group were represented as being both attached to the α-carbon atom, but Friedmann [10]

has more recently shown that this is not the case, but that the amino-group occupies the α and the sulphur atom the β position. Cystin must therefore be regarded as di-α-amino-β-thio-lactic acid.

$$
\begin{array}{ll}
\text{CH}_2\text{S}\text{—}\text{SCH}_2 & \text{CH}_2.\text{SH} \\
| \qquad\qquad | & | \\
\text{CH.NH}_2 \ \ \text{CH.NH}_2 & \text{CH.NH}_2 \\
| \qquad\qquad | & | \\
\text{CO.OH} \ \ \ \text{CO.OH.} & \text{CO.OH.} \\
\quad\text{Cystin.} & \text{Cystein.}
\end{array}
$$

Since then the synthesis of this compound, which had repeatedly been attempted without success, has been accomplished by Erlenmeyer, jun.,[11] and its formula may be regarded as finally established.

The revised formula renders possible the existence of an isomeric cystin, with the composition of di-β-amino-α-thio-lactic acid :—

$$
\begin{array}{ll}
\text{CH}_2.\text{NH}_2 \ \ \text{CH}_2\text{NH}_2 \\
| \qquad\qquad\quad | \\
\text{CH.S — SCH} \\
| \qquad\qquad | \\
\text{CO.OH.} \quad \text{CO.OH.}
\end{array}
$$

And this substance also has been synthetically prepared by Gabriel.[12]

Carl Neuberg and P. Meyer [13] have put forward the view that some cystin calculi consist wholly or in part of this isomeric cystin, although they found that the cystin of sediments and that which is present in solution in the urines which they examined was of the ordinary kind. This would imply that some cystinurics excrete one and some the other cystin, for it cannot be supposed that urinary calculi are formed of a material which is not present in the urine of the patients who produce them. Some calculi which Neuberg examined when dissolved in ammonia yielded on evaporation of the ammonia acicular crystals which he believed to consist of iso-cystin. This view has not met with acceptance from other investigators. The great majority of the calculi are undoubtedly composed of ordinary cystin, which is deposited in hexagonal plates from ammoniacal solutions. That the materials obtained from proteins and from such calculi are identical in their chemical and physical properties, including their effect upon the polarised ray, has been proved by Rothera,[14] Abderhalden,[15] and Gaskell.[16] Moreover, Emil Fischer and Zuzuki,[17] who examined some of the same calculus material which Neuberg used, found that it gave a pronounced red colour with Millon's reagent, and they suggest that the acicular crystals described consisted of tyrosin. I have never obtained the acicular crystals from any cystin calculus which I have examined, although slender hexagonal prisms were not infrequently obtained. It is obvious that the occurrence of iso-cystin as a metabolic product is as yet very far from proven.

There are a few scattered records, of the last century, of the detection of cystin in animal tissues. Cloetta [18] found it in the kidneys of an ox but failed to obtain it from other ox kidneys. Scherer [19] isolated cystin from the liver of a man who had died from typhoid fever, and Kulz [20] obtained it on one occasion among the products of pancreatic digestion *in vitro*. Such observations lent support to the view which had been held all along that the sulphur of proteins was the parent of the sulphur of cystin, which was regarded as an intermediate product of sulphur metabolism.

The year 1899 was marked by a great advance in our knowledge of the origin of cystin, for in that year K. A. H. Mörner [21] showed that cystin could be obtained in abundance by the hydrolysis of hair. Soon afterwards G. Embden [22] obtained it in like manner by the hydrolysis of serum- and egg-albumins. Not only was there thus provided a ready source of supply of this compound, which had previously only been obtainable from the rare cystin calculi and from the no

---

* Lectures I. and II. were published in THE LANCET of July 4th (p. 1) and 11th (p. 73), 1908, respectively.
[1] Philosophical Transactions of the Royal Society, 1810, vol. c., p. 223.
[2] Traité de Chimie, 1833, vol. vii., p. 424.
[3] Comptes Rendus de l'Académie des Sciences, Paris, 1838, vol. vi., p. 897.
[4] Annals of Philosophy, 1824, vol. viii., p. 146.
[5] On Stomach and Urinary Diseases, second edition, 1825, p. 166.
[6] Comptes Rendus de l'Académie des Sciences, Paris, 1837, vol. v., p. 394.
[7] Annalen der Chemie (Liebig's), 1838, vol. xxvii., p. 197.
[8] Baudrimont and Malaguti state (Journal de Pharmacie, 1838, vol. xxiv., p. 633) that they presented a complete analysis to the Académie des Sciences in 1837, but I can find no published account of this. The Comptes Rendus for that year contain only a brief note of their paper, in which the presence of sulphur in cystin is mentioned.
[9] Zeitschrift für physiologische Chemie, 1888, vol. xii., p. 254.
[10] Hofmeister's Beiträge zur chemischen Physiologie und Pathologie, 1902, vol. iii., p. 1.

[11] Berichte der deutschen chemischen Gesellschaft, 1903, vol. xxxvi., p. 2720.
[12] Ibid., 1905, vol. xxxviii., p. 637.
[13] Zeitschrift für physiologische Chemie, 1905, vol. xliv., p. 472.
[14] Journal of Physiology, 1905, vol. xxxii., p. 175.
[15] Zeitschrift für physiologische Chemie, 1907, vol. li., p. 391.
[16] Journal of Physiology, 1907, vol. xxxvi., p. 142.
[17] Zeitschrift für physiologische Chemie, 1905, vol. xlv., p. 405.
[18] Liebig's Annalen, 1856, vol. xcix., p. 299.
[19] Jahresbericht für Chemie, 1857, vol. v., p. 561.
[20] Zeitschrift für Biologie, 1884, vol. xx., p. 1.
[21] Zeitschrift für physiologische Chemie, 1899, vol. xxviii., p. 395.
[22] Ibid., 1901, vol. xxxii., p. 94.

less rare urinary sediments, but also it was shown that cystin has a place among the numerous a-amino-acids of which the complex molecules of proteins are built up. Like other primary protein fractions it is contained in widely different proportions in different members of the protein group.

Whether or no cystin is the sole protein fraction which contains sulphur, and is responsible for the entire sulphur contents of proteins, save the small quantity of sulphate which they hold apparently in some sort of combination, cannot be regarded as finally decided. Mörner's work shows that all the sulphur of some proteins is in this form, but it is not yet proved that other sulphur compounds obtained from certain substances of the group are derivatives of cystin. Thus the a-thio-lactic acid which has frequently been obtained offers obvious difficulties in this respect, for in it the sulphur atom occupies the a position, whereas in cystin it is attached to the β carbon atom.

Some account of what is known concerning sulphur metabolism forms a necessary preliminary to any discussion of the disturbance which it undergoes in cystinuria. However, our knowledge concerning it is far less advanced than that of the metabolism of nitrogenous compounds; the processes at work are evidently of considerable complexity and it appears certain that the cystin of the proteins broken down in the body does not all follow the same catabolic path. Almost the whole of the sulphur of the proteins of food and of tissues ultimately appears in the urine, for little of it is excreted by way of the alimentary canal. A small portion is present in the saliva and gastric contents as sulphocyanide, but as the sulphocyanide grouping is not represented in protein molecules it must be supposed that it is formed by synthesis within the organism. This fraction is ultimately reabsorbed from the alimentary canal and is excreted, at least in part, unchanged in the urine. A more considerable sulphur fraction goes to form the taurin of the bile and this portion too appears in the urine, probably as a part of the neutral sulphur. The bulk of the sulphur takes a more direct route and is excreted as sulphates. Of the urinary sulphur, by far the greater part is in such fully oxidised form and only some 14-20 per cent. is contained in a number of unoxidised or incompletely oxidised constituents which go to make up what is known as neutral sulphur. Of the sulphates, the output of which may be swelled by sulphates absorbed as such from the alimentary canal, some portion is in combination as aromatic or ethereal sulphates; the greater part, some 90 per cent. of the total, is in the form of simple salts. The formation of the aromatic sulphates is usually ascribed to the working of a protective mechanism, by means of which aromatic substances of harmful nature, such as are apt to be absorbed from the alimentary canal, are rendered harmless and inert. The ratio of aromatic to simple sulphates is therefore regarded as affording a valuable index of the amount of protein decomposition brought about by the bacteria which inhabit the alimentary canal. Folin has recently thrown doubt upon this prevalent view, and seeing that the output of aromatic sulphate is little affected by a change from a diet rich in proteins to one poor in such constituents, he classes these compounds among the products of tissue metabolism as distinguished from that of the proteins of the food. However, the large amount of evidence available upon the other side—the increased output of aromatic sulphates in cases in which intestinal decomposition is abnormally active, the conversion of the whole of the sulphate into aromatic sulphate in some cases of carboluria, and the effects of intestinal disinfection in limiting the output—appear to me to call for the production of stronger evidence than has yet been brought forward before the accepted view need be given up. As F. G. Hopkins[23] has suggested, the comparative uniformity of the excretion of these compounds upon widely different diets may well be due to the want of conspicuous variations in the bacterial activity in the intestine, a factor which is not known to be dependent upon the amount of protein contained in the food.

That variations in the amount of protein in the diet, although they influence very conspicuously the output of sulphates in the urine, have little effect upon that of neutral sulphur is a well-established fact, from which Folin concludes that the latter is made up of products of tissue

metabolism as distinguished from that of protein foods. He compares the neutral sulphur with the endogenous uric acid and kreatinin among nitrogenous waste products, whereas the sulphates may be compared to urea. Abstention from food causes relative increase of neutral sulphur and, it is said, an absolute increase also, and so does the increased breakdown of tissue proteins which is brought about by certain toxic substances, such as chloral and chloroform.

Of the materials which go to make up the so-called neutral sulphur our knowledge is as yet very incomplete. Some is probably derived from taurin; the cystin-like compound found by Baumann and Goldmann may contribute a fraction; minute quantities of sulphocyanide have already been mentioned, and among the sulphur-containing substances must be reckoned certain little-known acids of high molecular weight, but which do not yield the reactions of proteins, to which the names of uroproteic and uroferric acids have been assigned. According to Dombrowski[24] the essential yellow pigment of urine, urochrome, is also a sulphur compound. In the urine of animals there have been met with also products of the decomposition of cystin in the alimentary canal, such as thio-sulphates, methyl mercaptan, and ethyl sulphide.

Some interesting experiments carried out by Baumann and Preusse[25] and Jaffe,[26] at a time when it was not yet known that cystin is one of the primary fractions of proteins, were of much importance as showing that it is present in the animal organism as an intermediate product of protein metabolism. These investigators found that when monochlorbenzene or monobrombenzene was administered by the mouth to a dog there appeared in the urine of the animal a compound known as a mercapturic acid, which is excreted in combination with glycuronic acid. The process which leads to the formation of the mercapturic acid may be classed among the chemical protective processes, cystein being taken into combination with the halogen-benzene, just as glycocoll combines with benzoic acid to form hippuric acid. Mester[27] afterwards found that when a halogen benzene was taken by a man hardly any excretion of mercapturic acid resulted. When it was shown that the rational formula of cystin required revision the force of this evidence was apparently destroyed, supposing that the formula of the mercapturic acid as given by Baumann were correct; but Friedmann[28] followed up his work on cystin itself by a reinvestigation of the mercapturic acids and showed that their formulæ also required correction in a similar sense. Consequently the excretion of mercapturic acid by dogs after the administration of brombenzenes was reinstated as a true experimental cystinuria, such as Baumann and Jaffe originally believed it to be.

Of the cystin formed in the breaking down of proteins within the body, a portion, which has been estimated at some 30 per cent. of the total amount, is set apart for the formation of taurin, which is found in the bile in combination with cholalic acid, as taurocholic acid. Taurin stands in a simple chemical relationship to cystin as the following formulæ show:—

$$\begin{array}{llll}
\text{CH}_2\text{S}\!-\!\text{SCH}_2 & \text{CH}_2.\text{SO}_2.\text{OH} & \text{CH}_2\text{SO}_2.\text{OH} \\
| \qquad\quad | & | & | \\
\text{CH.NH}_2 \quad \text{CH.NH}_2 & \text{CH.NH}_2 & \text{CH.NH}_2. \\
| \qquad\quad | & | \\
\text{CO.OH} \quad \text{COOH.} & \text{CO.OH.} \\
\text{Cystin.} & \text{Cysteic acid.} & \text{Taurin.}
\end{array}$$

and it has actually been obtained from it, in vitro, by Friedmann.[29] It is obvious that the cystin destined to follow this metabolic path must be set apart at a very early stage of catabolism before either the sulphur atom or the amino-group has been removed, and there is evidence that the removal of both of these is early effected.

The question whether the taurin, as such, combines directly with cholalic acid to form taurocholic acid, or whether the combination first occurs with cystin itself or with some other precursor of taurin, taurocholic acid being afterwards produced by oxidation, is still an open one. In favour of the latter view may be quoted an observation made by L. Blum[30] that when cystin is injected into the portal circulation of an

[23] Guy's Hospital Gazette, 1907, vol. xxi, p. 424.

[24] Bulletin de l'Academie des Sciences de Cracovie, 1907, p. 777.
[25] Zeitschrift für physiologische Chemie, 1879, vol. vii., p. 159.
[26] Berichte der deutschen chemischen Gesellschaft, 1879, vol. xii., 1093.
[27] Zeitschrift für physiologische Chemie, 1889, vol. xiv., p. 109.
[28] Hofmeister's Beiträge zur chemischen Physiologie und Pathologie, 1903, vol. iv., p. 486.
[29] Ibid., 1902, vol. iii., p. 1.
[30] Ibid., 1904, vol. v., p. 1

animal with a biliary fistula the sulphur of the bile is thereby increased, but that the additional sulphur is neither in the form of cystin nor in that of taurin.

It seems probable that the formation of taurocholic acid is not limited merely by the quantities of cystin and cholalic acid available, although these factors appear to play a part. Thus von Bergmann[31] found that when cystin is given by the mouth to a dog with a biliary fistula the taurocholic acid of its bile is not appreciably increased, whereas if cholalic acid be administered at the same time a conspicuous increase is obtained. On the other hand, sodium cholate when given alone caused an obvious but transitory increase of taurocholic acid. This suggests that no excess of cholalic acid is available in the canine organism but that an excess of cystin is available which is, however, quickly exhausted. Only when an excess of both cystin and cholalic acid were swallowed was taurocholic acid continuously formed in abnormal quantities. In rabbits, on the other hand, Wohlgemuth[32] found that the administration of cystin alone suffices to cause an increased formation of taurocholic acid. Rothera has shown that in man the administration of cholalic acid alone, or of this acid and of cystin together, does not bring about any decrease of the sulphates of the urine such as might be expected to result if an excess of taurin were formed, always supposing that the taurin fraction finally appears in the urine as neutral sulphur. When cystin was swallowed with cholalic acid it was burnt completely to sulphate just as was the case when no cholalic acid was given with it. Goldmann's[33] investigations leave no doubt that the excretion of mercapturic acid which follows the administration of halogen-benzenes to dogs is at the expense of the sulphates of the urine and not at that of the neutral sulphur. In other words, the cystin which is in combination in the mercapturic acid is some of that which in ordinary circumstances would have been fully burnt to sulphuric acid. However, Blum[34] states that when an experimental cystinuria is induced in dogs with biliary fistulæ, presumably by the halogen-benzene method, although this is not stated, the taurin disappears almost completely from their bile, and this would suggest that the mercapturic cystin is that which ordinarily goes to form taurin. If this be so the inference is not to be avoided that under normal conditions the sulphur of taurin is mainly excreted as sulphate. However, there is not a little evidence which points in a contrary direction and which tends to show that the taurin sulphur is excreted as neutral sulphur. Thus Salkowski[35] found that in dogs, and in man also, taurin introduced into the stomach does not increase the output of sulphates, but is largely excreted as taurin-carbamic acid. In rabbits, on the other hand, and probably in the vegetivora in general, the administration of taurin by the mouth leads to an increase of the urinary sulphates. Kunkel's[36] observation that the establishment of a biliary fistula in a dog and the withdrawal of bile from the alimentary canal conspicuously diminished the neutral sulphur in the animal's urine points in the same direction, but Kunkel himself declines to base any generalisation upon an experiment upon a single dog.

The question of the ultimate fate of the sulphur which goes to form taurin is clearly one which calls for further investigation. When cystin itself is introduced into the stomach of animals, even in very large doses, no unchanged cystin is excreted in their urine. In addition to an increase of the sulphates an increase of neutral sulphur has been observed in animals so treated, and this has been largely in the form of thio-sulphates, which are probably formed by the decomposition of cystin in the intestine, and thence absorbed. When doses of a few grammes of cystin are swallowed by a normal man the cystin is wholly burnt to sulphate and no increase of the neutral sulphur has been observed.[37] Poly-peptides containing cystin, such as di-alanyl-cystin and di-leucyl-cystin, were found by Abderhalden and Samuely[38] to be dealt with by normal man in the same way as free cystin is. By injecting cystin into the systemic veins of animals L. Blum[39] caused cystin to appear in their urine, but when the injection was made into the branches of the portal vein it was apparently destroyed in the liver and no cystinuria resulted. However, neither Blum nor Rothera,[40] who also tried the experiment, was able to bring about the destruction of cystin by adding it to crushed hepatic tissue. Obviously the whole of the cystin ingested in protein foods, as distinguished from that derived from the tissues, is not burnt straightway to sulphates and excreted as such. Some, as we have seen, is probably utilised in the formation of taurin, and some must escape direct destruction and be built up into the tissue proteins, the cystin content of which must necessarily be maintained.

We may now turn to the consideration of the derangement of sulphur metabolism in cystinuria, in virtue of which the subjects of that anomaly excrete some part of their cystin as such in their urine, often to their serious disadvantage. That cystinuria is less rare than alkaptonuria I am convinced, although I cannot bring forward any statistical evidence in support of this statement, which is based upon personal experience acquired during a search for examples of both anomalies during the past ten years. That a far larger number of cases of cystinuria than of alkaptonuria are on record counts for little, for the former has been known for well-nigh a century, whereas our knowledge of the latter does not extend back more than 60 years. Moreover, the surgical consequences of cystinuria are such as compel attention. To the greater frequency of the condition in males than in females, a characteristic which it shares with other inborn chemical errors ; to the evidence of its congenital occurrence and its persistence, and to its so frequent transmission from generation to generation in a family, I referred at some length in my first lecture.

The liability to the formation of calculi composed of cystin, and to other urinary disorders such as cystitis, gives to cystinuria a practical as well as a theoretical importance, as its study may possibly lead to the discovery of some means of averting its evil consequences. It may even happen that cystin is deposited in crystalline form in the organs and tissues of a cystinuric, as sodium biurate is in those of a victim of gout. Only one case of the kind is as yet on record. An infant, described by Abderhalden,[41] died at the age of 21 months with symptoms of inanition. The child was one of a family which included a number of cystinurics. At the necropsy the internal organs showed innumerable white specks, visible to the naked eye, which were found to be deposits of cystin. From the spleen cystin was readily extracted by ammonia and was deposited from the solution in hexagonal plates, the nature of which was fully confirmed by chemical methods. In the few other available records of post-mortem examinations of cystinurics no mention is made of such deposits in the tissues.

The urine of cystinurics has been described by Golding-Bird and others[42] as possessing an odour not unlike that of sweet-briar and as assuming a greenish tint on standing and acquiring an odour like that of putrid cabbage when decomposition sets in. Bence Jones has even described one specimen which assumed a bright apple-green tint. I cannot say that I have ever noticed a smell recalling that of sweet-briar from such urines nor any distinctly green tint, but the odour of sulphuretted hydrogen which is given off in decomposition is sufficiently obvious. The deposits of crystalline cystin which are thrown down from such urines are not very abundant but may often be seen with the naked eye to consist of glancing crystals. Under the microscope these appear as hexagonal plates, the edges of which usually show paired inequalities of length. Hexagons of different sizes are often superposed upon each other and with the plates longer or shorter hexagonal prisms are occasionally intermixed. In some specimens the crystals show a radiating striation and jagged edges but they still retain roughly hexagonal forms. Fresh crystals are deposited after the urine has been passed, especially on addition of acetic acid, and Delépine[43] came to the conclusion that in a case which he investigated their deposition was favoured by the presence of an organism, probably one of the blastomycetes, which

[31] Ibid., 1903, vol. iv., p. 192.
[32] Zeitschrift für physiologische Chemie, 1903, vol. xl., p. 81.
[33] Ibid., 1885, vol. ix., p. 260.
[34] La Semaine Médicale, 1906, vol. xxvi., p. 554.
[35] Virchow's Archiv, 1877, vol. lviii., p. 460.
[36] Archiv für die gesammte Physiologie (Pflüger's), 1877, vol. xiv., p 344.
[37] Rothera: loc. cit., sub. 14.
[38] Zeitschrift für physiologische Chemie, 1905, vol. xlvi., p. 187.

[39] Hofmeister's Beiträge zur chemischen Physiologie und Pathologie, 1903, vol. v., p. 1.
[40] Loc. cit., sub. 14.
[41] Zeitschrift für physiologische Chemie, 1903, vol. xxxviii., p. 557.
[42] Urinary Deposits, London, 1851, p. 181.
[43] Proceedings of the Royal Society, 1890, vol. xlvii., p. 198.

could be separated by filtration. No comfirmation of this has yet come from other observers.

The ready solubility of cystin in ammonia affords a ready means of identifying the crystals, which are insoluble in acetic acid. In case of doubt the following test may be applied, which was described by Wollaston in his original paper on Cystic Oxide. Some crystals are dried upon a glass slide and covered with a cover-slip. A drop of strong hydrochloric acid is then allowed to flow over the deposit and as each crystal is bathed in the acid there springs from it a stellate cluster of delicate prisms which grow rapidly under the eye and which are composed of cystin hydrochlorate. If now a drop of water be allowed to dilute the acid the prisms melt away as rapidly as they were formed. The urinary crystals which are most likely to be mistaken for the hexagons of cystin are the almost colourless plates of uric acid which are sometimes deposited from feebly pigmented urines and which may assume roughly hexagonal forms.

The cessation of deposits must not be taken as evidence that a patient has ceased to excrete cystin, for when no crystals are found in the untreated urine, and especially in such as is alkaline in reaction, the addition of acetic acid may cause their deposition. Even in acid urine cystin is by no means insoluble. However, there is reason to believe that the excretion of cystin may be actually suspended, at least for a time. Thus Loewy and Neuberg[44] failed to detect any either as sediment or in solution in the urine of a medical man who had undoubtedly been cystinuric at a former period, and Lewis and Simon[45] have described a case in which cystin could no longer be found although cadaverin was isolated from the urine. It is noteworthy that in both cases the urine yielded hydrogen sulphide very readily, and this was attributed by Loewy and Neuberg to an excessive output of neutral sulphur. This suggests that possibly the error of sulphur metabolism persists although the form of its manifestation is changed, and that in such cases some other intermediate product than cystin itself may come to excretion. It may be hoped that further investigations will throw fresh light upon such intermittent or temporary cystinuria. In some of the records of temporary cystinuria the evidence adduced does not suffice to establish a fact of so much importance.

In the fæces of cystinurics no cystin can be found. Dewar and Gamgee[46] state positively that cystin is present in the sweat in some cases and that silver coins carried in the pockets of the patients are apt to become blackened. In a single case in which I examined some collected sweat I failed to find any evidence of the presence of cystin therein.

The error of metabolism of which cystinuria is a manifestation is clearly a far more complex one than that which underlies alkaptonuria and far less uniform in character. Although the excretion of cystin in the urine is apparently a constant feature, and is that by which it has hitherto always been recognised, cystin is by no means the only protein fraction which is implicated and in some cases other amino-acids or their derivatives are to be found in the urine. The commonest of these are the diamines cadaverin and putrescin, which stand in intimate chemical relationship to the diamino-acids, lysin and ornithin respectively, of which the former is a primary constituent of the protein molecule, whilst the latter enters into the composition of another such fraction, arginin. More rarely leucin and tyrosin are excreted unchanged, and it is probable that other amino-acids, less easily detected, will be found to be excreted in some cases. Thus it comes about that cases of cystinuria differ widely among themselves, not only in the number of protein fractions which are implicated but also in their behaviour as regards the individual fractions, so that what is true of one cystinuric may be quite untrue of another.

The cases hitherto investigated admit of classification upon the following lines : 1. In some no diamines and no primary protein fractions other than cystin have been found in the urine. 2 In some the urine has contained cadaverin or putrescin or both, in addition to cystin, but the excretion of diamines is apt to occur in an intermittent manner. 3. In a very few cases leucin or tyrosin or both have been excreted with or without diamines. 4. In a single case, that of Loewy and Neuberg,[47] cystin was present in the urine, but no

diamine, leucin, or tyrosin. When diamino-acids were given by the mouth the patient excreted the corresponding diamines, and when tyrosin and aspartic acid were given these were excreted unchanged. In other cases in which similar feeding experiments have been tried no such results have followed the swallowing of diamino- or monamino-acids, and this patient was also exceptional in his method of dealing with cystin introduced by the mouth.

The investigation of the pathogeny of cystinuria has been greatly impeded by the lack of a simple and reliable method for the estimation of cystin in urine, and the methods employed by different observers have been of very different values. The naphthalene-chloride sulphon method employed by Abderhalden[48] is not easy of application. Concentration of the urine under reduced pressure, with the addition of acetic acid, gives results which are presumably reliable, and J. Gaskell's[49] method, recently described, in which the deposition of cystin is aided by the addition of acetone to the urine acidified with acetic acid, is easily carried out, and is satisfactory, save for the detection of very small quantitative variations, in connexion with which the degree of dilution of the urine has a disturbing effect upon the results. The mere addition of acetic acid, without concentration, does not suffice, for, as with Heintz's old method of estimating uric acid, the results obtained are capricious and quite unreliable. The method of Mester,[50] which has been extensively employed, is based upon the assumption that an increase of the ratio of neutral sulphur to total sulphur, beyond the average ratio for normal urines is due to cystin excreted as such. The objections to this indirect method are obvious, in view of the conspicuous effect of diet upon the excretion of sulphates, although, as has been already pointed out, the output of neutral sulphur is little affected by the amount of protein in the food. Alsberg and Folin[51] adopt a different plan of reckoning, by which any increase of neutral sulphur above the average normal output, which they estimate at a lower figure than most other observers, is reckoned as cystin sulphur. This plan avoids the error due to fluctuations of the sulphate excretion, but involves the assumption that the cystin is excreted wholly at the expense of the sulphates, the normal neutral sulphur remaining intact.

As regards the disturbance of sulphur metabolism in cystinuria it is at any rate certain that only a portion of the total cystin of the proteins broken down is excreted as such, and that the error is in no sense complete. Cystinurics always excrete sulphates, and neutral sulphur other than that contained in the cystin. In no single recorded instance has any approach been made to a maximal excretion, which would attain to some five grammes per diem. The figures for the average daily output given by different observers have been obtained by methods of very different degrees of reliability, but if the various records are compared, and especially if one has had opportunities of personal study of a series of cases, it is borne in upon one as highly probable that the excretion of cystin in different cases makes some approach to uniformity and that it averages some 0·3 to 0·5 gramme in the 24 hours.

The figures included in the following table are collected from the published records and from the notes of several cases which have recently been investigated by T. S. Hele[52] in the laboratory under my charge at St. Bartholomew's Hospital. In most instances averages have been calculated from the published figures, and in Caracciolo's case the amount of cystin has been worked out by Mester's method from the figures given in his paper.

It would be unwise to lay too much stress upon the uniformity of many of the figures in view of the differences in the methods employed and their unequal values, but the figures are certainly suggestive, and with the more satisfactory methods of estimation there is more uniformity of results. As regards the large excretion estimated by Mester in his case it may be mentioned that in one of Hele's cases Mester's method of estimation gives an output of about one gramme. If further work with reliable methods should show that the uniformity here foreshadowed is an actual fact

44 Biochemische Zeitschrift, 1907, vol. ii., p. 438.
45 American Journal of the Medical Sciences, 1902, vol. cxxiii., p. 838.
46 Journal of Anatomy and Physiology, 1871, vol. v., p. 412.
47 Zeitschrift für physiologische Chemie, 1904, vol. xliii., p. 338.

Loc. cit., sub. 41
49 Loc. cit., sub. 16
50 Loc. cit., sub. 27
51 American Journal of Physiology, 1905, vol. xiv., p. 54.
52 I am much indebted to Mr. Hele for allowing me to make use of these and other unpublished figures from his work. See epitome, Brit Med. Jour., 1908, vol. i. (supplement), p. 440.

and that the quantity of cystin excreted in different cases of cystinuria does not exceed a certain limit, one would be driven to suppose that some definite fraction of the cystin of the proteins broken down escapes its usual fate—such a fraction, for example, as is normally diverted to the formation of taurin.

There is a consensus of opinion among those who have worked at the subject that the unchanged cystin is excreted at the expense of the sulphates rather than at that of the neutral sulphur. The high ratios of neutral sulphur to total sulphur in the urine of cystinurics, which far exceed the normal ratios, bear witness to this, and it is upon the assumption that the neutral sulphur is not implicated that the indirect methods of estimating cystin suggested by Mester and by Alsberg and Folin are based. That the output of sulphates is diminished is a fact which is established beyond question, but the immunity of the normal neutral sulphur does not appear to me to be so well established. Loewy and Neuberg obtained in their case normal values for the neutral sulphur after removal of cystin by concentration of the urine acidified with acetic acid, but Thiele,[53] who employed the same method for the removal of cystin, obtained unduly high values of neutral sulphur after its removal. In a few cases in which cystin excretion has ceased, or nearly ceased, for a time the excretion of neutral

The next question which calls for discussion—namely, whether or no the excretion of cystin is influenced by the amount of protein in the diet—is one of much interest but which cannot yet be regarded as definitely settled. That the output is in part endogenous and derived from the tissue proteins cannot be doubted, for Alsberg and Folin found cystin in the urine of their patient after he had been for 13 days upon a diet which was practically protein-free. They found, however, that although the ratio of neutral to total sulphur was naturally far higher on a protein-free diet the actual quantity of neutral sulphur excreted was considerably greater on a diet rich in protein. As the excretion of neutral sulphur remains almost constant in normal individuals, even on such extremes of dietary, they infer that the increase was in the form of cystin and that the output of that substance, which averaged 0·5 gramme on a nitrogen-free diet, rose to 1 gramme on one rich in protein. In support of this view they adduce the fact that whereas upon the latter diet there was an abundant deposit of crystals, these almost ceased to be deposited when protein was withheld. Further observations, with such extremes of diet, maintained as in the case referred to over considerable periods, and including direct estimations of cystin as well as of total and neutral sulphur, are greatly to be desired. At present there are no other estimations carried out during a period of nitrogen-free

Table Showing Daily Excretion of Cystin in 19 Cases of Cystinuria.

No.	Sex and age in years of patient.	Daily output of cystin. Calculated average unless otherwise stated.	Method of estimation employed.	Observers and references.
1	M., 18	0·520 grm.	Precipitation by acetic acid.	Niemann : Deutsches Archiv für klinische Medicin, 1876, vol. xviii., p. 232.
2	M., adult.	0·425 ,,	,,	Loebisch : Liebig's Annalen, 1876, vol. clxxxii., p. 231.
3	M., 25	0·241 ,,	,,	Ebstein : Deutsches Archiv. 1879, vol. xxiii., p. 138.
4	F., 23	0·430 ,,	,,	Ebstein : Ibid., 1881, vol. xxx., p. 594.
5	M., 13	0·186 ,,	,,	Stadthagen : Virchow's Archiv, 1885, vol. c., p. 416.
6	F., 41	0·1402 ,,	,,	Leo : Zeitschrift für klinische Medicin, 1889, vol. xvi., p. 325.
7	F., 29	0·19–0·3 ,,	,,	Piccini e Conti : Lo Sperimentale, 1891, vol. xlv., p. 359.
8	F., 65	0·464 ,,	,,	Marriott and Wolf : American Journal of the Medical Sciences, 1906, vol. cxxxi., p. 197.
9	M., adult.	Circ. 1·0 ,,	Mester's method.	B. Mester : Zeitschrift für physiologische Chemie, 1889, vol. xiv., p. 110.
10	F., 50	0·423 ,,	,,	Percival : Archivio Italiano di Clinica Medica, 1902, vol. xli., p. 50.
11	F., adult.	0·454 ,,	,,	Morcigne : Archives de Médecine Expérimentale et d'Anatomie Pathologique, 1899, vol. xl., p. 254.
12	M., 28	0·469 ,, (calculated).	,,	Caracciolo : Riforma Medica, 1907, p. 174.
13	M., adult.	0·471 grm.	Concentration and acetic acid.	Loewy and Neuberg : Zeitschrift für physiologische Chemie, 1904, vol. xliii. p. 338.
14	M., adult.	0·561 ,,	,,	Thiele : Journal of Physiology, 1907, vol. xxxvi., p. 68.
15	M., adult.	0·4 ,,	,,	Abderhalden and Schittenhelm : Zeitschrift für physiologische Chemie, 1905, vol. xlv., p. 468.
16	M., adult.	On protein-rich diet, 1 grm.; on protein-free diet, 0·5 grm.	Alsberg and Folin's method.	Alsberg and Folin : American Journal of Physiology, 1905, vol. xiv., p. 54.
17	F., 23	0·314 grm.	Gaskell's method.	T. S. Hele : Case I.
18	M., 63	0·412 ,,	,,	,, Case II.
19	M., adult.	0·456 ,,	,,	,, Case III.

sulphur has remained much above the normal. This was so in Loewy and Neuberg's second case and the ready formation of hydrogen sulphide in the case recorded by Lewis and Simon suggests that the same was true for it. Bödtker[54] also obtained ratios of neutral to total sulphur as high as in any investigated case of cystinuria, even when the urine of his patient no longer deposited cystin, even with acetic acid, and when the presence thereof could only be detected by benzoylation after concentration of the urine. Unless we are prepared to admit that precipitation methods may wholly fail, even when cystin is abundantly present in urine, the inference cannot be avoided that under certain conditions cystinurics excrete a sulphur compound which is not free cystin but a compound or unoxidised derivative thereof. If this be so, estimations of cystin cannot be safely based upon variations, either relative or absolute, of the neutral sulphur.

diet. Thiele,[55] whose observations were unavoidably restricted to isolated days, and are therefore open to the objection that an unequal rate of excretion of nitrogen and cystin may come into play, concluded that the output of cystin by his patient was not affected by the nature of the food taken. The cystin extracted by concentration and addition of acetic acid hardly varied in amount on a day of complete abstinence from food on account of an operation with a nitrogen excretion of 5 grammes, on a day of diet poor in protein on which 9 grammes of nitrogen were excreted, and on one of protein-rich diet on which the nitrogen mounted to 17·29 grammes. Moreover, the residual neutral sulphur, excluding that of the separated cystin, showed very slight variations amounting to 0·359 gramme SO_3 on the fasting day and to 0·428 on that of protein-rich diet. Abderhalden and Schittenhelm[56] give two

---

[53] Journal of Physiology, 1907, vol. xxxvi.: p. 68.
[54] Zeitschrift für physiologische Chemie, 1905, vol. xlv., p. 392.

[55] Loc. cit., sub. 53.
[56] Zeitschrift für physiologische Chemie, 1905, vol. xlv., p. 468.

estimations of cystin by the same method in a case described by them. With an excretion of 16·6 grammes of nitrogen the extracted cystin amounted to 0·31 gramme with 12 grammes of nitrogen to 0·42 gramme. Of earlier observers Leo and Mester expressed the belief that the excretion of cystin in cystinuria is not influenced by diet. In a case under my care, which was investigated by Hele, it was not possible to give any extremes of diet and the capricious appetite of the patient even prevented the maintenance of a constant scale of feeding, but the results obtained pointed to a hardly appreciable augmentation of cystin when the protein of the food was increased.

If the excretion of cystin were directly dependent upon the breaking down of food and tissue proteins alike the ratio $\frac{cystin}{nitrogen}$, the C : N ratio, should not vary with the quantities of the same proteins taken but only with the quality of the food proteins—i.e., their richness or poorness in the cystin fraction. In other words, it should behave as does the H : N ratio in alkaptonuria. If, on the other hand, the output of cystin be independent of, or only slightly influenced by, the amount of protein in the food the C : N ratio should fall conspicuously as the protein food is increased, seeing that the nitrogen will be increased thereby but not the cystin.

The following table, calculated from the figures recorded for a number of cases of cystinuria, shows that such a variation of the ratio is constantly observed, but it is not possible to say how far this is due to differences in the quality rather than in the quantity of the proteins taken. The very unequal variations in different cases suggest that qualitative differences may have an important share in bringing about the results.

Table showing Variations of the C : N Ratio in Cystinuria.

Names of observers.	Total N in 24 hours.	C : N ratio.	Method of estimating cystin.
Moreigne ...	4·13 grm. 16·8 ,,	7·4 : 100 4·8 : 100	Mester's method.
Percival ...	3·98 ,, 15·1 ,,	5·9 : 100 3·1 : 100	,, ,,
Alsberg and Folin ...	Mean 5·19 ,, Mean 14·84 ,,	9·6 : 100 6·7 : 100	Absolute increase of neutral sulphur above normal average.
Abderhalden and Schittenhelm.	12·0 ,, 16·6 ,,	3·3 : 100 1·08 : 100	Concentration and acetic acid.
Thiele ...	5·15 ,, 9·1 ,, 17·29 ,,	9·4 : 100 6·2 : 100 2·9 : 100	,,
Hele : Case I.	4·30 ,, 11·16 ,,	5·2 : 100 4·2 : 100	Gaskell's method.
,, Case II.	6·52 ,, 16·90 ,,	5·52 : 100 3·37 : 100	,, ,,

Even in Alsberg and Folin's case the C : N ratio was higher on a nitrogen-free diet, which shows that even if the increase of neutral sulphur was all due to increase of cystin this increase was not proportional to the amount of protein added to the diet. The balance of evidence appears to be in favour of the view that the excretion of cystin is not materially influenced by diet ; but it must not be forgotten that cases of cystinuria differ widely in their behaviour, and it is conceivable that in one case protein food may augment the output of cystin and in another may not do so. Some observers [57] have noted differences in the quantities of cystin in the day and night urines respectively, but their results have been contradictory, and Hele was unable to find any constant difference in the cases which he investigated.

Much interest attaches to the effects of the administration of cystin, as such, to cystinurics. As we have seen, a normal man burns cystin so given to sulphate, and Alsberg

[57] Beale : Kidney Diseases, Urinary Deposits, and Calculous Disorders, third edition, 1869, p. 385. Ebstein : Deutsches Archiv, 1881, vol. xxx., p. 594. Piccini e Conti : Lo Sperimentale, 1891, vol. xlv., p. 353. Bartels : Virchow's Archiv, 1863, vol. xxvi., p. 419.

and Folin found that their patient behaved in this respect as does a normal man. Two doses of cystin, obtained by hydrolysis of hair, amounting to 1·2 and 6 grammes respectively, were given by the mouth at different times. The neutral sulphur output was unchanged and the increased excretion of sulphur was wholly as sulphates. In one of our cases Hele obtained a similar result from the administration of five grammes of cystin from hair. Thiele gave 4·6 grammes of the patient's own cystin thoroughly purified. The cystin excreted was estimated by concentration of the urine and addition of acetic acid and showed no increase worthy of mention. The quantities for the day before the cystin day, for the cystin day itself, and for the day following were 0·577, 0·604, and 0·561 gramme respectively. However, on the cystin day there was a very conspicuous increase of neutral sulphur as well as of sulphate, and the total increase of sulphur of both kinds corresponded to 3·526 grammes of cystin, of which increase 35 per cent. was in the form of neutral sulphur. It will be remembered that in this case the output of neutral sulphur, above the normal, was always higher than was accounted for by the cystin extracted from the urine. On Alsberg and Folin's reckoning this patient would have been taken as having excreted some of the cystin given as such. In Loewy and Neuberg's case, which stands wholly apart from all others hitherto investigated, the administration of cystin by the mouth produced far more remarkable results. When five grammes of cystin from hair were so taken the excreted cystin, estimated, as in Thiele's case, by concentration with acetic acid, rose from 0·388 gramme per diem to 7·04 grammes in the collected urine of the three days following the administration, an increase which corresponds to the whole of the cystin swallowed. In other words, this cystinuric was quite unable to burn protein cystin given by the mouth, and simply added it to his ordinary daily output. Whether increase of protein food influenced the excretion in this case we are not told. Still more extraordinary was the result when cystin obtained from a calculus was given. The taking of 3·52 grammes of this material caused no increase of cystin in the urine, but there resulted a conspicuous increase of the neutral sulphur, mainly in the form of thio-sulphate. Why in this case cystin obtained from hair and from a calculus respectively should have been dealt with in such wholly different ways by this patient remains a mystery ; no solution of which can be suggested unless these authors are right in regarding them as different substances. The excretion of thio-sulphate suggests the possibility that some of the cystin taken by the mouth may sometimes undergo decomposition in the intestine, and that this may account for the increase of neutral sulphur other than cystin in the urine, which has sometimes followed its administration.

The fact that in ordinary cases of cystinuria swallowed cystin is burned to sulphate, as by normal persons, certainly lends support to the view that the cystin which the patients excrete unchanged has its origin in the breaking down of the tissue proteins and not of those of the food. As Thiele showed even the cystin which the patients himself has excreted can be dealt with when it enters the organism by way of the alimentary canal. It is evident that, as the limitation of the error to a part only of the catabolised cystin also indicates, only a certain path of sulphur metabolism is interfered with in cystinuria. If this be so Loewy and Neuberg's patient must be the victim of a distinct anomaly from that manifested in ordinary cases and, as we shall see, there is much other evidence to support this view.

Before leaving the subject of the disturbance of sulphur metabolism in cystinuria certain observations upon the effects of administration of cholalic acid must be referred to. Von Bergmann's observations upon the effects of such administration to dogs with biliary fistulæ suggested to Simon and Campbell, [58] and independently to Alsberg, [59] that the excretion of cystin in the urine might possibly be due to deficient formation of cholalic acid and that the cystin set apart for the formation of taurocholic acid might on this account fail to be utilised and be excreted unchanged. If this were so the administration of cholalic acid might have the effect of restricting or of abolishing the output of urinary cystin. This leads me to mention that, as Blum has

[58] Johns Hopkins Hospital Bulletin, 1904, vol. xv., p. 164.
[59] Journal of Medical Research, 1904, vol. xiii., p. 105.

pointed out, what is most of all to be desired, as an observation likely to throw valuable light upon the nature of the error in cystinuria, is an examination of the bile of patients who exhibit this anomaly. No such examination has yet been recorded. Possibly in these days of biliary surgery an opportunity may present itself, but even an analysis of the bile obtained from the gall-bladder at a necropsy should throw important light upon the question whether cystinurics form taurocholic acid in the ordinary way. If it be correct that much of the neutral sulphur of urine is derived from taurin, the fact that cystinurics apparently excrete normal quantities of neutral sulphur in addition to cystin suggests that the formation of taurocholic acid is not interfered with, but more direct evidence would be of far greater value.

Alsberg showed that when cholalic acid is administered to cats protein catabolism is increased, as is evidenced by an increased excretion of nitrogen, as well as of total sulphur both in the form of neutral sulphur and of sulphate. Rothera[60] found the same to be true for man, and the taking of cholalic acid by the mouth caused no diminution of the urinary sulphate, such as might be expected if the taurin fraction is excreted as neutral sulphur and the cholalic acid causes an increased formation of taurin.

[60] Loc. cit., sub. 14.

There are obviously great difficulties in arriving at any definite conclusions as to the influence of cholalic acid upon taurocholic formation in the absence of a biliary fistula, and for this reason the observations of Simon and Campbell are not conclusive. They could obtain no evidence that in their cystinuric the taking of cholalic acid had any influence upon the output of cystin, but this was not estimated by any direct method and the fluctuations of the excretion of neutral sulphur during the period over which the observations extended were very wide, so that it is not easy to gauge the effect of the cholalic acid upon the ratio of neutral sulphur to sulphate.

All that we know of the pathology of cystinuria renders it highly improbable that the formation of cholalic acid is primarily at fault, but these observations are of special interest as embodying a rational attempt to influence by treatment a condition which may give rise to grave morbid events.

As I have already pointed out, the excretion of cystin in the urine is after all only one of the manifestations of the metabolic anomaly which we know as cystinuria. Other protein fractions than cystin may also be implicated in the error, with the result that cadaverin and putrescin, leucin and tyrosin may be excreted side by side with cystin, but the consideration of these further aspects of the subject must be deferred to my concluding lecture.

The Croonian Lectures

ON

INBORN ERRORS OF METABOLISM.

Delivered before the Royal College of Physicians of London on June 18th, 23rd, 25th, and 30th, 1908,

BY ARCHIBALD E. GARROD, M.A., M.D.
OXON., F.R.C.P. LOND.,

ASSISTANT PHYSICIAN TO, AND LECTURER ON CHEMICAL PATHOLOGY AT, ST. BARTHOLOMEW'S HOSPITAL; SENIOR PHYSICIAN, HOSPITAL FOR SICK CHILDREN, GREAT ORMOND STREET.

LECTURE IV.*

Delivered on June 30th.

CYSTINURIA (*continued*).

MR. PRESIDENT AND FELLOWS,—The foundation of the wider conception of cystinuria as an error of protein metabolism of which the excretion of cystin in the urine, although the most constant, is only one of the manifestations, was laid by Udranszky and Baumann[1] in 1888. These investigators found the diamines, cadaverin and putrescin, both in the urine and fæces of a cystinuric patient under their observation, and the significance of this find was greatly enhanced when shortly afterwards Stadthagen and Brieger[2] detected cadaverin in the urine of two other cystinurics. In Baumann's case the diamines were readily isolated from the urine of 24 hours by benzoylation in the presence of sodium hydrate,[3] and the daily yield varied between 0·2 and 0·4

gramme of benzoyl-diamines. The major part of the urinary diamine was cadaverin, whereas in the fæces putrescin was the more abundantly present. In 1893, when Garcia[4] had the same patient under investigation, putrescin was alone found in the urine, but in 1897, nine years after their original discovery, both diamines were once more excreted as at first.

In no condition other than cystinuria have these diamines been found in urine in quantities which could be detected by the ordinary methods in the excretion of 24 hours. By evaporating down as much as 100 litres of normal urine Dombrowski[5] was able to demonstrate the presence of traces of cadaverin therein, and after evaporation of large volumes of the urine of patients with pernicious anæmia William Hunter[6] obtained small yields of benzoyl-diamines. Roos[7] also found cadaverin in the fæces of a patient with dysentery and putrescin in those of a sufferer from cholera nostras, but the urine was not examined for diamines. During the past ten years I have myself benzoylated the urine of 24 hours of large number of patients suffering from various maladies, but such searches for diamines have invariably proved fruitless save in cases of cystinuria.

Even in cystinuric cases the search for diamines in the urine and fæces is far from being uniformly successful. In a number of cases they have been found by Bödtker,[8] C. E. Simon, Riegler, Marriott and Wolf, Thiele, Cammidge, Schölberg, and myself, but in others Cohn,[9] Baumann,[10] Alsberg and Folin, Loewy and Neuberg, Hurtley, Hele, and I have been unable to detect their presence, although in

that of benzoyl-putrescin 176° C. Two other methods have been employed for the detection of diamines in urine—viz., the picric acid method of Stadthagen and Brieger (loc. cit., sub. 2) and the phenyl-isocyanate method of Loewy and Neuberg (Zeitschrift für physiologische Chemie, 1904, vol. xliii., p. 355).

* Lectures I., II., and III. were published in THE LANCET of July 4th (p. 1), 11th (p. 73), and 18th (p. 142) respectively.
[1] Zeitschrift für physiologische Chemie, 1889, vol. xiii., p. 562.
[2] Berliner klinische Wochenschrift, 1889, vol. xxvi., p. 344.
[3] 1500 cubic centimetres of urine are shaken with 200 cubic centimetres of 10 per cent. sodium hydrate solution and from 20 to 25 cubic centimetres of benzoyl-chloride, till the smell of the last disappears. The precipitate is filtered off, washed, and treated with hot alcohol. The filtered alcoholic extract is thrown into excess of water. If diamines be present their benzoyl compounds will in a short time separate in minute crystals. The melting-point of benzoyl-cadaverin is 129° C.,
[4] Zeitschrift für physiologische Chemie, 1893, vol. xvii., p. 577.
[5] Archives Polonaises des Sciences biologiques et médicales, 1903, vol. ii.
[6] Transactions of the Medical Society of London, 1890, vol. xiii., p. 386.
[7] Zeitschrift für physiologische Chemie, 1892, vol. xvi., p. 192.
[8] Norsk Magazin for Lægevidenskaben, 1892, vol. liii., p. 1220; Zeitschrift für physiologische Chemie, 1905, vol. xlv., p. 393.
[9] Berliner klinische Wochenschrift, 1899, vol. xxxvi., p. 503.
[10] Pfeiffer: Centralblatt für Krankheiten der Harn- und Sexual-Organe, 1897, vol. viii., p. 173.

some instances the examinations were persisted with over considerable periods.

My own experience relates to nine cases of cystinuria. In four of these diamines were found in the urine at one time or another and in one the fæces also ; in several cases no opportunity of examining the fæces presented itself. My impression is that the likelihood that diamines will be detected in any given specimen of cystin urine is comparatively small, but that if in any case the examinations be continued over sufficiently long periods they are likely to be found eventually.

Of the two diamines cadaverin has been much the more frequently found in the urine, whereas there is evidence that putrescin is more often present in the fæces. In no other case have cadaverin or putrescin, or either of them, been shown to be present so continuously, in quantities easy of detection both in urine and fæces, as in the original one of Udranszky and Baumann, but even in that case the quantities and relative proportions of the diamines present varied considerably and at one time they were almost absent from the urine for several days. In the case investigated by Cammidge and myself [11] cadaverin was found in the urine of two days only out of 41, and putrescin in the fæces at one of six examinations. In another case, in which specimens of urine were sent up to us in gallon jars, one such specimen was rich in cadaverin which was readily extracted by benzoylation from each separate fraction treated, but in no other sample from this patient was any diamine found. From the urine of yet a third patient Schölberg and I [12] got putrescin and cadaverin on several occasions, but five years later, when the same patient was under continuous observation for several weeks, Hurtley and I [13] were uniformly unsuccessful in our search for diamines in his urine and fæces.

These results, taken in conjunction with those of other observers, indicate that in some cases of cystinuria the presence of diamines, in quantities which can be detected in the urine of 24 hours, is a very intermittent phenomenon which may only be manifested at long intervals. It is evident that it cannot safely be asserted that a given cystinuric never excretes them, even when they cannot be found in the urine for days or even weeks together. It is possible that the apparent intermittence is merely due to varying amounts, for a number of experiments in which cadaverin was added to normal urine in different proportions, have convinced me that failure to detect diamines by the methods in use cannot be held to exclude their presence in quantities, small indeed, but much greater than the normal traces found by Dombrowski.

The substance known as cadaverin is penta-methylene-diamine, and putrescin is tetra-methylene-diamine. They are two members of a series of such compounds of which other members are known.

$$\begin{array}{ll} CH_2.NH_2 & CH_2.NH_2 \\ | & | \\ CH_2 & CH_2 \\ | & | \\ CH_2 & CH_2 \\ | & | \\ CH_2 & CH_2.NH_2. \\ | & \\ CH_2.NH_2. & \\ \text{Cadaverin.} & \text{Putrescin.} \end{array}$$

An obvious explanation of the occurrence of these two particular members of the diamine series, both in the excreta of cystinurics and when proteins undergo decomposition, under the influence of bacteria, is afforded by the fact that they are intimately related to two of the protein fractions, the diamino-acids lysin and ornithin.

Lysin is converted into cadaverin by the elimination of carbon dioxide, and in like manner ornithin, which enters into the composition of the important protein fraction arginin, yields putrescin by a similar change.

$$\begin{array}{llll} CH_2.NH_2 & CH_2.NH_2 & CH_2.NH_2 & CH_2.NH_2 \\ | & | & | & | \\ CH_2 & CH_2 & CH_2 & CH_2 \\ | & | & | & | \\ CH_2 -CO_2 = & CH_2 & CH_2 -CO_2 = & CH_2 \\ | & | & | & | \\ CH_2 & CH_2.NH_2. & CH.NH_2 & CH_2.NH_2. \\ | & & | & \\ CH.NH_2 & & CO.OH. & \\ | & & & \\ CO.OH. & & & \\ \text{Lysin.} & \text{Cadaverin.} & \text{Ornithin.} & \text{Putrescin.} \end{array}$$

[11] Journal of Pathology and Bacteriology, 1900, vol. vi., p. 327.
[12] THE LANCET, August 24th, 1901, p. 526.
[13] Journal of Physiology, 1906, vol. xxxiv., p. 217.

When Udranszky and Baumann wrote our knowledge of the structure of protein molecules was far more imperfect than it now is, and, as the known source of the diamines in question was the bacterial decomposition of proteins, they naturally suggested that their presence in the urine and fæces of cystinurics might result from a similar decomposition carried out in the alimentary canal, and that cystinuria itself was probably indirectly due to an intestinal infection. The absence of cystin from the fæces weighed strongly against its being itself a product of intestinal decomposition, but it was thought that the diamines thence absorbed might in some way preserve the cystin from its usual fate. However, the administration of diamines to dogs was found not to produce cystinuria in them, intestinal disinfection was found to have no influence upon the excretion of cystin or diamines, and cultures from the fæces of cystinurics have failed to reveal the presence of abnormal bacteria having the power of forming diamines from proteins. In the absence of any positive evidence in its favour the infective theory of cystinuria, which at one time met with wide acceptance, has now been abandoned and opinion has veered round to the view that the diamines which cystinurics excrete are themselves products of metabolism derived from the lysin and arginin of proteins broken down in the organism, a view which was independently advanced by Moreigne [14] in France and by C. E. Simon [15] in America.

This hypothesis receives very strong support from the observations of Loewy and Neuberg [16] upon their exceptional patient already frequently referred to. Although no diamines could at ordinary times be detected in his urine, when lysin was administered to him by the mouth large quantities of cadaverin in large quantities, and when arginin was so administered he excreted putrescin. In two other cases of cystinuria Hurtley and Hele and I [17] could find no putrescin in the urine after the administration of five-gramme doses of arginin carbonate, although one of the patients had spontaneously excreted putrescin five years previously, sometimes alone and sometimes in association with cadaverin. As we shall see, these results are fully in keeping with those obtained with other protein fractions in Loewy and Neuberg's case and other cases of cystinuria respectively.

It would be very interesting to know whether in Loewy and Neuberg's case the taking of diamino-acids by the mouth was followed by the appearance of diamines in the fæces as well as in the urine, for, although the diamines have comparatively seldom been found in the fæces of cystinurics their occurrence therein is beyond question, and if in cystinuria they are products of metabolism we must assume that they are in part excreted by way of the alimentary canal.

It does not necessarily follow that the diamines present in the fæces in cases of grave intestinal infection, such as Roos examined, have the same origin as those excreted in cystinuria, for it is quite possible that they are formed in the intestine by bacterial action, and even that the normal traces in urine have such an origin ; but in Udranszky and Baumann's case the abundant diamine yield from the fæces was undoubtedly connected with the urinary output and with the underlying cystinuria, for diamines are not to be detected in normal fæces, and their patient had no intestinal disorder. It is worthy of note that in the case at which Cammidge and I worked the urine of the only day on which putrescin was found in the fæces yielded no diamine.

It is conceivable that the change from diamino-acids to diamines might occur after excretion, both in urine and fæces, but some observations which we have made on specimens of urine kept for considerable times lend no support to such a view. Specimens which yielded no diamine when fresh equally failed to yield any when kept or even when decomposing. T. S. Hele allows me to quote the interesting fact from his as yet unpublished results that when he recently examined large specimens of urine from a cystinuric already referred to, who at one time excreted cadaverin, he failed to obtain, by the method employed by Wohlgemuth for their detection in cases of phosphorus poisoning, any evidence of the presence therein of lysin or arginin. There is therefore no evidence yet forthcoming to

[14] Archives de Médecine Expérimentale et d'Anatomie Pathologique, 1899, vol. xi., p. 254.
[15] American Journal of the Medical Sciences, 1900, vol. cxix., p. 39.
[16] Zeitschrift für physiologische Chemie, 1904, vol. xliii., p. 355.
[17] Loc. cit., sub. 13, p. 220.

support the tempting hypothesis that diamino-acids are excreted as such by cystinurics.

Garcia[18] thought that the excretion of diamines varied with the amount of protein in his patient's diet, and Thiele[19] has recently expressed the same opinion. The latter observer obtained a far larger yield of benzoyl-cadaverin from the urine of a day of excessive meat diet than in that of a day of abstinence or of a diet rich in carbohydrates. However, his observations are too few to allow any definite conclusion to be drawn from them, seeing how widely the diamine excretion varies apart from changes of diet. In the case which Cammidge and I investigated the diet of the days in which diamines were excreted did not differ from that of other days.

To sum up, the excretion of cadaverin and putrescin in some cases of cystinuria, both in the urine and fæces, in quantities such as have not been found in any other condition, is an established fact. The easier explanation which ascribes their presence to an intestinal infection offers no adequate explanation of their association with the excretion of cystin and receives no support save from its inherent plausibility. The alternative theory which regards the diaminuria as an outcome of the same error of metabolism as cystinuria has much more in its favour, although it presents certain obvious difficulties, especially as other protein fractions than cystin are undoubtedly present in the urine of some of the subjects of this anomaly.

Leucin and tyrosin have been found in several cystin urines. Thus in 1891 Piccini and Conti[20] detected crystals of tyrosin, together with those of cystin and of uric acid, in the urinary sediment in their case. Percival[21] speaks confidently of the presence of both leucin and tyrosin in small amounts in the urine of his cystinuric patient. Moreigne[22] also described the presence of tyrosin, but the method which he recommends for its detection suggests that he mistook for crystals of that substance the prismatic crystals of cystin hydrochlorate. The presence of these amino-acids in the urine of one cystinuric at least has been established beyond all possibility of doubt by Abderhalden and Schittenhelm,[23] who extracted from the urine of their patient both tyrosin and leucin in considerable quantities, and fully proved their nature by the ultimate analysis of the tyrosin which separated after evaporation of the urine, and of the naphthylene-sulphon derivate of leucin obtained therefrom. Emil Fischer and Zuzuki[24] also detected tyrosin in a cystin calculus. Leucin and tyrosin would appear to be far rarer constituents of such urines than are the diamines. In no case which I have had under observation have they been detected, although sought for. Millon's reagent lends valuable aid in the detection of tyrosin, for although all urines yield some pink colour with this reagent, even in the cold, the presence of tyrosin in any considerable quantity conspicuously intensifies the colour on heating. It may indeed safely be concluded that any urine which does not yield more than the normal reaction with Millon's reagent does not contain tyrosin in appreciable amount. The urine of Loewy and Neuberg's patient contained no tyrosin, but when tyrosin was given to him by the mouth he excreted it unchanged and almost quantitatively. After a dose of 6·2 grammes of tyrosin no less than 4·82 grammes were recovered from the urine, and its nature was confirmed by ultimate analysis. On the other hand, C. E. Simon,[25] Alsberg and Folin,[26] Thiele,[27] Hele, Hurtley, and I[28] have failed to find any tyrosin in the urine of other cystinurics after the administration of similar doses. Hurtley and I obtained from the urine of our patient, on the days on which tyrosin was given, a benzoyl compound which melted at 253° C. The yield was small and its nature has not been determined. Thiele obtained no such product by benzoylation of the urine of the tyrosin day in his case. Loewy and Neuberg also found that aspartic acid, another protein fraction, was excreted by their patient when administered by the mouth, but in Alsberg and Folin's case aspartic acid so

given did not reappear in the urine. Hurtley and I obtained on many occasions on benzoylating the urine of our patient small quantities of a benzoyl compound which, after repeated recrystallisation from alcohol, melted at 205° C. The collected yields only amounted to 0·22 gramme of substance, and a single analysis, which was alone possible, gave figures which suggested that it was probably the benzoyl compound of a derivative of tryptophane. However, we are not prepared to make any definite suggestion as to the nature of this substance, which we have not obtained from any other cystin urine.

The question arises whether the implication of the several protein fractions follows any definite sequence, according to the extent of the error, or whether in different cases now one and now another fraction is implicated. To this question no certain reply can yet be given. It may be that the excretion of cystin in all cases hitherto examined is due to the fact that it is by the presence of cystin in the urine that the anomaly has always been recognised hitherto; but the cystin fraction appears alone to escape destruction in so many of the cases that it is highly probable that it is the first to be involved in the error.

Not a few cystinurics have excreted cystin and diamines but no leucin or tyrosin, but Percival, who found leucin and tyrosin in the urine of his patient, failed to detect any diamine, and Abderhalden and Schittenhelm make no mention of the presence of diamines in their case in which leucin and tyrosin were undoubtedly excreted. The evidence available suggests that the incidence of the error upon the several protein fractions is capricious rather than that the amino- and diamino-acids are involved in any definite order.

The nitrogenous metabolism of cystinurics deviates from the normal in certain ways not fully explained. When diamines and amino-acids are present an excess of undetermined nitrogen in the urine—i.e., of nitrogen not contained in urea, uric acid, kreatinin, and ammonia—is to be expected, but Alsberg and Folin, whose observations upon this aspect of the subject are the most complete yet carried out, found an abnormally high nitrogenous residue, although the urine of their patient contained no diamine nor amino-acid other than cystin. This excess could not be ascribed to the small quantity of cystin excreted. Marriott and Wolf also found an excess of undetermined nitrogen, but their patient excreted no diamine. Alsberg and Folin suggest that the excess of undetermined nitrogen in the urine in their case, which was conspicuously increased when cystin was given by the mouth, may have been due to an incomplete destruction of cystin, the sulphur of which was excreted as sulphate, whereas the nitrogen was apparently not eliminated in urea.

It will be clear, from all which has gone before, that we are still far from being in a position to formulate a satisfactory theory of cystinuria. Before this can be done it will be necessary to accumulate many more data by patient investigation of individual cases, and, above all, quantitative data. Obviously the anomaly is a very complex one, of different range in different cases and even of distinct natures. No theory which will explain the ordinary cases of cystinuria can apply to such a case as that studied by Loewy and Neuberg, in which any protein fraction given by the mouth appeared in the urine, the monamino-acids as such and the diamino-acids as diamines, whereas these same fractions, when given in combination as polypeptides or in the more complex form of proteins, were dealt with in the ordinary way. This raises important side issues relating to the degree of disintegration which proteins normally undergo in the alimentary canal, and suggests that if they were broken down into their component amino- and diamino-acids before absorption this patient should have excreted the fractions unchanged, as, indeed, he did excrete cystin and the mixed products of advanced tryptic digestion. When the patient took glycocoll by the mouth some 20 per cent. of the quantity administered was apparently burned, whereas when glycyl-glycin was given, a dipeptide which is not split up by the pancreatic ferment but is split by the intestinal juice, only 10 per cent. escaped destruction. Polypeptides were completely burnt.[29]

The varying extent of the error, as regards the number of protein fractions involved in cases of cystinuria, suggests that it is manifested at an early stage of the catabolic series

18 Loc. cit., sub. 4.
19 Journal of Physiology, 1907, vol. xxxvi., p. 68.
20 Lo Sperimentale, 1891, vol. xlv., p 353.
21 Archivio Italiano di Clinica Medica, 1902, vol. xli., p. 50.
22 Loc. cit., sub. 14.
23 Zeitschrift für physiologische Chemie, 1905, vol. xlv., p. 468.
24 Ibid., 1905, vol. xlv., p. 405.
25 Zeitschrift für physiologische Chemie, 1905, vol. xlv., p. 357.
26 American Journal of Physiology, 1905, vol. xiv., p. 54.
27 Loc. cit., sub. 19.   28 Loc. it., sub. 13.

29 Biochemische Zeitschrift, 1907, vol. ii., p. 438.

and concerns a mechanism which deals with a number of amino- and diamino-acids in common. In this respect cystinuria stands in conspicuous contrast to alkaptonuria, which involves a late stage in the catabolism of two closely allied protein fractions, a stage so late, indeed, that the tyrosin and phenyl-alanin derived from the breaking down of food and tissues are alike implicated. The fact that the abnormal substances excreted retain their amino-groups intact points in the same direction, for there is strong evidence that desamination occurs at an early stage of the breaking down of proteins. It is a process to which all the protein fractions are normally subjected and, as Lang[30] has shown, desaminating enzymes are widely distributed in the tissues. Moreover, it would appear that the several amino-acids are desaminated with various degrees of ease and that different members of the group are specially attacked in different organs and tissues. Lang found that phenyl-alanin was the most difficult of desamination of the protein fractions and under some conditions cystin and tyrosin were refractory. The maximum yield of ammonia, by which the change was estimated, was obtained when glycocoll or leucin was exposed to the action of the liver or the pancreas.

This complex process of desamination carried out in a variety of tissues, by the action of enzymes which appear to differ somewhat in their modes of action, may possibly afford a clue to the differences observed in individual cases of cystinuria. Thus the observation of Lang that the power of removing the amino-group is specially conspicuous in the intestinal mucous membrane on the one hand, and in the liver on the other, gives a hint of a possible explanation of such different types of cystinuria as are exemplified in the ordinary cases and in Loewy and Neuberg's respectively, an explanation based upon absence or inhibition of the enzyme which effects the change in the one situation or in the other. It would seem, also, that when for any reason the amino-acids escape this early change no alternative mechanism is available for dealing with them and they are excreted unchanged, or in the case of lysin and arginin as diamines. Thus cystinuria, like alkaptonuria, may be classed as an arrest rather than as a perversion of metabolism.

PENTOSURIA.

Our knowledge of the sugars of the pentose group is only of recent date, and dates back only a few years further than the discovery of pentosuria, the last of the inborn errors of metabolism to be considered in these lectures. It was in the year 1887 that Kiliani[31] showed that arabinose, the sugar of gum arabic, which had previously been classed as a hexose, was a member of a different class of sugars from any which had up to then been thoroughly investigated and that its molecule contained only five carbon atoms, its formula being $C_5H_{10}O_5$. Not long afterwards Wheeler and Tollens[32] found that the wood sugar, xylose, was a second member of the group, and since then other pentoses have been recognised or obtained, including a series of methyl-pentoses, of which rhamnose is the best known. However, the original pair, arabinose and xylose, remain the important members of the class from the point of view of physiology.

Like other sugars the pentoses exist in dextro- and lævo-rotatory forms and also in optically inactive, racemic, combinations. The structures of the isomeric arabinoses and xyloses are represented by the following formulæ:—

COH	COH	COH	COH
HOCH	HCOH	HOCH	HCOH
HCOH	HOCH	HCOH	HOCH
HCOH	HOCH	HOCH	HCOH
CH₂.OH.	CH₂.OH.	CH₂.OH.	CH₂.OH.
d. arabinose.	l. arabinose.	d. xylose.	l. xylose.

The names l. and d. arabinose and xylose express their affinities to the l. and d. series of hexose sugars respectively, and not their own optical properties. As a matter of fact, both l. arabinose and l. xylose, the forms usually met with in nature, are dextro-rotatory. The pentoses are abundantly represented in vegetable structures, not as such, but in the form of complex anhydrides known as pentosanes.

In 1892 Salkowski and Jastrowitz[33] described the case of a young man, neurasthenic and the victim of the morphia habit, whose urine gave the reduction tests for sugar, did not ferment with yeast, was optically inactive, but yielded an osazone the melting point of which was 159° C. The melting point of the osazone suggested that the sugar present was a pentose, although no compound of that class had previously been recognised as a product of animal metabolism nor as a constituent of animal tissues. Further investigation confirmed this conjecture and thus was laid the foundation of our knowledge of pentosuria. Soon afterwards Hammarsten demonstrated the presence of a pentose in the pancreas, and since then the presence of such a substance in the nucleo-proteins of various tissues and organs has been practically proven, although the demonstration is not so complete as in the case of the pancreas. Neuberg has identified the pancreatic pentose as l. xylose.

The power of the normal human organism to destroy pentoses is very limited, and quite small doses, such as a gramme of arabinose or even less, lead to the appearance of some of the sugar in the urine. Different observers have found that the fractions excreted varied widely, and the destruction may be in part effected in the alimentary canal before absorption.

Seeing that many vegetable foods are rich in pentosanes it is not surprising that some degree of alimentary pentosuria is occasionally induced by their free consumption, but the quantities excreted after the eating of such fruits as plums and cherries, even in abundance, is very small, and delicate tests are required for their detection. Such an alimentary pentosuria, which is wholly distinct from the so-called essential pentosuria, is little likely to lead to diagnostic errors, but von Jaksch[34] has found that when the unfermented fruit juices, which are popular beverages on the continent, are taken, in such quantities as a litre or more, the next passed urine yields Trommer's and Nylander's tests and also the special tests for pentoses. Under such conditions the sugar excreted is the l. arabinose contained in fruits and the urine rotates the polarised ray to the right. Hence it is evident that the presence of a pentose in urine in quantities easy of detection is not necessarily due to an error of metabolism, but may be of accidental alimentary origin.

Again, Kulz and Vogel[35] found that from some diabetic urines there may be obtained by means of phenyl-hydrazine, in addition to glucosazone, an osazone soluble in hot water, which has the melting point and nitrogen content of a pentosazone. Only small yields of this product were obtained when quantities of urine amounting to several litres were so treated, and other observers have failed to detect pentose in diabetic urines, and among them Bendix,[36] who employed the same method as was used by Kulz and Vogel. It would appear, indeed, that the presence of pentose in diabetic urines is a rare phenomenon and that it has no direct bearing upon the pathology of essential pentosuria. On the other hand, it should be mentioned that Kulz and Vogel found a pentose in the urine of dogs rendered diabetic by removal of the pancreas, but the actual nature of the 5-carbon sugar excreted in such circumstances, and in human diabetes, has not been determined.

The case studied by Salkowski and Jastrowitz[37] has not remained an isolated one, and of quite recent years such observations have been multiplied, so that some 30 cases of essential pentosuria are now on record. The great majority of the recorded cases have been met with in Germany, especially at watering-places to which patients with the milder forms of diabetes are wont to resort, and the three cases recorded in America have been in patients of German or Russian birth.[38] In this country no case has yet been described. There can be little doubt that pentosuria is a very rare anomaly. An assiduous search carried out over several years by myself, and by others at my instigation, has failed to reveal a case, and eloquent testimony to its rarity is afforded by the fact, to which Dr. Pavy kindly allows me

[30] Hofmeister's Beiträge, 1904, vol. v., p. 321.
[31] Berichte der deutschen chemischen Gesellschaft, 1887, vol. xx., p. 339.
[32] Annalen der Chemie und Pharmacie (Liebig's), 1889, p. 254.

[33] Centralblatt für die medicinischen Wissenschaften, 1892, vol. xxx., p. 337.
[34] Centralblatt für Innere Medicin, 1906, vol. xxvii., p. 145.
[35] Zeitschrift für Biologie, 1895, vol. xxxii., p. 185.
[36] Die Pentosurie, Stuttgart, 1903, p. 55.
[37] Loc. cit., sub. 33.
[38] T. Janeway: American Journal of the Medical Sciences, 1906, vol. cxxxii., p. 423. Kaplan: New York Medical Journal, 1906, vol. lxxxiv., p. 233.

to refer, that he has never met with an authentic example of the condition, and that those of his cases in which there have been grounds to suspect its presence have all been found on further investigation to be of other natures.

However, the rarity of pentosuria amòng us cannot be ascribed to any immunity of the Anglo-Saxon race, for there is evidence of a special frequency of the anomaly in Jews, and among the large Jewish population in our midst, both alien and native, it cannot fail to occur. Of 26 recorded examples no less than five are definitely stated to have been in members of the Hebrew race, and it is probable that other patients whose race is not mentioned would swell the proportion. Erben[39] quotes von Jaksch as saying that in his experience the majority of pentosurics are of Jewish origin. The tendency of pentosuria to occur in several collaterals of a family, and the great preponderance of males among its subjects, have already been referred to, and also the fact that the youngest recorded patient was a boy aged 15 years.[40] At least two patients have been over 60 years of age.

Pentosuria is not necessarily associated with any morbid symptoms. In some instances the detection of sugar at an examination for life assurance has been the first indication of anything amiss, but this is also true of not a few cases of mild diabetes. There is a growing opinion among those whose experience enables them to speak with authority on the subject that this anomaly is in itself harmless and calls for no treatment. The most recent writers suggest that it should be included in the class of which cystinuria and alkaptonuria are members. In a considerable proportion of cases there have been symptoms of neurasthenia, and if any morbid condition can be looked upon as specially associated with the urinary peculiarity it is this. However, race and racial temperament may here play no unimportant part, and the diagnosis of diabetes, usually made, may contribute to the result. Most pentosurics have been supposed to have diabetes and have been for a time treated accordingly.

The question of the existence of any relationship of essential pentosuria to diabetes is of considerable interest. There can be no doubt that from the metabolic standpoint the two are absolutely distinct, despite the fact already referred to that some diabetics excrete small quantities of pentose. It has been demonstrated that the pentosuric is as capable of dealing with ingested glucose as an ordinary individual; only when it is given in doses which suffice to overtax the power to destroy glucose which is possessed by a normal man does that sugar appear in his urine.

However, when we turn to the clinical evidence their independence of each other is less obvious. A certain number of the observed pentosurics have excreted glucose as well as pentose at times. The original patient of Salkowski and Jastrowitz did so, but in him the morphia habit may have been the exciting cause of the temporary glycosuria. One of Blumenthal's[11] patients was glycosuric, and so also was one of af Klerker's.[12] Some pentosurics have been members of diabetic families. Kj. O. af Klerker's patients were two brothers, Jews, whose father and another brother had diabetes. Schüler's[43] patient had a brother and two sisters who were diabetic, and in one of the cases described by Rosenfeld,[44] who lays special stress upon these points, the patient, whose father and brother had diabetes, developed pentosuria after being in a railway accident. The evidence of its recent development was afforded by the fact that an examination of his urine a year and three-quarters previously had revealed nothing abnormal. The pentosuria diminished as time went on. In this connexion also race may come into play, seeing that members of the Jewish race show a special liability to diabetes, but of those mentioned above af Klerker's patients are stated to have been Jews.

Clearly the correct diagnosis of cases of pentosuria is a matter of real practical importance. If it be not made the patient is not only subjected to the distress which the diagnosis of diabetes entails, but is subjected to irksome dietary restrictions which in his case are wholly uncalled for, are quite useless, and when not necessary are certainly undesirable. On the other hand, it is certainly better that a few persons who exhibit an anomaly so rare as pentosuria should be erroneously classed as diabetic than that a diagnosis of pentosuria on inadequate grounds should lead to the omission of treatment in a number of cases of true diabetes. In pentosuria the urine is not excessive in quantity, nor are the symptoms so often prominent in diabetic cases, such as thirst and undue appetite, met with in connexion therewith.

The specific gravity of the urine usually varies from 1.025 to 1.035 and when it is boiled with Fehling's solution it reduces, as does a diabetic urine containing some 0.5 per cent. of glucose. Many observers have described the reduction as delayed and as occurring suddenly after the test-tube has been removed from the flame. Urine containing small percentages of glucose may behave in this way, and Bial,[45] who has had an exceptionally large experience in this matter, states that, when fresh, pentose urines may show no such peculiarity of behaviour and that the reduction may occur before the boiling point is reached. He suggests that the delayed but sudden reduction so often observed is due to the urine having been kept for some time, with the addition of preservative substances, such as toluene or chloroform. The only specimen of pentose urine which I have had the opportunity of examining, for which I am indebted to the kindness of Professor von Jaksch, behaved in this manner, but it had been preserved for some long time under toluene. Pentose urines also yield Moore's test for sugar and that of Nylander, although not very strikingly, for the amount of the sugar present is never large. They also yield the saffranin test.

With the yeast test no fermentation occurs and after standing with yeast in a warm place for 24 hours the reducing power of the urine is not impaired. Kj. O. af Klerker lays special stress on this as affording a ready means of detecting pentose when present in association with glucose. In every case hitherto observed, with the exception of one of Luzzato's,[16] the urine had been optically inactive, unless glucose were also present. In Luzzato's case, in which it was clearly established that the sugar present was arabinose, the urine was dextro-rotatory.

With the phenyl-hydrazine test a crystalline osazone is obtained, but as it is to some extent soluble in hot water it is only thrown out in crystalline form after the liquid has cooled. In appearance the crystals closely resemble those of phenyl-glucosazone, but whereas the latter after recrystallisation melts at 205° C., the melting point of pentosazone is much lower, between 156° and 160° C. This melting point supplies one of the most important means of diagnosis of pentosuria, but to insure complete certainty as to its nature the nitrogen content of the osazone must be estimated. The theoretical amount for a pentosazone is 17.07 per cent. N.

The special tests which are available for the detection of pentoses are based upon the property which such sugars possess of yielding furfurol when heated with mineral acids. The phloroglucin test is performed as follows: a small quantity of phloroglucin is dissolved in 5 or 6 cubic centimetres of fuming hydrochloric acid, a slight excess remaining undissolved. To one portion of the solution half a cubic centimetre of the urine to be tested is added in a test-tube, and to the other portion an equal quantity of normal urine as control. Both tubes are then placed in a beaker of boiling water. If pentose be present the test containing it quickly assumes a deep red colour from the surface downwards, whereas the normal specimen shows little change. The colour can be readily extracted by shaking with amylic alcohol, and the amylic extract, suitably diluted, shows a spectroscopic absorption band between the Fraunhofer lines D and E. More satisfactory for diagnostic purposes is the orcin test. A specimen of urine is warmed with an equal volume of concentrated hydrochloric acid in which enough orcin to cover a knife-point has been dissolved. A green colour quickly develops if pentose be present and the liquid becomes turbid by the formation of a bluish-green flocculent precipitate. After cooling the precipitate may be taken up with amyl alcohol, which acquires a rich green colour and shows a characteristic absorption band between

[39] Prager medicinische Wochenschrift, 1906, vol. xxxi., p. 301.
[40] Chobola: Centralblatt für innere Medicin, Abstract, 1907, vol. xxviii., p. 864.
[41] Berliner klinische Wochenschrift, 1895, vol. xxxii., p. 567.
[42] Nordiskt medicinskt Arkiv, 1905, Afd. ii., Heft i., pp. 1 and 53.
[43] Münchener medicinische Wochenschrift, 1905, vol. lii., p. 1657.
[44] Medizinische Klinik, 1906, vol. ii., p. 1041.

[45] Berliner Klin k, 1907, Heft 2?6.
[46] Hofmeister's Beitrage, 1904, vol. vi., p. 87

the solar C and D lines. A second band nearer to the red end of the spectrum and a fainter band in green, which are often seen, are not of diagnostic importance.

Bial's [47] modification of the orcin test is very useful for clinical purposes. It is based upon the effect of the addition of ferric chloride in quickening the reaction. The reagent is prepared as follows: one gramme of orcin is dissolved in 500 cubic centimetres of hydrochloric acid of specific gravity 1·151 and to the solution 25 drops of a 10 per cent. solution of ferric chloride are added. The liquid when kept in an amber-coloured bottle will preserve its activity for a long period. Five cubic centimetres of the reagent are boiled in a test-tube and after the tube has been removed from the flame five drops of the urine to be tested are allowed to fall upon the surface of the liquid from a pipette. If pentose be present a green ring appears at the junction of the liquids, and if the tube be gently shaken the green colour spreads through the liquid. With the spectroscope the characteristic absorption band is seen. Care is required that the correct proportions be employed in mixing the reagent and the hydrochloric acid must be of the prescribed strength. If a pentose urine be not available the efficiency of the reagent may be tested with a dilute solution of gum arabic. I can testify, from my own experience, that the reaction so obtained is a very striking and characteristic one. By its means it is easy to exclude pentosuria in doubtful cases and it is easily applied in ordinary clinical work. Bial and also Kraft maintain that, if performed in the manner described, the test is diagnostic and is yielded by none but pentose urines. However, for the diagnosis of pentosuria one should not rely upon any single reaction and confirmatory tests should be applied.

The discrimination between pentosuria and glycosuria should present no real difficulty as glucose does not yield the above colour tests. It is in connexion with glycuronic acid that the risk of error comes in. Glycuronic acid has obvious chemical relationships to the pentoses as their respective formulæ show. By splitting off carbon dioxide

$$
\begin{array}{ll}
\text{COH} & \text{COH} \\
\text{HCOH} & \text{HCOH} \\
\text{HOCH} & \text{HOCH} \\
\text{HCOH} & \text{HCOH.} \\
\text{HCOH} & \text{CH}_2.\text{OH.} \\
\text{CO.OH.} & \\
\text{d. glycuronic acid.} & \text{l. xylose.}
\end{array}
$$

from it a pentose results and the acid itself yields the furfurol reactions. Urine containing glycuronic acid may yield the phloroglucin test, but as the acid is almost always in combination in urine, as compound glycuronates, the reactions are not obtained until the combination is loosened. However, some compound glycuronates, such as the menthol compound, undergo spontaneous breaking up in urine.

It is not easy to make sure from the writings of those who have had most experience of these reactions whether in clinical work the risk of mistaking a urine containing glycuronic acid for a pentose urine is really great. It is stated that the orcin test is not yielded by compound glycuronates unless heating be unduly prolonged, and although Bial's reagent, if boiled with a urine containing a compound glycuronate, gives the green colour and the absorption band, it is claimed that no risk of confusion exists if the urine be added after the test-tube has been removed from the flame. With free glycuronic acid Paul Mayer [18] obtained crystalline phenyl-hydrazine compounds, one of which had a melting point like that of a pentosazone, but this is not yielded by compound glycuronates, and in case of doubt an estimation of the nitrogen in the crystalline product will settle the point.

Again, the polarimeter may help in the diagnosis, for all known compound glycuronates are lævo-rotatory, although the free acid is dextro-rotatory, whereas pentose urine is almost always optically inactive. Lastly, patients who excrete glycuronic acid in appreciable quantities are almost always taking some drug which is known to be excreted as a compound glycuronate. Nevertheless there is some doubt whether in certain cases [49] which have been described as

examples of pentosuria the reducing property of the urine was not really due to glycuronic acid.

The most remarkable fact of all in regard to pentosuria is the optical inactivity of the excreted sugar. In one case Neuberg [50] succeeded in isolating the pentose from the osazone obtained from a large volume of urine treated with di-phenyl-hydrazine, and was able to identify it as racemic arabinose. Thus the pentose of urine stands out as the sole example of a racemic sugar occurring as such in the animal organism, and forms an exception to the rule that such organisms are built up of optically inactive materials. When racemic arabinose is administered by the mouth to a normal man it does not appear in the urine unchanged. Neuberg and Wohlgemuth [51] found that when 15 grammes of this sugar were so taken, the urine, which had previously contained no reducing substance, had been optically inactive and had failed to yield the orcin test, acquired after the lapse of four hours strongly reducing properties, was optically active, and gave the pentose reactions. Of the excreted arabinose, to which these properties were due, no less than two-thirds was the lævo-rotatory d. arabinose, and it was evident that the individual in question had a much greater power of destroying the ordinary l. arabinose than the other form. This observation only makes it the more remarkable that in pentosuria the racemic arabinose is excreted as such and alone. It is certain that essential pentosuria is not alimentary. This is fully proved by its persistence when all pentosane is excluded from the diet, and moreover the arabinose present in vegetable foodstuffs is always the dextro-rotatory l. arabinose and not the racemic sugar.

Nor can it be derived from the pentose contained in the nucleo-proteins of the food and body tissues. This is an entirely different sugar l. xylose, which cannot be supposed to become converted into racemic arabinose. Moreover, the quantity present in the body is too small to serve as a source of supply. Grund [52] estimated the total quantity in the human body at ten grammes and Bendix [53] makes it only slightly more. Lastly, Bial and Blumenthal [54] found no increase of pentose in the urine after feeding with 500 grammes of calf's thymus, and the uric acid and phosphate excretion of pentosurics affords no evidence of an abnormal breaking down of nucleo-proteins.

Our knowledge of the quantities of pentose excreted by pentosurics is very imperfect. The ordinary Fehling's method of estimation is not applicable to such urines because the cuprous-hydrate precipitate does not separate satisfactorily and other methods have to be resorted to, such as Allihn's, Knapp's, or weighing the phloroglucin precipitate. The estimated amounts in individual cases have differed widely—from one to seven grammes in the 24 hours—and Neuberg [55] states that all such estimates are too low, sometimes by 100 per cent., because much of the pentose is in combination with urea as a ureide, and the portion so combined does not reduce until the ureide is broken up by heating with an acid. In view of the uncertainty which surrounds this matter it is obviously impossible as yet to draw any conclusions from the published figures obtained by various methods in different cases, as to the constancy or otherwise of the output. The quantity estimated—viz., the uncombined arabinose—may represent no constant fraction of the total amount. Neuberg [56] mentions from 30 to 36 grammes as the figure which may be reached by the day's excretion, which is far more than twice as much as any other recorded estimate with which I am acquainted.

Exclusion of carbohydrate from the diet has been shown to have no influence upon the pentose output, as estimated by reduction tests, nor is the pentosuric less able to burn dextrose and lævulose than a normal subject. Still more remarkable is the fact, ascertained by Bial and Blumenthal [57] that even when five grammes of l. arabinose were administered to a pentosuric by the mouth, the amount of the optically active arabinose excreted was no greater than when a like dose is administered to a normal man. Just as a

[47] Deutsche medicinische Wochenschrift, 1902, vol. xxviii., p. 253.
[48] Zeitschrift für physiologische Chemie, 1900, vol. xxix., p. 59.
[49] Caporelli: Riforma Clinica e Terapeutica, 1896, i.; Colombini: Monatshefte der praktischen Dermatologie, 1897, vol. xxiv., p. 129.

[50] Berichte der deutschen chemischen Gesellschaft, 1900, vol. xxxiii., p. 2243.
[51] Zeitschrift für physiologische Chemie, 1902, vol. xxxv., p 41.
[52] Zeitschrift für physiologische Chemie, 1902, vol. xxxv., p 111.
[53] Die Pentosurie, Stuttgart, 1903, p. 20.
[54] Deutsche medicinische Wochenschrift, 1901, vol. xxvii., p. 349.
[55] Pathologie des Stoffwechsels (von Noorden), second edition, 1907, vol. ii., p. 220.
[56] Ibid., p. 221.
[57] Loc. cit., sub. 54.

cystinuric destroys cystin given by the mouth so the pentosuric is able to destroy the usual proportion of a dose of l. arabinose when so given. From this it is clear that, even if any treatment of pentosuria is called for, no good purpose is served by restriction of the carbohydrate nor even of the pentose intake.

Neither is there conclusive evidence that the excretion of pentose, as estimated by reduction tests, is influenced by protein food. However, Kj. O. af Klerker, in the course of an elaborate series of observations, found that the output varied much during the day and that the pentose curve from hour to hour followed that of nitrogen. One of his Jewish patients excreted only half his average output of pentose during the day following the rigid fast of the Day of Atonement, to which an abnormally low nitrogen on the same day also bore witness. Bial and Blumenthal found that the blood of a pentosuric patient gave the orcin reaction, and concluded that pentose was present therein. Thus a renal origin is apparently excluded, and their patient also reacted in the ordinary manner to the phlorhizin test of renal efficiency.

The conclusion appears to be inevitable that the racemic arabinose of pentose urines is derived from some substance formed within the organism, and that glucose is not this parent substance. C. Neuberg has suggested [58] that the most likely parent substance is d. galactose, but himself points out that the evidence in favour of this hypothesis is purely circumstantial, and that no direct proof of its correctness is forthcoming.

By a series of changes with glycuronic acid as an intermediate stage, the pentose of nucleo-proteins, l. xylose, can be derived from d. glucose, and l. arabinose stands in a like relation to d. galactose. Moreover, Neuberg points out that on account of the symmetry of the groupings upon its four middle carbon atoms d. galactose is readily converted into optically inactive derivatives, mucic acid by oxidation, and dulcite by reduction. By converse treatment these products are reconverted into galactose but into the racemic form. On this account Neuberg finds it easier to suppose

that this particular sugar may be the parent substance of the racemic arabinose of pentosuria. The formation of d. galactose in the animal body is an undoubted fact; lactose, the disaccharid formed from glucose and galactose, is abundantly formed during lactation, and galactose has also been shown by Thierfelder [59] to be the sugar yielded by cerebrin.

So far no conclusive evidence is forthcoming that galactose administered as such by the mouth, or in lactose, has any appreciable effect upon the output of arabinose by a pentosuric, nor are there any recorded observations upon a female pentosuric during lactation. Blumenthal and Bial gave to their patient 100 grammes of galactose by the mouth, not with any idea of testing Neuberg's hypothesis, which had not then been put forward, but observed no conspicuous increase of urinary pentose. Kj. O. af Klerker, [60] who gave 100 grammes of lactose to one of his patients, observed a distinct increase of the hourly output of pentose six or seven hours later, but the total day's excretion was in no way excessive. Tintemann [61] observed a slight increase of pentose in the urine after 50 grammes of galactose given on an empty stomach.

Obviously much further patient research is necessary before our knowledge of this remarkable anomaly of metabolism can be placed upon a satisfactory footing, and we may hope that to this research British workers will, in due course, contribute their quota.

And now the time allotted to me is exhausted, and it only remains to thank you, Mr. President, and my other hearers, for the kind attention accorded to me. The learned Dr. Croon, in memory of whom these lectures were established, placed upon the title-page of his tract, "De Ratione Motus Musculorum," some words of Aristotle, "ἐν πᾶσι τοῖς φυσικοῖς ἔνεστί τι θαυμαστόν." [62] These words I would borrow as a motto for the present course; for if it be true that in every phenomenon of nature there is something of the marvellous, surely that factor is nowhere more in evidence than in the workings of the metabolic processes in living things.

58 Ergebnisse der Physiologie, 1904, vol. iii., 1 Abtheilung, p. 426.

59 Zeitschrift für physiologische Chemie, 1890, vol. xiv., p. 209.
60 Loc. cit., sub. 12.
61 Zeitschrift für klinische Medicin, 1906, vol. lvii., p. 190.
62 Περὶ Ζώων Μορίων, I. 5.

17

Reprinted from *Science* 110:543–548 (1949)

Sickle Cell Anemia, a Molecular Disease[1]

Linus Pauling, Harvey A. Itano,[2] S. J. Singer,[2] and Ibert C. Wells[3]

Gates and Crellin Laboratories of Chemistry,
California Institute of Technology, Pasadena, California[4]

THE ERYTHROCYTES of certain individuals possess the capacity to undergo reversible changes in shape in response to changes in the partial pressure of oxygen. When the oxygen pressure is lowered, these cells change their forms from the normal biconcave disk to crescent, holly wreath, and other forms. This process is known as sickling. About 8 percent of American Negroes possess this characteristic; usually they exhibit no pathological consequences ascribable to it. These people are said to have sicklemia, or sickle cell trait. However, about 1 in 40 (4) of these individuals whose cells are capable of sickling suffer from a severe chronic anemia resulting from excessive destruction of their erythrocytes; the term sickle cell anemia is applied to their condition.

The main observable difference between the erythrocytes of sickle cell trait and sickle cell anemia has been that a considerably greater reduction in the partial pressure of oxygen is required for a major fraction of the trait cells to sickle than for the anemia cells (11). Tests *in vivo* have demonstrated that between 30 and 60 percent of the erythrocytes in the venous circulation of sickle cell anemic individuals, but less than 1 percent of those in the venous circulation of sicklemic individuals, are normally sickled. Experiments *in vitro* indicate that under sufficiently low oxygen pressure, however, all the cells of both types assume the sickled form.

The evidence available at the time that our investigation was begun indicated that the process of sickling might be intimately associated with the state and the nature of the hemoglobin within the erythrocyte. Sickle cell erythrocytes in which the hemoglobin is combined with oxygen or carbon monoxide have the biconcave disk contour and are indistinguishable in

that form from normal erythrocytes. In this condition they are termed promeniscocytes. The hemoglobin appears to be uniformly distributed and randomly oriented within normal cells and promeniscocytes, and no birefringence is observed. Both types of cells are very flexible. If the oxygen or carbon monoxide is removed, however, transforming the hemoglobin to the uncombined state, the promeniscocytes undergo sickling. The hemoglobin within the sickled cells appears to aggregate into one or more foci, and the cell membranes collapse. The cells become birefringent (11) and quite rigid. The addition of oxygen or carbon monoxide to these cells reverses these phenomena. Thus the physical effects just described depend on the state of combination of the hemoglobin, and only secondarily, if at all, on the cell membrane. This conclusion is supported by the observation that sickled cells when lysed with water produce discoidal, rather than sickle-shaped, ghosts (10).

It was decided, therefore, to examine the physical and chemical properties of the hemoglobins of individuals with sicklemia and sickle cell anemia, and to compare them with the hemoglobin of normal individuals to determine whether any significant differences might be observed.

EXPERIMENTAL METHODS

The experimental work reported in this paper deals largely with an electrophoretic study of these hemoglobins. In the first phase of the investigation, which concerned the comparison of normal and sickle cell anemia hemoglobins, three types of experiments were performed: 1) with carbonmonoxyhemoglobins; 2) with uncombined ferrohemoglobins in the presence of dithionite ion, to prevent oxidation to methemoglobins; and 3) with carbonmonoxyhemoglobins in the presence of dithionite ion. The experiments of type 3 were performed and compared with those of type 1 in order to ascertain whether the dithionite ion itself causes any specific electrophoretic effect.

Samples of blood were obtained from sickle cell anemic individuals who had not been transfused within three months prior to the time of sampling. Stroma-free concentrated solutions of human adult hemoglobin were prepared by the method used by Drabkin (3). These solutions were diluted just before use with the

[1] This research was carried out with the aid of a grant from the United States Public Health Service. The authors are grateful to Professor Ray D. Owen, of the Biology Division of this Institute, for his helpful suggestions. We are indebted to Dr. Edward R. Evans, of Pasadena, Dr. Travis Winsor, of Los Angeles, and Dr. G. E. Burch, of the Tulane University School of Medicine, New Orleans, for their aid in obtaining the blood used in these experiments.
[2] U. S. Public Health Service postdoctoral fellow of the National Institutes of Health.
[3] Postdoctoral fellow of the Division of Medical Sciences of the National Research Council.
[4] Contribution No. 1333.

appropriate buffer until the hemoglobin concentrations were close to 0.5 grams per 100 milliliters, and then were dialyzed against large volumes of these buffers for 12 to 24 hours at 4° C. The buffers for the experiments of types 2 and 3 were prepared by adding 300 ml of 0.1 ionic strength sodium dithionite solution to 3.5 liters of 0.1 ionic strength buffer. About 100 ml of 0.1 molar NaOH was then added to bring the pH of the buffer back to its original value. Ferrohemoglobin solutions were prepared by diluting the

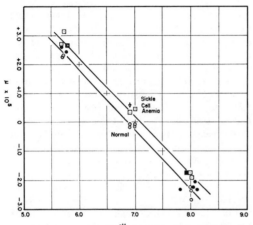

FIG. 1.   Mobility(μ)-pH curves for carbonmonoxyhemoglobins in phosphate buffers of 0.1 ionic strength. The black circles and black squares denote the data for experiments performed with buffers containing dithionite ion. The open square designated by the arrow represents an average value of 10 experiments on the hemoglobin of different individuals with sickle cell anemia. The mobilities recorded in this graph are averages of the mobilities in the ascending and descending limbs.

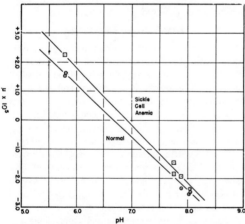

FIG. 2.   Mobility(μ)-pH curves for ferrohemoglobins in phosphate buffers of 0.1 ionic strength containing dithionite ion. The mobilities recorded in the graph are averages of the mobilities in the ascending and descending limbs.

concentrated solutions with this dithionite-containing buffer and dialyzing against it under a nitrogen atmosphere. The hemoglobin solutions for the experiments of type 3 were made up similarly, except that they were saturated with carbon monoxide after dilution and were dialyzed under a carbon monoxide atmosphere. The dialysis bags were kept in continuous motion in the buffers by means of a stirrer with a mercury seal to prevent the escape of the nitrogen and carbon monoxide gases.

The experiments were carried out in the modified Tiselius electrophoresis apparatus described by Swingle (*14*). Potential gradients of 4.8 to 8.4 volts per centimeter were employed, and the duration of the runs varied from 6 to 20 hours. The pH values of the buffers were measured after dialysis on samples which had come to room temperature.

RESULTS

The results indicate that a significant difference exists between the electrophoretic mobilities of hemoglobin derived from erythrocytes of normal individuals and from those of sickle cell anemic individuals. The two types of hemoglobin are particularly easily distinguished as the carbonmonoxy compounds at pH 6.9 in phosphate buffer of 0.1 ionic strength. In this buffer the sickle cell anemia carbonmonoxyhemoglobin moves as a positive ion, while the normal compound moves as a negative ion, and there is no detectable amount of one type present in the other.[4] The hemoglobin derived from erythrocytes of individuals with sicklemia, however, appears to be a mixture of the normal hemoglobin and sickle cell anemia hemoglobin in roughly equal proportions. Up to the present time the hemoglobins of 15 persons with sickle cell anemia, 8 persons with sicklemia, and 7 normal adults have been examined. The hemoglobins of normal adult white and negro individuals were found to be indistinguishable.

The mobility data obtained in phosphate buffers of 0.1 ionic strength and various values of pH are summarized in Figs. 1 and 2.[5]

Occasionally small amounts (less than 5 percent of the total protein) of material with mobilities different from that of either kind of hemoglobin were observed in these uncrystallized hemoglobin preparations. According to the observations of Stern, Reiner, and Silber (*12*) a small amount of a component with a mobility smaller than that of ox hemoglobin is present in human erythrocyte hemolyzates.

[5] The results obtained with carbonmonoxyhemoglobins with and without dithionite ion in the buffers indicate that the dithionite ion plays no significant role in the electrophoretic properties of the proteins. It is therefore of interest that ferrohemoglobin was found to have a lower isoelectric point in phosphate buffer than carbonmonoxyhemoglobin. Titration studies have indicated (*5, 6*) that oxyhemoglobin (similar in electrophoretic properties to the carbonmonoxy compound) has a lower isoelectric point than ferrohemoglobin in

The isoelectric points are listed in Table 1. These results prove that the electrophoretic difference between normal hemoglobin and sickle cell anemia hemoglobin

TABLE 1
ISOELECTRIC POINTS IN PHOSPHATE BUFFER, $\mu = 0.1$

Compound	Normal	Sickle cell anemia	Difference
Carbonmonoxyhemoglobin	6.87	7.09	0.22
Ferrohemoglobin	6.87	7.09	0.22

exists in both ferrohemoglobin and carbonmonoxyhemoglobin. We have also performed several experiments in a buffer of 0.1 ionic strength and pH 6.52 containing 0.08 M NaCl, 0.02 M sodium cacodylate, and 0.0083 M cacodylic acid. In this buffer the average mobility of sickle cell anemia carbonmonoxyhemoglobin is 2.63×10^{-5}, and that of normal carbonmonoxyhemoglobin is 2.23×10^{-5} cm/sec per volt/cm.[6]

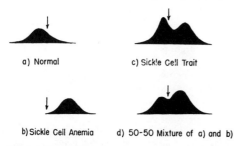

a) Normal

b) Sickle Cell Anemia

c) Sickle Cell Trait

d) 50-50 Mixture of a) and b)

FIG. 3. Longsworth scanning diagrams of carbonmonoxyhemoglobins in phosphate buffer of 0.1 ionic strength and pH 6.90 taken after 20 hours' electrophoresis at a potential gradient of 4.73 volts/cm.

These experiments with a buffer quite different from phosphate buffer demonstrate that the difference between the hemoglobins is essentially independent of the buffer ions.

Typical Longsworth scanning diagrams of experiments with normal, sickle cell anemia, and sicklemia carbonmonoxyhemoglobins, and with a mixture of the first two compounds, all in phosphate buffer of pH 6.90 and ionic strength 0.1, are reproduced in Fig. 3. It is apparent from this figure that the sicklemia material contains less than 50 percent of the anemia component. In order to determine this quantity accurately some experiments at a total protein concentra-

the absence of other ions. These results might be reconciled by assuming that the ferrous iron of ferrohemoglobin forms complexes with phosphate ions which cannot be formed when the iron is combined with oxygen or carbon monoxide. We propose to continue the study of this phenomenon.

[6] The mobility data show that in 0.1 ionic strength cacodylate buffers the isoelectric points of the hemoglobins are increased about 0.5 pH unit over their values in 0.1 ionic strength phosphate buffers. This effect is similar to that observed by Longsworth in his study of ovalbumin (7).

tion of 1 percent were performed with known mixtures of sickle cell anemia and normal carbonmonoxyhemoglobins in the cacodylate-sodium chloride buffer of 0.1 ionic strength and pH 6.52 described above. This buffer was chosen in order to minimize the anomalous electrophoretic effects observed in phosphate buffers (7). Since the two hemoglobins were incompletely resolved after 15 hours of electrophoresis under a potential gradient of 2.79 volts/cm, the method of Tiselius and Kabat (16) was employed to allocate the

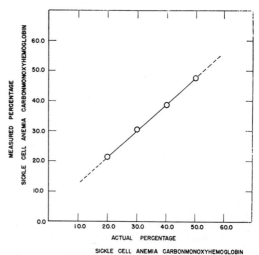

FIG. 4. The determination of the percent of sickle cell anemia carbonmonoxyhemoglobin in known mixtures of the protein with normal carbonmonoxyhemoglobin by means of electrophoretic analysis. The experiments were performed in a cacodylate sodium chloride buffer described in the text.

areas under the peaks in the electrophoresis diagrams to the two components. In Fig. 4 there is plotted the percent of the anemia component calculated from the areas so obtained against the percent of that component in the known mixtures. Similar experiments were performed with a solution in which the hemoglobins of 5 sicklemic individuals were pooled. The relative concentrations of the two hemoglobins were calculated from the electrophoresis diagrams, and the actual proportions were then determined from the plot of Fig. 4. A value of 39 percent for the amount of the sickle cell anemia component in the sicklemia hemoglobin was arrived at in this manner. From the experiments we have performed thus far it appears that this value does not vary greatly from one sicklemic individual to another, but a more extensive study of this point is required.

Up to this stage we have assumed that one of the two components of sicklemia hemoglobin is identical with sickle cell anemia hemoglobin and the other is identical with the normal compound. Aside from the genetic evidence which makes this assumption very probable (see the discussion section), electrophoresis experiments afford direct evidence that the assumption is valid. The experiments on the pooled sicklemia carbonmonoxyhemoglobin and the mixture containing 40 percent sickle cell anemia carbonmonoxyhemoglobin and 60 percent normal carbonmonoxyhemoglobin in the cacodylate-sodium chloride buffer described above were compared, and it was found that the mobilities of the respective components were essentially identical.[7] Furthermore, we have performed experiments in which normal hemoglobin was added to a sicklemia preparation and the mixture was then subjected to electrophoretic analysis. Upon examining the Longsworth scanning diagrams we found that the area under the peak corresponding to the normal component had increased by the amount expected, and that no indication of a new component could be discerned. Similar experiments on mixtures of sickle cell anemia hemoglobin and sicklemia preparations yielded similar results. These sensitive tests reveal that, at least electrophoretically, the two components in sicklemia hemoglobin are identifiable with sickle cell anemia hemoglobin and normal hemoglobin.

DISCUSSION

1) *On the Nature of the Difference between Sickle Cell Anemia Hemoglobin and Normal Hemoglobin*: Having found that the electrophoretic mobilities of sickle cell anemia hemoglobin and normal hemoglobin differ, we are left with the considerable problem of locating the cause of the difference. It is impossible to ascribe the difference to dissimilarities in the particle weights or shapes of the two hemoglobins in solution: a purely frictional effect would cause one species to move more slowly than the other throughout the entire pH range and would not produce a shift in the isoelectric point. Moreover, preliminary velocity ultracentrifuge[8] and free diffusion measurements indicate that the two hemoglobins have the same sedimentation and diffusion constants.

The most plausible hypothesis is that there is a difference in the number or kind of ionizable groups in the two hemoglobins. Let us assume that the only groups capable of forming ions which are present in carbonmonoxyhemoglobin are the carboxyl groups in the heme, and the carboxyl, imidazole, amino, phenolic hydroxyl, and guanidino groups in the globin. The number of ions nonspecifically adsorbed on the two proteins should be the same for the two hemoglobins

[7] The patterns were very slightly different in that the known mixture contained 1 percent more of the sickle cell anemia component than did the sickle cell trait material.

[8] We are indebted to Dr. M. Moskowitz, of the Chemistry Department, University of California at Berkeley, for performing the ultracentrifuge experiments for us.

under comparable conditions, and they may be neglected for our purposes. Our experiments indicate that the net number of positive charges (the total number of cationic groups minus the number of anionic groups) is greater for sickle cell anemia hemoglobin than for normal hemoglobin in the pH region near their isoelectric points.

According to titration data obtained by us, the acid-base titration curve of normal human carbonmonoxyhemoglobin is nearly linear in the neighborhood of the isoelectric point of the protein, and a change of one pH unit in the hemoglobin solution in this region is associated with a change in net charge on the hemoglobin molecule of about 13 charges per molecule. The same value was obtained by German and Wyman (5) with horse oxyhemoglobin. The difference in isoelectric points of the two hemoglobins under the conditions of our experiments is 0.23 for ferrohemoglobin and 0.22 for the carbonmonoxy compound. This difference corresponds to about 3 charges per molecule. With consideration of our experimental error, sickle cell anemia hemoglobin therefore has 2–4 more net positive charges per molecule than normal hemoglobin.

Studies have been initiated to elucidate the nature of this charge difference more precisely. Samples of porphyrin dimethyl esters have been prepared from normal hemoglobin and sickle cell anemia hemoglobin. These samples were shown to be identical by their x-ray powder photographs and by identity of their melting points and mixed melting point. A sample made from sicklemia hemoglobin was also found to have the same melting point. It is accordingly probable that normal and sickle cell anemia hemoglobin have different globins. Titration studies and amino acid analyses on the hemoglobins are also in progress.

2) *On the Nature of the Sickling Process*: In the introductory paragraphs we outlined the evidence which suggested that the hemoglobins in sickle cell anemia and sicklemia erythrocytes might be responsible for the sickling process. The fact that the hemoglobins in these cells have now been found to be different from that present in normal red blood cells makes it appear very probable that this is indeed so.

We can picture the mechanism of the sickling process in the following way. It is likely that it is the globins rather than the hemes of the two hemoglobins that are different. Let us propose that there is a surface region on the globin of the sickle cell anemia, hemoglobin molecule which is absent in the normal molecule and which has a configuration complementary to a different region of the surface of the hemoglobin molecule. This situation would be somewhat analogous to that which very probably exists in antigen-antibody reactions (9). The fact that sickling occurs only when the partial pressures of oxygen and carbon monoxide are low suggests that one of these sites is very near to the iron atom of one or more of the hemes, and that when the iron atom is combined with either one of these gases, the complementariness of the two structures is considerably diminished. Under the appropriate conditions, then, the sickle cell anemia hemoglobin molecules might be capable of interacting with one another at these sites sufficiently to cause at least a partial alignment of the molecules within the cell, resulting in the erythrocyte's becoming birefringent, and the cell membrane's being distorted to accommodate the now relatively rigid structures within its confines. The addition of oxygen or carbon monoxide to the cell might reverse these effects by disrupting some of the weak bonds between the hemoglobin molecules in favor of the bonds formed between gas molecules and iron atoms of the hemes.

Since all sicklemia erythrocytes behave more or less similarly, and all sickle at a sufficiently low oxygen pressure (*11*), it appears quite certain that normal hemoglobin and sickle cell anemia hemoglobin coexist within each sicklemia cell; otherwise there would be a mixture of normal and sickle cell anemia erythrocytes in sicklemia blood. We might expect that the normal hemoglobin molecules, lacking at least one type of complementary site present on the sickle cell anemia molecules, and so being incapable of entering into the chains or three-dimensional frameworks formed by the latter, would interfere with the alignment of these molecules within the sicklemia erythrocyte. Lower oxygen pressures, freeing more of the complementary sites near the hemes, might be required before sufficiently large aggregates of sickle cell anemia hemoglobin molecules could form to cause sickling of the erythrocytes.

This is in accord with the observations of Sherman (*11*), which were mentioned in the introduction, that a large proportion of erythrocytes in the venous circulation of persons with sickle cell anemia are sickled, but that very few have assumed the sickle forms in the venous circulation of individuals with sicklemia. Presumably, then, the sickled cells in the blood of persons with sickle cell anemia cause thromboses, and their increased fragility exposes them to the action of reticulo-endothelial cells which break them down, resulting in the anemia (*1*).

It appears, therefore, that while some of the details of this picture of the sickling process are as yet conjectural, the proposed mechanism is consistent with experimental observations at hand and offers a chemical and physical basis for many of them. Furthermore, if it is correct, it supplies a direct link between the existence of "defective" hemoglobin molecules and the pathological consequences of sickle cell disease.

3) *On the Genetics of Sickle Cell Disease*: A genetic basis for the capacity of erythrocytes to sickle was recognized early in the study of this disease (*4*). Taliaferro and Huck (*15*) suggested that a single dominant gene was involved, but the distinction between sicklemia and sickle cell anemia was not clearly understood at the time. The literature contains conflicting statements concerning the nature of the genetic mechanisms involved, but recently Neel (*8*) has reported an investigation which strongly indicates that the gene responsible for the sickling characteristic is in heterozygous condition in individuals with sicklemia, and homozygous in those with sickle cell anemia.

Our results had caused us to draw this inference before Neel's paper was published. The existence of normal hemoglobin and sickle cell anemia hemoglobin in roughly equal proportions in sicklemia hemoglobin preparations is obviously in complete accord with this hypothesis. In fact, if the mechanism proposed above to account for the sickling process is correct, we can identify the gene responsible for the sickling process with one of an alternative pair of alleles capable through some series of reactions of introducing the modification into the hemoglobin molecule that distinguishes sickle cell anemia hemoglobin from the normal protein.

The results of our investigation are compatible with a direct quantitative effect of this gene pair; in the chromosomes of a single nucleus of a normal adult somatic cell there is a complete absence of the sickle cell gene, while two doses of its allele are present; in the sicklemia somatic cell there exists one dose of each allele; and in the sickle cell anemia somatic cell there are two doses of the sickle cell gene, and a complete absence of its normal allele. Correspondingly, the erythrocytes of these individuals contain 100 percent normal hemoglobin, 40 percent sickle cell anemia hemoglobin and 60 percent normal hemoglobin, and 100 percent sickle cell anemia hemoglobin, respectively. This investigation reveals, therefore, a clear case of a change produced in a protein molecule by an allelic change in a single gene involved in synthesis.

The fact that sicklemia erythrocytes contain the two hemoglobins in the ratio 40:60 rather than 50:50 might be accounted for by a number of hypothetical schemes. For example, the two genes might compete for a common substrate in the synthesis of two different enzymes essential to the production of the two different hemoglobins. In this reaction, the sickle cell gene would be less efficient than its normal allele. Or, competition for a common substrate might occur at some later stage in the series of reactions leading to the synthesis of the two hemoglobins. Mechanisms of this sort are discussed in more elaborate detail by Stern (*13*).

The results obtained in the present study suggest that the erythrocytes of other hereditary hemolytic anemias be examined for the presence of abnormal hemoglobins. This we propose to do.

Based on a paper presented at the meeting of the National Academy of Sciences in Washington, D. C., in April, 1949, and at the meeting of the American Society of Biological Chemists in Detroit in April, 1949.

References

1. BOYD, W.   *Textbook of pathology.*   (3rd Ed.)   Philadelphia : Lea and Febiger, 1938.   P. 864.
2. DIGGS, L. W., AHMANN, C. F., and BIBB, J.   *Ann. int. Med.*, 1933, **7**, 769.
3. DRABKIN, D. L.   *J. biol. Chem.*, 1946, **164**, 703.
4. EMMEL, V. E.   *Arch. int. Med.*, 1917, **20**, 586.
5. GERMAN, B. and WYMAN, J., JR.   *J. biol. Chem.*, 1937, **117**, 533.
6. HASTINGS, A. B. *et al.*   *J. biol. Chem.*, 1924, **60**, 89.
7. LONGSWORTH, L. G.   *Ann. N. Y. Acad. Sci.*, 1941, **41**, 267.
8. NEEL, J. V.   *Science*, 1949, **110**, 64.
9. PAULING, L., PRESSMAN, D., and CAMPBELL, D. H.   *Physiol. Rev.*, 1943, **23**, 203.
10. PONDER, E.   *Ann. N. Y. Acad. Sci.*, 1947, **43**, 579.
11. SHERMAN, I. J.   *Bull. Johns Hopk. Hosp.*, 1940, **67**, 309.
12. STERN, K. G., REINER, M. and SILBER, R. H.   *J. biol. Chem.*, 1945, **161**, 731.
13. STERN, C.   *Science*, 1948, **108**, 615.
14. SWINGLE, S. M.   *Rev. sci. Inst.*, 1947, **18**, 128.
15. TALIAFERRO, W. H. and HUCK, J. G.   *Genetics*, 1923, **8**, 594.
16. TISELIUS, A. and KABAT, E.   *J. exp. Med.*, 1939, **69**, 119.

18

Reprinted from *J. Biol. Chem.* **199**:661–667 (1952)

GLUCOSE-6-PHOSPHATASE OF THE LIVER IN GLYCOGEN STORAGE DISEASE

By GERTY T. CORI AND CARL F. CORI

(*From the Department of Biological Chemistry, Washington University School of Medicine, St. Louis, Missouri*)

(Received for publication, June 27, 1952)

In the preceding paper (1) it was shown that the structure of liver glycogen in ten cases of von Gierke's disease was within the normal range of variations in eight and definitely abnormal in two cases. That the liver glycogen which accumulates in excessive amounts is not readily available for blood sugar formation is indicated by the tendency to hypoglycemia, the frequently flat blood sugar curve after injection of epinephrine, and the persistence of the glycogen post mortem.

The following considerations led to an investigation of the glucose-6-phosphatase activity of the liver. The enzymes required for the conversion of glycogen to glucose in the liver are (1) phosphorylase, (2) amylo-1,6-glucosidase, (3) phosphoglucomutase, and (4) glucose-6-phosphatase. For glycogen synthesis from glucose there is needed, in addition to (1) and (3), hexokinase and branching enzyme. Therefore, if the hepatic type of glycogen storage disease were due to an enzyme deficiency, (2) and (4) might be implicated. It is shown in this paper that in two fatal cases there was an almost complete lack of the specific glucose-6-phosphatase of the liver.

EXPERIMENTAL

Livers obtained at autopsy were frozen within 4 hours after death, while samples obtained at biopsy were frozen in the operating room. In no case were the samples allowed to thaw before use. When shipped, they were packed in dry ice and they arrived frozen. The livers could be kept for several months at $-18°$ without significant change in phosphatase activity.

For the experiments small pieces (0.3 to 4 gm.) were chiseled off the frozen livers, weighed, and homogenized immediately in ice-cold water in a cold stainless steel homogenizer, or ground in a cold mortar without sand. For 1 part of liver, 2 parts of water were used. The homogenates were filtered through gauze to remove small amounts of fibrous tissue, yielding a filtrate which could be pipetted without difficulty. Several dilutions of the same homogenate were incubated with glucose-6-phosphate in order to test proportionality to enzyme concentration. In some, but not all cases less enzyme had more than the proportional effect. The composition of the

243

reaction mixtures is given in Fig. 1 and Table I. Initial and incubated samples were deproteinized with trichloroacetic acid and the filtrates analyzed for inorganic phosphate by the method of Fiske and Subbarow (2). Each experiment included samples incubated without the addition of glucose-6-phosphate. The amount of inorganic phosphate found after incubation without substrate was deducted from that formed in the presence of substrate. In all instances only small amounts of inorganic P were formed on incubation without substrate.

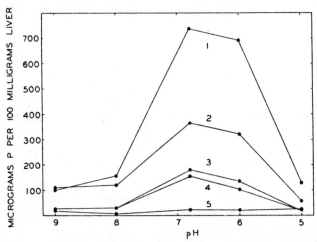

FIG. 1. Action of liver phosphatases on glucose-6-phosphate at different pH levels. The experimental conditions were the same as in Table I, except that 0.1 M Veronal was used at pH 8 and 9 and 0.1 M acetate at pH 5. Each point represents an average of two to six experiments. Curve 1, normal liver from a 4 month-old infant, accidental death (P. J.); Curve 2, cirrhotic liver, autopsy (R. S.); Curve 3, von Gierke's disease, biopsy (C. H.); Curve 4, von Gierke's disease, biopsy (Dennis B.); Curve 5, von Gierke's disease, autopsy (P. S.).

Glucose-6-phosphatase—In order to establish the range of variation in the activity of this enzyme, homogenates were prepared from normal livers as well as from livers with pathological lesions other than storage disease. The results obtained at different pH are summarized in Fig. 1. In normal liver (Curve 1) and a cirrhotic liver (Curve 2) the activity in the pH range 6 to 7 with glucose-6-phosphate as substrate is much greater than at pH 5 or 9, the optimal pH range of the (unspecific) acid and alkaline phosphatases, respectively. These curves show that the human glucose-6-phosphatase, like that of other species (3–5), has its pH optimum near 6.8; they also show that the activity of the specific phosphatase exceeds many times that of the acid or alkaline phosphatases. Curves 3, 4, and 5 were obtained with homogenates from 3 cases of von Gierke's disease, the former two relatively mild and the latter fatal at an early age. There was practically no phosphatase activity at any pH value in the severe case, and a marked reduction in the two mild cases.

It seemed possible that the homogenate obtained from the severe case contained an inhibitor, and, in order to test this, it was mixed with the homogenate from normal liver and incubated at different pH values with glucose-6-phosphate. No significant inhibition could be detected over the pH range 5 to 9. The low activity in the severe case was apparently not due to a lack of cofactors, since addition of a boiled juice prepared from normal liver had no effect.

Although the materials from the three autopsy cases in Fig. 1 were all frozen within the same length of time, there was the danger that a more rapid postmortem change in enzyme activity occurred in the case with von Gierke's disease than in the control cases. It was possible, however, to determine phosphatase activity in a biopsy and autopsy specimen from the same patient (M. A. S., Table I). In both specimens, secured 12 days apart, the glucose-6-phosphatase activity measured at pH 6.8 was practically absent. In Table I is recorded an experiment in which in a mixture with a homogenate from another liver there was no significant inhibition of the glucose-6-phosphatase activity.

Table I also includes measurements on four other patients with von Gierke's disease, all of them living and with relatively mild clinical symptoms. In two of these the glucose-6-phosphatase activity was within the range found in other liver diseases or even in normal liver. In case C. H., Table I, the glycogen of liver and muscle was found to differ from normal glycogen by having very short outer chains, and it was pointed out that a deficiency of amylo-1,6-glucosidase might result in this type of structural abnormality (1). The liver material available for biopsy permitted only glucose-6-phosphatase activity to be measured, and it may be seen that the value obtained was below normal.

Acid and Alkaline Phosphatases—In Fig. 1 it would appear as if there were also a deficiency in acid and alkaline phosphatases in cases of von Gierke's disease. It seemed possible, however, in view of the strong activity of glucose-6-phosphatase in normal liver, that there was some residual activity of this enzyme left even at pH 5 and 9. In this case the reduction in activity at these pH values in Curves 3, 4, and 5 could be due in part to the loss of specific phosphatase rather than to a decrease in acid and alkaline phosphatases. In order to test this point, the homogenates from liver shown in Curves 1, 4, and 5 (Fig. 1) were incubated for 1 hour at 30° with 0.03 M sodium glycerophosphate (Mallinckrodt). At pH 5 the amounts of inorganic P liberated (in micrograms per 100 mg. of liver) were 135, 127, and 122, respectively, while at pH 9 the values were 48, 17, and

39. In view of the unspecific action of these phosphatases, the amount of inorganic P liberated on incubation without substrate has not been deducted. At pH 5 this makes little difference, while at pH 9 the values (after deduction) were 39, 7, and 26, respectively. In normal liver there was considerably more hydrolysis of glycerophosphate in the pH range 6 to 8 than in two cases of von Gierke's disease. Thannhauser *et al.* (6) had

TABLE I

Glucose-6-Phosphatase Activity in Liver Homogenates

The liver specimens were obtained by biopsy except as noted. The reaction mixture contained, in 1 ml., 0.3 ml. of 0.1 M K citrate, pH 6.8, 0.5 ml. of 0.01 M K glucose-6-phosphate, pH 6.8, or 0.5 ml. of H_2O, and 0.2 ml. of homogenate (equivalent to 4 to 33 mg. of liver). After 1 hour at 30°, 1 ml. of 10 per cent trichloroacetic acid was added and inorganic P determined in 1 ml. of filtrate. The values recorded represent the difference in inorganic P formed during incubation with and without glucose-6-P.

Case	No. of experiments	P liberated from glucose-6-P by 100 mg. liver	Remarks
		γ	
K. M.	2	273	10 yrs. old; abdominal tumor
D. S.	4	327	8 yrs. old; liver cirrhosis
C. S.	2	429	1.5 mos. old; normal liver
M. A. S.	2	22	Von Gierke's disease,* Case 2
P. S.†	4	23	" " " 1
Dennis B.	2	155	" " " 5
C. H.	5	180	" " " 9
S. S.	3	282	" " " 8
D. B.	2	402	" " " 7
D. S.		312	See text for explanation
M. A. S.†		15	
D. S. + M. A. S.		294	

* The history of these cases is given in Table II of the preceding paper (1).
† Autopsy specimen.

previously reported that the alkaline phosphatase was decreased in von Gierke's disease. At any rate, Curve 5, Fig. 1, shows that the action of the unspecific phosphatases on glucose-6-phosphate is too weak to contribute significant amounts of glucose to the blood in severe cases of the disease.

Phosphoglucomutase—This enzyme was measured in the three control cases in Table I and in one patient with von Gierke's disease (M. A. S., biopsy sample). The reaction mixture was the same as in Table I, except that glucose-1-phosphate was substituted for glucose-6-phosphate and that the solution also contained 0.003 M cysteine and 0.01 M $MgCl_2$ to insure maximal activity of the mutase. Inorganic P and easily hydrolyzable P were determined before and after incubation, with and without addition of glucose-1-phosphate. From the values obtained could be calculated how much glucose-1-phosphate had been converted to glucose-6-phosphate and how much of the latter had been hydrolyzed to glucose and inorganic phosphate.

It was found that the amount of inorganic P formed from glucose-1-phosphate was practically the same as that formed in a parallel experiment in which glucose-6-phosphate was added as substrate. This indicates that the phosphoglucomutase reaction was not limiting the rate of glucose-6-phosphatase. To give two examples, the homogenate from case C. S., Table I, formed (per 100 mg. of liver per hour) 440 γ of inorganic P from glucose-1-phosphate, as compared to 429 γ from glucose-6-phosphate. In case M. A. S. the amount of inorganic P formed from glucose-1-phosphate was 18 γ, similar to the value of 22 γ reported in Table I with glucose-6-phosphate as substrate. In both cases some of the glucose-1-phosphate which disappeared was found as glucose-6-phosphate at the end of incubation, but the amounts were strikingly different. In the subject with normal phosphatase activity there were present 29 γ of glucose-6-phosphate per ml. of reaction mixture, while in the one with deficient phosphatase 102 γ were present. The effect of concentration of glucose-6-phosphate on the breakdown of glycogen is discussed below.

DISCUSSION

The sequence of reactions, glucose → glucose-6-P $\rightleftharpoons$ glucose-1-P $\rightleftharpoons$ glycogen, is obviously not disturbed in von Gierke's disease. Glucose-6-phosphate can also be formed from non-carbohydrate sources (lactic acid, glycerol, amino acids, etc.). Blood sugar formation in the liver, with lactic acid as an example of a non-carbohydrate source, can then be illustrated as follows:[1]

$$\begin{array}{c} \text{Glycogen} \\ \Updownarrow \\ \text{Glucose + ATP} \rightarrow \text{glucose-6-phosphate} \rightleftharpoons \text{lactic acid} \\ \downarrow \\ \text{Glucose + phosphate} \end{array}$$

If glucose-6-phosphatase activity were deficient, the over-all equilibria of these reactions would favor glycogen deposition, and both glycogen and non-carbohydrate sources would be unavailable for blood sugar formation. The main pathway by which glycogen could disappear would be through the chain of glycolytic reactions, glucose-6-P → pyruvate → lactate, or

[1] Where only one arrow is shown, the equilibrium of the reaction is far to one side.

pyruvate $\rightarrow CO_2 + H_2O$. Another pathway for the disposal of glucose-6-phosphate would be its oxidation to 6-phosphogluconate. It would be of interest to investigate these enzyme reactions in von Gierke's disease.

In severe cases which are fatal at an early age the deficiency in glucose-6-phosphatase is so marked that it could be regarded as the primary cause of the symptoms. The mild cases, in which hepatomegaly, low fasting blood sugar, and abnormally high liver glycogen may persist for many years without causing serious clinical symptoms, are more difficult to evaluate. Some of these show a rise in blood sugar after epinephrine injection, while others do not. The glucose-6-phosphatase activity, except in two cases, was not below that found in cirrhosis of the liver, in which there is no accumulation of glycogen. It is impossible to estimate to what levels glucose-6-phosphatase activity has to be reduced before glycogen begins to accumulate, because this will be relative to the rates at which glycogen is formed from glucose and from other sources. The disparity between these rates might be much greater in von Gierke's disease of the liver than, for example, in cirrhosis of the liver, in which all enzymes, including hexokinase, might be expected to show decreased activity.

It is not possible to explain abnormal storage of glycogen in skeletal muscle or heart muscle by a lack of glucose-6-phosphatase, since this enzyme is normally absent from these tissues. Here one would have to look for a different enzymatic lesion, and, should the opportunity arise, it is suggested that the activity of amylo-1,6-glucosidase be measured for reasons that have been mentioned in a previous paper (1).

This investigation was made possible by the generous cooperation of a number of physicians whose names have been given (1) and to whom we wish again to express our appreciation.

SUMMARY

1. The glucose-6-phosphatase activity has been measured in liver homogenates from six cases of von Gierke's disease. In two fatal cases the enzyme was practically absent; this could not be explained by the presence of an inhibitor or by the lack of cofactors. The activity of the acid phosphatase at pH 5 and the alkaline phosphatase at pH 9, with glycerophosphate as substrate, was not appreciably diminished. Owing to the absence of the specific phosphatase which has its optimum at pH 6.8, it could be shown that the unspecific phosphatases had practically no action on glucose-6-phosphate in the physiological pH range. The phosphoglucomutase reaction was not rate-limiting, since glucose-6-phosphate accumulated when glucose-1-phosphate was added to the homogenates.

2. In two out of four surviving cases of von Gierke's disease with mild

symptoms the activity of the specific phosphatase was below the level found in other liver diseases.

3. The pathogenesis of the disease is discussed in terms of a specific enzymatic lesion.

BIBLIOGRAPHY

1. Illingworth, B., and Cori, G. T., *J. Biol. Chem.*, **199**, 653 (1952).
2. Fiske, C. H., and Subbarow, Y., *J. Biol. Chem.*, **66**, 375 (1925).
3. Fantl, P., and Rome, N., *Australian J. Exp. Biol. and Med. Sc.*, **23**, 121 (1945).
4. Swanson, M. A., *J. Biol. Chem.*, **184**, 647 (1950).
5. de Duve, C., Berthet, J., Hers, H. G., and Dupret, L., *Bull. Soc. chim. biol.*, **32**, 20 (1950).
6. Thannhauser, S. J., Sorkin, S. Z., and Boncoddo, N. F., *J. Clin. Invest.*, **19**, 681 (1940).

19

Reprinted from *Am. J. Obst. Gynec.* **42**:925–937 (1941)

THE ROLE OF ISO-IMMUNIZATION IN THE PATHOGENESIS OF ERYTHROBLASTOSIS FETALIS*

PHILIP LEVINE, M.D., NEWARK, N. J., LYMAN BURNHAM, M.D.,
ENGLEWOOD, N. J., E. M. KATZIN, M.D., NEWARK, N. J., AND
PETER VOGEL, M.D., NEW YORK, N. Y.

(From the Division of Laboratories of the Newark Beth Israel Hospital, and the Woman's Hospital of New York)

STUDIES on the cause of intra-group transfusion accidents associated with pregnancy have established the importance of the concept of iso-immunization of the mother by blood factors in the fetus transmitted from the father.[1-5] More recently it was found that the same theory of iso-immunization may serve as the basis for a theory on the pathogenesis of erythroblastosis fetalis, the well-described familial hemolytic disease of the newborn.[3, 4, 6]

The data to be presented indicate that erythroblastosis fetalis results from (1) iso-immunization of the mother by dominant hereditary blood factors in the fetus, as evidenced by the production of immune intra-group agglutinins and (2) the subsequent passage of these maternal agglutinins through the placenta and their continuous action on the susceptible fetal blood. In the great majority of the cases the blood factor involved has been shown to be either identical with or related to the Rh (Rhesus) agglutinogen first described by Landsteiner and Wiener with the aid of rabbit sera prepared by injection of Rhesus blood.[7] In other words, the rabbit anti-Rhesus immune sera and the sera of pregnant women suffering from intra-group transfusion accidents gave almost identical agglutination reactions on all human bloods tested. Accordingly, a pregnant woman whose blood does not contain the Rh factor (Rh–, occurring in about 15 per cent of the general population) if married to an Rh husband (85 per cent in the random population), may produce anti-Rh agglutinins as a result of immunization with the Rh fetal blood. Should these agglutinins penetrate the placenta in suitable concentration they may serve as the source of the intrauterine hemolysis of fetal blood, the characteristic feature of erythroblastosis fetalis.[4]

The term iso-immunization denotes immunization within the same species, i.e., the individual being immunized and source of the antigenic (immunizing) stimulus belong to the same species. It is obvious that patients receiving repeated blood transfusions may be subjected to iso-

*Aided by a grant from the Blood Betterment Association of New York City.

immunization. A prerequisite condition is an antigenic difference in the bloods of recipient and donor. In each instance, this difference is expressed by the presence of a particular blood factor in the donor and its absence in the recipient. Although this condition is frequently satisfied in many cases of repeated transfusions, immune agglutinins are rarely produced because many human blood factors which immunize animals are not antigenic in man. An exception to this rule is the Rh factor which has recently been shown to be responsible for transfusion accidents in Rh– recipients who in the course of several transfusions produced anti-Rh agglutinins.[5, 8, 9] In short, the Rh factor is a good antigen for Rh– mothers as well as for Rh– recipients.

Iso-immunization in pregnancy as the cause of the production of atypical agglutinins responsible for an intra-group transfusion accident was suggested by Levine and Stetson[1] in 1939. In this case the patient (Group O), who harbored a dead fetus for two months, was transfused with her husband's blood (Group O) after the delivery of a macerated fetus (October, 1937). This transfusion was followed by an immediate severe reaction resulting in jaundice, anuria, and ultimate recovery. It was later shown that this patient's blood contained an atypical agglutinin which agglutinated about 80 per cent of Group O bloods. As was to have been expected, this agglutinin gradually diminished in titer and entirely disappeared from the blood at the end of one year. In 1940, or three years after the transfusion accident, it could be shown that the patient was Rh– and her husband Rh+. There is still further evidence that the atypical agglutinin in this patient was anti-Rh in its specificity since the same donors selected as compatible* for subsequent uneventful transfusions in 1937 were later (1940) found to be Rh– in tests with a human anti-Rh serum.

Recent studies have shown that intra-group transfusion accidents previously reported in the literature[5] and additional cases recently observed by Levine, Katzin, and Burnham,[2, 3, 10] occurred frequently in conditions associated with pregnancy. It is characteristic of this group of cases that the accident occurs at the first transfusion. Accordingly, the immune agglutinin which must have been present prior to the transfusion but could not be demonstrated with the usual cross-matching procedure, was probably induced by the pregnancy.

The atypical agglutinins in some of these cases were shown to have the unusual property of greater activity at 37° C. than at 20° C. (low room temperature). For this reason, the term "warm agglutinins" was applied to these antibodies.[10] This observation offers a possible explanation for the frequent failure of the cross-matching tests to detect this form of intra-group incompatibility. It is therefore recommended that in performing the compatibility test, the patient's serum and donor's blood cell suspension be incubated at 37° C. for thirty minutes before the mixture is centrifuged (one minute at 500 r.p.m.) for sedimentation and subsequent resuspension.[4, 11, 12]

The relationship of iso-immunization and certain pathologic states in the pregnant woman or in the fetus was pointed out by Levine and

---

*These tests were carried out by Dr. E. M. Katzin.

Katzin.[2] In an analysis of 12 intra-group transfusion accidents associated with pregnancy in which atypical agglutinins were demonstrated, these authors observed a high incidence of toxemia, spontaneous abortions and miscarriages, and stillbirths in the past obstetric histories. Accordingly, it was assumed that "there does appear to be a correlation of the complications with the incidence of atypical agglutinins, and one can speculate as to their relationship."

This observation made it possible to enlarge the source of the material to be studied instead of limiting it to the comparatively rare transfusion accidents in pregnancy.

Shortly thereafter, one of us (L. B.) observed a patient who suffered from a severe transfusion accident following the delivery of an infant in whom a diagnosis of erythroblastosis fetalis was established. The obstetric history in this case (R. C.)* and in still another case (J. L.) not transfused, were so striking as to suggest a theoretical basis for the pathogenesis of erythroblastosis fetalis.[3, 4] One of these mothers (R. C.) had three pregnancies, the first and third of which resulted in infants with erythroblastosis fetalis and the second pregnancy terminated in a macerated fetus. In the other case (J. L.), there had been 10 pregnancies, the first of which resulted in a normal infant; there were 3 spontaneous abortions and the remaining 6 pregnancies resulted in infants who survived one to three days. In at least 3 of these neonatal deaths there was sufficient evidence to support a diagnosis of erythroblastosis fetalis. This woman's blood was tested in the eighth month of her eleventh pregnancy. Since atypical agglutinins were already present, it was anticipated that the baby to be born would be affected. Actually, the infant suffered from anemia of the newborn, one of the several manifestations of erythroblastosis fetalis.

The two cases mentioned were among the seven patients with 37 pregnancies which formed the basis for the preliminary observation that the pathogenesis of erythroblastosis fetalis depends on iso-immunization of the mother by the fetus. The findings in these 37 pregnancies are reproduced in Table I.

TABLE I. OUTCOME OF 37 PREGNANCIES IN 7 PATIENTS

(Modified After Levine, Katzin and Burnham[3])

Normal babies	10
Babies with erythroblastosis	7
Neonatal deaths	3
Stillbirths	5
Abortions or miscarriages (at least 6 spontaneous)	10
No data available	2

In 6 of these 7 women in whose blood atypical agglutinins were demonstrated, there were indications that the specificity of the antibodies corresponded to the anti-Rh. Obviously, the blood cells of these women

*Another patient of Dr. Burnham (G. B.) who died from a transfusion anuria following the delivery of an infant with fetal hydrops, was mentioned in previous papers.[2, 3, 10] These two cases were the first to show the relationship of intra-group transfusion accidents and erythroblastosis fetalis. A more detailed discussion is given in a paper by Burnham.[13]

did not contain the Rh factor and were therefore Rh–. These considerations suggested a statistical study of the bloods of. women known to have given birth to infants in whom a diagnosis of erythroblastosis fetalis was established. If iso-immunization with the Rh factor plays a significant role in the pathogenesis of this disease, one should expect to find (1) a high incidence of Rh– reactions in this group of selected mothers and (2) a high incidence of anti-Rh agglutinins in their sera.

Whenever possible, the bloods of the fathers and the affected children were also tested, for, if the iso-immunization theory is correct, then 100 per cent of the fathers and the affected children in the series of Rh– mothers should be Rh+.

In each case, suspensions of red blood cells were tested with potent human anti-Rh agglutinins to determine whether the blood was Rh– or Rh+. Since the Rh factor is a constant immutable hereditary property of the red blood cells, it is obvious that such tests could be made at any interval after the last delivery of an affected infant.

The presence or absence of atypical agglutinins was determined by testing each serum with blood suspensions of at least 10 individuals of Group O, of which at least 1 was Rh–. It is obvious that the possibility of demonstrating atypical agglutinins was better if the serum was studied soon after birth of the affected infant.

EXPERIMENTAL

Selection of the Cases.—This study is based chiefly on blood tests of 153 mothers who delivered one or more infants suffering from one of three clinical forms of erythroblastosis fetalis, i.e., fetal hydrops, icterus gravis, or anemia of the newborn. Of these, 115 were under observation either by one of the authors or by other collaborators.* Thirty-eight mothers were referred to us from several sources mainly outside of the metropolitan area of New York. The diagnosis in the affected infants in the latter group could be accepted on the basis of a history submitted by the physicians referring the case. In each case, some or all characteristic features were present, such as a significant obstetrical history of previous miscarriages or stillbirths, or infants with erythroblastosis fetalis—and the accepted diagnostic criteria such as severe icterus at or shortly after birth, enlarged liver and spleen, excess of normoblasts in the peripheral circulation, progressive anemia, and in fatal cases post-mortem findings of extramedullary hematopoiesis.

In addition to the 153 mothers of infants in whom a diagnosis of erythroblastosis fetalis was established, data will be presented below (p. 933) in a smaller number of women whose pregnancies terminated in habitual abortion, miscarriages, stillbirths, or macerated fetuses.

Technique of the Agglutination Tests.—*Diagnosis of Rh+ and Rh– blood:* One or more drops of anti-Rh serum are mixed in small test tubes (75 × 10 mm.) with 2 drops of a washed 1 per cent to 2 per cent cell suspension (this corresponds to a suspension made by adding one drop of whole blood to 4 c.c. saline). The cell suspension is preferably prepared from the clot.

*Drs. L. Goldman, C. Javert, S. Polayes, and H. Schwartz.

The tubes are shaken and incubated in a water-bath at 37° C. for one hour, at the end of which period each tube is properly identified and all tubes are centrifuged at low speed (500 r.p.m.) for one minute. After replacing the tubes in the rack, the sedimented cells are resuspended by gentle shaking and readings are recorded. Those mixtures in which no gross agglutination is visible are examined microscopically (low magnification) by withdrawing with the aid of a glass rod some of the mixture onto a slide. Bloods showing no agglutination are Rh−.

Detection of Anti-Rh in Human Sera.—Two drops of the serum to be tested are added to each of 10 small test tubes and each tube receives two drops of washed cell suspension of ten different Group O bloods. As a rule, at least one Rh− blood is included. The tests are incubated at 37° C. for one hour, and readings are made after centrifuging the tests and resuspension of the sediments as indicated above.

All cases were tested with one or more of three potent anti-Rh sera. Since the first serum employed (December, 1940) was of Group A (patient M. F.) all bloods of Groups O and A could be tested directly. In order to test bloods of Groups B and AB, it was necessary to treat this serum with an Rh− blood of Group B, in order to remove the normal iso-agglutinin anti-B. In April, 1941, the second potent anti-Rh serum was found (Patient E. B.). Since this serum belonged to Group AB, it could be used directly on bloods of all groups. In May, 1941, the third anti-Rh serum (M. S.) of Group O became available.

In the latter part of these studies, bloods of all groups could be tested in parallel with three anti-Rh sera. This became possible ˙because of the availability of Witebsky's group specific soluble Substances A and B,[15*] the addition of which in small quantities inhibited the action of agglutinin anti-A and anti-B without affecting the activity of anti-Rh.

In any event, each of the three sera gave distinct agglutination reactions so that there was never any doubt as to the diagnosis of Rh+ or Rh−, in any blood tested with a particular serum.

A brief obstetric history of these patients is given below:

1. *M. F. (Group A):* A patient of Dr. P. Vogel had 7 pregnancies of which 6 were full term and one terminated at six months. A diagnosis of erythroblastosis fetalis was made in the last two infants, the most recent one in December, 1940. Serum obtained several weeks post-partum contained potent anti-Rh agglutinins. The agglutinins remained very active though they were somewhat weaker in the blood specimen drawn in May, 1941.

2. *E. B. (Group AB):* A patient of Dr. Freed had two full-term pregnancies. The first infant was very pale at birth and survived but one and one-half hours. No other data were available. The second infant, born April, 1941, showed classical symptoms of congenital anemia.

3. *M. S. (Group O):* A patient of Dr. Paul delivered a normal infant in 1931; this pregnancy was followed by a long interval of sterility. The second pregnancy (April, 1941) resulted in an infant suffering from icterus gravis.

Specificities of the Anti-Rh sera, M. F., E. B., and M. S.—A complicating feature in the behavior of the human anti-Rh sera is their failure to give entirely parallel reactions when tested with numerous

*Supplied by courtesy of Eli Lilly & Co.

bloods.[14, 17]   Thus these sera give varying incidences of Rh– reactions (Table II).

TABLE II.   RESULTS OF PARALLEL TESTS ON 334 RANDOM BLOODS OF ALL GROUPS

ANTI-Rh SERUM FROM PATIENT	PER CENT	
	Rh+	Rh–
E. B.	87	13
M. F.	85	15
M. S.	73	27

The greater incidence of Rh– reactions with serum M. S. is manifested by the fact that some bloods diagnosed as Rh+ in their reactions with sera E. B. or M. F. are not agglutinated by serum M. S.*   This fact is applicable to the bloods of only 3 of the 29 mothers which were tested with each of the three sera.

These serologic observations which are of more than academic interest, will be published in detail elsewhere, but reference will be made later (page 932) to the significance of these findings in connection with the iso-immunization by the fetus.

Statistical Data on the Rh Blood Factor in 153 Mothers of Infants with Erythroblastosis Fetalis.—This study reveals an incidence of about 90 per cent Rh– reactions in contrast to 15 per cent Rh– reactions in the random population.   The actual analysis is rendered somewhat complex because not all bloods were tested with each of the three anti-Rh sera (Table III).

TABLE III

ANTI-Rh SERUM	NUMBER OF MOTHERS' BLOODS TESTED	PER CENT	
		Rh+	Rh–
M. F.	142	10	90
E. B.	60	15	85
M. S.	28	7	93
Combined results	153	7	93

The higher incidence of Rh+ with serum E. B. is due to the fact that more mothers Rh+ with serum M. F. were tested with serum E. B. than with serum M. S.   At least one of the bloods Rh+ with serum E. B. can be considered Rh–, because this blood contains very active anti-Rh agglutinins.

Three mothers whose blood was Rh+ with serum M. F. or E. B. were Rh– with serum M. S. These three cases listed as Rh– could be considered as instances in which iso-immunization could occur, because in two instances the husband's blood, and in the third the affected child's blood were Rh+ with each of the three sera.

The striking difference in the incidence of the Rh factor in the selected population of mothers with erythroblastic babies and in the random population strongly supports the concept of iso-immunization.   Since the incidence of Rh– bloods is very high in these mothers, it could be anticipated on the basis of the iso-immunization theory that the affected infants and the fathers in the series of Rh– mothers are exclusively Rh+.   As will be shown such results were obtained in tests on 89 fathers and on 76 infants in whom a diagnosis of erythroblastosis fetalis was made (Table IV).

*Very rarely, the converse specific effect is obtained, i.e., blood Rh– with serum M. F. is Rh+ with serum E. B. or M. S.

TABLE IV. INCIDENCE OF RH+ AND RH− IN HUSBANDS AND AFFECTED INFANTS OF
THE 141 RH− MOTHERS

	NUMBER TESTED	Rh+	Rh−	EXPECTANCY OF Rh− IN RANDOM POPULATION
Husbands	89	89	0	13 (89 × 15%)
Affected infants	76	76	0	11 (76 × 15%)

The findings in this table are based on tests with serum M. F. which gives an
incidence of 15% Rh− reactions.

The findings in Tables III and IV strongly indicate the iso-immunization of the Rh− mother by the Rh factor in fetal blood. The final proof
for such iso-immunization can be supplied only by the demonstration of
anti-Rh agglutinins in the mother's blood. Obviously, the likelihood
of finding such agglutinins will be greater if the mother's blood is
tested soon after the delivery of an infant suffering from erythroblastosis fetalis. This is borne out by the findings presented in Table V.

TABLE V. INCIDENCE OF ANTI-RH AGGLUTININS IN 141 RH− MOTHERS

INTERVAL AFTER LAST DELIVERY OF AN AFFECTED INFANT	AGGLUTININS PRESENT	AGGLUTININS NOT FOUND
2 months post partum	33	37
2 months to 1 year past partum	5	15
1 year or longer post partum	2	39
During next pregnancy	2	5
No data	0	3
Total	42	99

The failure to detect anti-Rh agglutinins in many cases tested shortly
after the delivery of an affected infant does not exclude their presence
at some previous period during the course of their pregnancy. Since
the course of antibody production in general is characterized by a gradual rise, a period of maximum activity, and gradual disappearance, it is
conceivable that in some cases anti-Rh agglutinins, after exerting their
lytic effect on the fetus, rapidly disappeared from the blood so that
none could be demonstrated at the time of delivery. It is therefore
indicated that the bloods of mothers known to have delivered babies
with erythroblastosis fetalis be studied at several intervals during course
of future pregnancies. Furthermore, it is conceivable that antibodies
capable of reacting in vivo cannot be demonstrated because of limitations in the sensitivity of the technique employed.

It is of great interest that the anti-Rh agglutinins may be demonstrable for such unusually long periods, as two years post partum.
Probably the long duration indicates an intense degree of iso-immunization which may perhaps influence the outcome of the next pregnancy
with an Rh+ fetus. More specifically, reference is made to the clinical
observation on the high incidence of spontaneous abortions and miscarriages in mothers of infants with erythroblastosis fetalis (p.
933).[3, 18, 19]

The results shown in Tables III, IV, and V indicate that a combination of Rh− mother, Rh+ father, and affected infant can be used as a
laboratory test to support a diagnosis of erythroblastosis fetalis. Such
results which are present in about 90 per cent of the cases should be

of value in the diagnosis of mild, atypical, or borderline cases of this condition.

The 11 Cases in Which Iso-immunization with Rh is Excluded.— When these exceptional cases were first observed, it was assumed that blood factors other than Rh may also induce iso-immunization. This explanation seemed plausible since Levine and Polayes[20] had already demonstrated a new blood factor by means of an atypical hemolysin found to be the cause of a post-partum transfusion reaction.* Indeed, this antibody seemed to differentiate the bloods of several of the Rh+ mothers from those of their husbands or affected children in a manner compatible with the concept of iso-immunization. However, the results were only suggestive and further study was hampered by the weak activity of the antibody.

More decisive results, however, were obtained with the reactions of an atypical agglutinin found in the serum of one of the 11 Rh+ mothers (patient K. F.).[21] This patient of Dr. Javert had one full-term, normal pregnancy; one infant with erythroblastosis fetalis (1937) and a missed abortion. When tested she was again several months pregnant. The antibody in this serum gives strong reactions almost exclusively on bloods which are Rh– with serum M. S. Accordingly, the blood factor identified by this atypical agglutinin must have some genetic relationship with Rh. Since this serum became available late in the course of this study, the bloods of the remaining mothers of this group must be retested with this new agglutinin.

Reference again is made (p. 931) to the bloods of 3 mothers of erythroblastic babies which were Rh+ with two anti-Rh sera but Rh– with the third serum M. S. These cases were included in the Rh– series because in one instance the husband and in the other the affected child were Rh+ with each of the three anti-Rh sera.

The Theory of Heterospecific Pregnancy.—The Distribution of Blood Factors A, B, and Rh in Red Blood Cells, Tissue Cells, and Body Fluids.—It is of interest that a difference in the blood groups of mother and infant was the basis of an older theory on the pathogenesis of familial icterus gravis.†[22, 23] In the literature, this concept has been referred to as "heterospecific pregnancy." The theory, since abandoned for lack of evidence, does not differ in principle from the iso-immunization theory in which the Rh factor plays such a prominent role. According to the concept of heterospecific pregnancy, given a mother of Group O and an infant of Group A (or Group B), the maternal anti-A (or anti-B) agglutinins are theoretically capable of acting on the Group A (or Group B) fetal blood. Actually, there is now evidence that in the example cited the mother's normal iso-agglutinin anti-A (or anti-B) is increased in titer as a result of iso-immunization with the A (or B) blood of the fetus.[24, 25] Nevertheless, the maternal agglutinins are specifically inhibited from acting on the fetal blood because of the wide distribution of the A and B factors in tissues and body fluids.

However, this applies to about 80 per cent of all individuals (secretors)[26] and if a fetus of Group A belongs to the class of nonsecretors (20 per cent), it is conceivable that the maternal iso-agglutinin anti-A may serve as the source of the intrauterine hemolytic process.

---

*This mother had 4 miscarriages and 8 full-term, presumably normal infants.

†At that time (1923) the more comprehensive term, erythroblastosis fetalis, was not yet in use.

Accordingly, the older theory of heterospecific pregnancy may have to be invoked at least for selected cases of erythroblastosis fetalis in which the Rh or other blood factors fail to indicate iso-immunization by the fetus.

From these considerations on heterospecific pregnancy and in view of the established importance of the Rh factor in the pathogenesis of erythroblastosis fetalis, it can be assumed that the Rh factor is probably not present in tissue cells or body fluids, but rather is limited to red blood cells only. Otherwise, the maternal anti-Rh agglutinins would be specifically inactivated and therefore incapable of inducing the hemolytic action on the Rh+ fetal blood. Tests made by Levine and Katzin[27] with numerous specimens of saliva, a few specimens of sperm cells, and seminal fluid, indicate that the Rh factor is not present in the material tested.*

Iso-immunization in Habitual Abortion and Stillbirth.—Reports from the literature indicate that the obstetric history of mothers of infants with erythroblastosis fetalis reveals a high incidence of abortions, miscarriages, and stillbirths. Macklin[18] and Darrow[19] stated that the mechanism responsible for the pathogenesis of erythroblastosis fetalis applies also for these abortions and stillbirths. This view is supported by the results in the present study. That iso-immunization by the fetus and subsequent action of maternal agglutinins on fetal blood may be the mechanism of these abortions and stillbirths is indicated in the data presented in Table I.[3, 4]

Another group of women were investigated because of their history of habitual abortion and stillbirths, but these women had no infants with erythroblastosis fetalis. Many of these women were Rh– but a satisfactory statistical analysis similar to the study on erythroblastosis fetalis is more difficult because habitual abortions and stillbirths may be manifestations of many conditions. Nevertheless, there is sufficient evidence in at least five† cases of this group to include them with the mothers of erythroblastic infants, because iso-immunization by the Rh factor in the fetus could be demonstrated. In each of these cases, anti-Rh agglutinins were observed. Three of these patients suffered from intragroup transfusion accidents following an abortion or stillbirths.

Two of these 5 patients had just delivered presumably normal infants in spite of the presence of moderately active anti-Rh agglutinins. Unfortunately, hematologic and other clinical data were not obtained so that a mild form of erythroblastosis fetalis could not be entirely excluded. However, the obstetric histories reveal that one of these patients (H. H.) had three consecutive miscarriages and the other (L. L.) had two miscarriages and one premature infant who survived for ten days.

The association of intragroup transfusion accidents, in the presence of atypical agglutinins, in three patients following an abortion, a miscarriage, and a stillbirth, respectively, was observed by us in the analysis of reports by Parr and Krischner,[29] Johnson,[30] and Zacho.[31] Other instances of intragroup transfusion accidents following abortions, miscarriages, or stillbirths, in the absence of atypical agglutinins have been

*Similar observations on saliva were recently reported by Wiener and Forer.[28]
†Four of these patients were discussed in our previous papers.[1-3]

reported by several authors.[32, 33] Very probably the transfusion accidents reported in the five papers quoted are ultimately attributable to iso-immunization by the fetus.

In any event, it is the action of maternal immune agglutinins on the fetus which may cause its death at any stage of its development so that the same mechanism may be responsible for erythroblastosis fetalis in the surviving infant as well as some cases of habitual abortion.

DISCUSSION

The numerous theories on the pathogenesis of erythroblastosis fetalis were recently reviewed by Darrow[19] who, in a hypothetical discussion, anticipated the iso-immunization theory. This author suggested an antigen-antibody reaction based on differences in maternal and fetal hemoglobin or other constituents of red blood cells as the most plausible explanation. A similar view was previously mentioned by Ottenberg,[23] but neither author presented experimental data to support their theses.

The findings presented establish the significant role of iso-immunization of the mother by blood factors in the fetus in the pathogenesis of erythroblastosis fetalis. Of prime importance is the Rh factor in the fetal blood which is transmitted as a dominant mendelian gene from the father.[14] Statistical evidence was offered to prove that its presence in the blood of the father and affected infant is a prerequisite condition for the iso-immunization of the Rh– mother. But in the final analysis, the proof of iso-immunization by the Rh factor was the demonstration in many instances of anti-Rh agglutinins in the mother's blood. It is the continuous intrauterine action of anti-Rh agglutinins with the Rh positive fetal blood over a period varying from weeks to months which causes a progressive hemolysis of fetal blood.

The reaction of anti-Rh antibody and Rh blood takes the form of agglutination in the test tube but in the fetal circulation the end result is hemolysis. Attempts to demonstrate corresponding hemolytic reaction in the test tube by the addition of fresh human serum have so far failed.

One of the most striking features of erythroblastosis fetalis is the wide variety of clinical syndromes it embraces, such as, the extremely fatal form of fetal hydrops and the mild, frequently unrecognizable anemias of the newborn.[34] These clinical forms are probably the result of varying degrees and duration of iso-immunization during the course of the pregnancy. Nothing is as yet known concerning the exact time and conditions during the course of the pregnancy when iso-immunization begins. If some cases of habitual abortions are a manifestation of iso-immunization, then the process may start shortly after conception occurs. At any rate, it is conceivable that the prolonged action of immune iso-agglutinins on the susceptible fetal blood may induce more severe damage than the action of agglutinins produced very late in the course of a pregnancy.

The clinical observation has been made that some infants may be born apparently free from the condition, but in the course of a few days, severe anemia and jaundice make their appearance. It is difficult to correlate this fact with the iso-immunization theory since the infant after birth should be free from any further action by the maternal agglutinins. The source of the delayed hemolysis has been suspected to be the colostrum, but in a number of these cases the infant was not breast fed. An alternative explanation is the storage of mother's agglutinins by the tissue of the fetus so that their subsequent release may then induce the hemolysis several days after birth.

A few practical applications based on these studies are of importance to the obstetrician. In the first place the data presented make it clear why caution must be exercised in selecting compatible donors for transfusing mothers of infants with erythroblastosis fetalis or those with a history of habitual abortions, a stillbirth, or a neonatal death. It is of interest that 8 of the 141 Rh– mothers suffered from severe intragroup transfusion reactions. Such accidents can be prevented if Rh– donors are available for Rh– mothers[3] and if the modified compatibility (cross-matching) test be employed for the detection of the "warm" anti-Rh agglutinins (p. 927).[4, 11, 12]

Furthermore, there are indications from a small number of cases that the affected infant maintains higher levels of hemoglobin and red blood cell counts if he is transfused with Rh– blood instead of Rh+ blood. The rationale for this suggestion is drawn from the fact that the infant's own Rh+ blood is undergoing destruction.[35]

The hereditary nature of erythroblastosis fetalis, hitherto unknown, can now be stated in terms of the iso-immunization theory. In some families, every pregnancy but perhaps the first, terminates in either an abortion, a stillbirth, or an infant with erythroblastosis fetalis; while in other families, only one of several pregnancies results in an affected infant. Since the Rh factor is inherited as a simple mendelian dominant,[14] it is obvious from a genetic standpoint, that this striking difference in familial incidence of the disease is determined by the homozygosity (RhRh) or heterozygosity (Rhrh) of the father's blood.[36] The genetic details and evidence supporting this concept will be given elsewhere. It is, however, appropriate to state at this point, that the first born is frequently but not always spared because more than one pregnancy with an Rh+ fetus may be required before a sufficient degree of iso-immunization is attained.

It has been recorded by Javert[37] that erythroblastosis fetalis in one of its several forms, occurs once in 400 deliveries, but it is probable that this condition has a still higher frequency, especially if some cases of habitual abortion and stillbirths are manifestations of iso-immunization. Actually, the incidence of matings in which iso-immunization with Rh may occur (Rh+ husband and Rh– wife) is 85 by 15 or 13 per cent of all matings. Consequently, one would expect a much higher incidence of erythroblastosis fetalis. There are, however, a number of factors tending to reduce this incidence, such as, for example, the current tendency to small families and the inability of many Rh– women to respond to iso-immunization.

The data on erythroblastosis fetalis and the recent studies on the iso-immunization with the blood factors A and B in the absence of any pathologic conditions in the mother or the infant indicates that immunizing substances derived from the fetus make their way through the placental barrier into the maternal circulation. Since the Rh factor, in contrast to the A and B substances, is probably limited to only red blood cells, it is assumed that fetal blood in one form or another penetrates the villus in sufficient quantity to induce immunization in the mother.

SUMMARY AND CONCLUSIONS

1. In 93 per cent of the cases investigated, erythroblastosis fetalis results from the iso-immunization of the Rh– mother by the Rh factor in the red blood cells of the fetus.

2. In the remaining cases, blood factors other than Rh are responsible for the iso-immunization.

3. Agglutination tests for the Rh factor are of value as a laboratory aid in the diagnosis of erythroblastosis fetalis.

4. The pathologic manifestations of this disease are produced by the intrauterine action of maternal immune agglutinins on the susceptible red blood cells of the fetus.

5. It is probable that iso-immunization is also the cause of a certain proportion of habitual abortions and stillbirths.

6. Intra-group transfusion accidents associated with pregnancy can now be prevented by the use of Rh– donors and by means of modified cross-matching test.

The authors wish to express their appreciation to Dr. William Antopol, Pathologist and Director of Laboratories, Newark Beth Israel Hospital, for examination of the histologic sections and valuable suggestions, and to Miss Estelle Richardson for her technical assistance.

REFERENCES

(1) *Levine, Ph., and Stetson, R.:* J. A. M. A. 113: 126, 1939.   (2) *Levine, Ph., and Katzin, E. M.:* Proc. Soc. Exper. Biol. & Med. 45: 343, 1940.   (3) *Levine, Ph., Katzin, E. M., and Burnham, L.:* J. A. M. A. 116: 825, 1941.   (4) *Levine, Ph.:* Am. J. Obst. & Gynec. 42: 165, 1941.   (5) *Wiener, A. S., and Peters, H. R.:* Ann. Int. Med. 13: 2306, 1940.   (6) *Levine, Ph., Vogel, P., Katzin, E. M., and Burnham, L.:* Science 94: 371, 1941.   (7) *Landsteiner, K., and Wiener, A. S.:* Proc. Soc. Exper. Biol. & Med. 43: 223, 1940.   (8) *Levine, P., Katzin, E. M., Vogel, P., and Burnham, L.:* Symposium of the American Human Serum Association, June 2, 1941, Cleveland.   (9) *Wiener, A. S.:* Arch. Path. 32: 227, 1941.   (10) *Levine, P., Katzin, E. M., and Burnham, L.:* Proc. Soc. Exper. Biol. & Med. 45: 346, 1940.   (11) *Levine, P., Katzin, E. M., and Vogel, P.:* In preparation.   (12) *Levine, P.:* Cited by Wiener.[9]   (13) *Burnham, L.:* Am. J. Obst. & Gynec. 42: 389, 1941.   (14) *Landsteiner, K., and Wiener, A. S.:* J. Exper. Med. 74: 309, 1941.   (15) *Witebsky, E., Klendshoj, N. C., and Swanson, P.:* J. A. M. A. 116: 2654, 1941.   (16) *Levine, Ph.:* Unpublished observations.   (17) *Levine, Ph., and Katzin, E. M.:* Unpublished observations.   (18) *Macklin, M. T.:* Am. J. Dis. Child. 53: 1245, 1937.   (19) *Darrow, R. R.:* Arch. Path. 25: 378, 1938.   (20) *Levine, Ph., and Polayes, S. H.:* Ann. Int. Med. 14: 1903, 1941.   (21) *Levine, Ph., Javert, C., and Katzin, E. M.:* In preparation.   (22) *Hirszfeld, L.:* Konstitutionsserolgie und Blutgruppenforschung, 1928, Berlin, Julius Springer.   (23) *Ottenberg, R.:* J. A. M. A. 81: 295, 1923.   (24) *Jonsson, B.:* Acta path. et Microbiol. Scandinav. 13: 424, 1936.   (25) *Levine, Ph.:* Unpublished data.   (26) *Schiff, F., and Sasaki, H.:* Ztschr. f. Immunitätsforsch. u. exper. Therap. 77: 129, 1932.   (27) *Levine, Ph., and Katzin, E. M.:* Proc. Soc. Exper. Biol. & Med. (In press.)   (28) *Wiener, A. S., and Forer, S.:* Proc. Soc. Exper. Biol. & Med. 47: 215, 1941.   (29) *Parr, E. L., and Krischner, H.:* J. A. M. A. 98: 47, 1932.   (30) *Johnson, R. A., and Conway, J. F.:* Am. J. Obst. & Gynec. 26: 255, 1936.   (31) *Zacho, A.:* Ztschr. f. Rassenphysiol. 8: 1, 1936.   (32) *Goldring, W., and Graef, I.:* Arch. Int. Med. 58: 825, 1936.   (33) *Bernstein, A.:* Am. J. Obst. & Gynec. 39: 1045, 1940.   (34) *Diamond, L. K., Blackfan, K. D., and Baty, J. M.:* J. Pediat. 1: 269, 1932.   (35) *Katzin, E. M., Vogel, P., and Levine, P.:* Unpublished data.   (36) *Levine, Ph.:* In preparation.   (37) *Javert, C.:* Am. J. Obst. & Gynec. 34: 1042, 1937.

20

Galton Laboratory, University College, London

THE GENETICAL BACKGROUND OF COMMON DISEASES

By L. S. PENROSE

Introduction

A direct method of genetical investigation of common traits is to establish the fact of familial concentration by comparing the relatives of propositi, or index cases, with the general population. A raised incidence in relatives is then plausibly interpreted as indicating that the trait in question has some hereditary background. When the difference between the familial and population incidence is found to be large, say tenfold or more, the result is so striking that the probability of genetical causation appears to be high even in data compiled with little attention to detail. The surveys, carried out by *Luxenburger* [1936] to demonstrate the familial concentration of various kinds of mental illnesses, are cases in point.

However, in some conditions, the observed difference between the familial and population figures is relatively small so that, to ascertain its extent or even to find whether or not it really exists, great care must be exercised in the collection and analysis of the material. For example, data have to be compiled in such a way that ages and sexes can be separately classified. A good example of the precise application of this method is the study of peptic ulcer made by *Doll* and *Buch* [1950]. These investigators first set up a control based upon the observed incidence of the trait in the general population at each age, specifying the sexes separately. They were thus able to show convincingly that the risk of contracting the disease was about four times as great for parents and twice as great for sibs as for members of the population at large. Using a somewhat similar method, figures have been obtained by *Stecher* [1941] showing that, for females, the risk of developing *Heberden*'s nodes on the fingers was doubled in close relatives of affected cases. In the survey of mammary cancer by *Penrose*, *MacKenzie* and *Karn* [1948] the incidence was shown to be doubled in parents and trebled in sibs of affected cases. A study of the incidence of acute rheumatism in the sibs of affected propositi by *Roberts* [1951] indicated that the familial incidence was not as much as twice that in the general population. In this and most other surveys where sibships are each counted once, the crude incidence in sibs underrates the true value, which lies between the observed value and twice the observed value. When the true familial incidence in sibs is very low it closely approaches twice the observed value. Thus, in *Doll* and *Buch*'s data, the incidence in parents and sibs probably can be regarded as about equal.

In all these examples, the crucial value, the ratio of familial to population incidence, which can be called K, is small, varying between about 1½ and 4½. It is the purpose of this paper to examine some possible interpretations of these low K-values; they do not necessarily indicate the relative absence of genetical determination, as often supposed. Actually, a low value of K indicates that, if the cause of the trait is a gene, this gene must be very common in the population studied. However, if the value of K is actually indistinguishable from unity, it is reasonable to suppose that genetical differences play no appreciable part in causation in the population in question.

The Standard Formulae.

There are two simple traditional ways of interpreting familial concentration. The observed trait may be due to an autosomal recessive or an autosomal dominant gene. A third interpretation is that the gene is intermediate and, in the present instance, it will be assumed that this means that a trait is manifested in all cases in the homozygote but only in one half of those who are heterozygous. The manifestation of the trait in these heterozygotes could be determined by other genes but, in the present instance, it will be considered to be a threshold phenomenon determined non-genetically. The values of K which cover the range already mentioned are given in the left hand column of Table 1, and the corresponding frequencies of a gene interpreted as being the cause are set out.

Marked differences between parent-child and sib-sib relationships are shown in recessive traits. However, when the K-values are very low, indicating very common genes, the distinction between recessive and dominant hypotheses by this means may be difficult unless data are extensive. The intermediate hypothesis as defined here gives rise to values of q which are the same for both types of relationship and they do not differ much from those consistent with the single dominant gene hypothesis.

Table 1. Single gene frequency q, for different values of K, i.e. incidence in relative as compared with population incidence.

K	Recessive Gene		Intermediate Gene		Dominant Gene	
	Parent or child	Sib	Parent or child	Sib	Parent or child	Sib
1.0	1.000	1.000	1.000	1.000	1.000	1.000
1.5	0.667	0.690	0.333	0.333	0.262	0.281
2.0	0.500	0.547	0.200	0.200	0.170	0.178
3.0	0.333	0.406	0.111	0.111	0.100	0.103
5.0	0.200	0.288	0.059	0.059	0.056	0.056
10.0	0.100	0.188	0.027	0.027	0.026	0.026
Indefinitely large	$1/K$	$1/2K^{1/2}$	$1/4K$	$1/4K$	$1/4K$	$1/4K$
General formula	$qK=1$	$\begin{aligned}q(2K^{1/2}-1)\\=1\end{aligned}$	$\begin{aligned}q(4K-3)\\=1\end{aligned}$	$\begin{aligned}q(4K-3)\\=1\end{aligned}$	$\begin{aligned}(2-q)^2K\\=1+q-q^2\end{aligned}$	$\begin{aligned}4q(2-q)^2K\\=4+5q-6q^2+q^3\end{aligned}$

Trait and Disposition.

The great majority of traits or diseases thought to be strongly influenced by genes are also strongly influenced by environment. That is to say, quite apart from the effects of dominance, recessiveness and intermediacy, agreement between phenotype and genotype is not perfect. The trait for which there exists a genetical susceptibility or predisposition may not always be manifested, indeed it may be so only very rarely. In addition to this, the trait might sometimes occur in the absence of the predisposing genetical background. Such acquired cases, fresh mutations or phenocopies, are generally supposed to be very infrequent as compared with those occurring on the genetical basis; yet they must not be neglected in human data because cases are recorded when the specific outward effect is produced whatever the cause. The alternatives are set out in Table 2. There are four different classes, containing x, y, z and w people respectively. The total number affected is $(x + y)$ and the number genetically predisposed is $(x + z)$.

Table 2. Relationship between trait and hereditary disposition.

Class	Number of people	Description
(i)	x	Affected and also genetically predisposed (hereditary, endogenous or primary cases)
(ii)	y	Affected but not genetically predisposed (acquired, environmental, exogenous or secondary cases)
(iii)	z	Unaffected although genetically predisposed (latent, susceptible)
(iv)	w	Unaffected and not genetically predisposed

The proportion of predisposed cases, M, who are actually affected and show the trait, is $(x)/(x + z)$; this is sometimes termed penetrance though it is more correctly termed manifestation since, in human genetics, the distinction between penetrance and expressivity is invalid. The importance of the value of $(x)/(x + z)$ in the present discussion is that, provided y is a negligible quantity, x measures the incidence of the trait. Furthermore, when y is zero, the estimate of gene frequency in Table 1 is independent of the manifestation. The disposition frequency, which is the most significant factor for purposes of estimating eugenic prognosis or the value of preventive measures, can be ascertained directly from the gene frequency after the mode of inheritance, recessive, intermediate or dominant, has been determined.

In the case of the hypothesis of intermediacy, for example, we can suppose that, over and above the indeterminancy of the threshold effect in the heterozygote, there is an imperfect degree of manifestation. The supposition then would be that genotype AA would be always unaffected (given that $y = O$), genotype Aa would be affected in $M/2$ instances and aa would be affected in M instances. The relationship between parent and child (or between sib and sib) would then be as shown in Table 3. The gene frequency, in the intermediate case as here defined, is equal to the frequency of predisposed people, the incidence of those actually showing the trait being Mq.

Table 3. Correlation between parent and child for an incompletely manifested intermediate trait.

Child Parent		Trait		Total
		Absent	Present	
Trait Absent		$1-2Mq+\frac{1}{4}M^2q(1+3q)$	$Mq-\frac{1}{4}M^2q(1+3q)$	$1-Mq$
Present		$Mq-\frac{1}{4}M^2q(1+3q)$	$\frac{1}{4}M^2q(1+3q)$	Mq
Total		$1-Mq$	Mq	1

M = manifestation; q = gene frequency = predisposition frequency = F; $K = (1+3q)/4q$, independent of M.

The value of K, which is the proportion affected among relatives of affected propositi, i.e. $\frac{1}{4}M(1+3q)$, divided by the population incidence, i.e. Mq, is equal to $(1+3q)/4q$ and is independent of manifestation.

These considerations make it convenient to express Table 1 in a different, equivalent, form, as shown in Table 4, where the predisposition frequencies, F, are given instead of gene frequencies.

Multiple Gene Effects.

The intermediate gene hypothesis is similar in certain respects to the hypothesis of multiple additive genes. Gene frequency has no obvious meaning for characters due to multiple genes, except as a rough average value, but frequency of the predisposition is still a very significant quantity. According to the definition of intermediacy used here, a trait determined in this way would give parent-child and sib-sib correlations of 0.25 for all gene frequencies. Multiple alternative additive genes give rise to parent-child and sib-sib correlations of

Table 4. Predisposition frequency, *F*, for different values of *K*,

| K | Recessive Gene | | Intermediate Gene |
	Parent or child	Sib	Parent or child
1.0	1.000	1.000	1 000
1.5	0.444	0.476	0 333
2.0	0.250	0.299	0.200
3.0	0.111	0.165	0.111
5.0	0.040	0.083	0.059
10.0	0.010	0.035	0.027
Indefinitely large	$1/K^2$	$1/4K$	$1/4K$
General formula	$FK^2 = 1$	$F(2K^{1/2}-1)^2 = 1$	$F(4K-3) = 1$

0.5, when the character which they determine has a continuous distribution, but an apparent reduction is produced by the all-or-none classification of affected and unaffected people. Values of *K* corresponding to different degrees of *F*, on the assumption that the genetical background is continuous and the genotypical correlation 0.5, can be calculated from *Pearson*'s [1914] tables for tetrachoric functions. Within the range $K = 1.5$ to $K = 10$, the values of *F* obtained in this way differ little from those for the single intermediate gene, as shown in Table 5. The value of the observed crude correlation coefficient gradually diminishes as *K* increases but, at the value of $K = 3.2$ it is equal to 0.25.

The critical distinction between the hypothesis of single dominant or intermediate gene and multiple genes cannot be made unless other data than *K*-values are available. One kind of data concerns monozygotic twins. According to the present interpretation of the action

Table 5. Hypotheses of single intermediate gene and multiple additive alternative genes compared; values of *F* for different values of *K* in parents or sibs.

K	Single Intermediate Gene predisposition frequency F	Multiple Additive Genes predisposition frequency, F
1.0	1.000	1.000
1.5	0.333	0.384
2.0	0.200	0.239
3.0	0.111	0.115
5.0	0.059	0.048
10.0	0.027	0.015

i. e. incidence in relative as compared with population incidence.

| Intermediate Gene | Dominant Gene | |
Sib	Parent or child	Sib
1.000	1.000	1.000
0.333	0.455	0.483
0.200	0.311	0.324
0.111	0.190	0.195
0.059	0.109	0.109
0.027	0.052	0.052
1/4K	1/2K	1/2K
$F(4K-3) = 1$	$FK[1+(1-F)^{1/2}]$ $= F + (1-F)^{1/2}$	$4FK[1+(1-F)^{1/2}]$ $= 1+3F+3(1-F)^{1/2}+F(1-F)^{1/2}$

of an intermediate gene, the expected correlation between monozygotic twins would be exactly 0.5; and the incidence of the trait in the monozygotic twin of a propositus would be $F = 1/(2K - 1)$.

Assuming that the threshold effect producing the trait is a sharp one, when multiple additive genes are involved, this correlation would be higher than 0.5 in monozygotic twins and the corresponding F-value would approach unity. On the other hand, the more genes which are involved in the manifestation of a trait, the lower will be the familial incidence. Thus a relatively strong likeness between monozygotic twin pairs, together with a weak likeness between dizygotic pairs or ordinary sibs, is suggestive of multiple gene causation with a sharp threshold controlling manifestation. In making any practical use of this suggestion, however, the effects of incomplete manifestation due to environmental causes, both on twin pairs and other kinds of relatives, would have to be taken into account. The consequent reduction of all these incidence rates if M is less than unity may not always be proportional. But, as a general rule, if the incidence of a trait in monozygotic twins of propositi manifesting a trait is more than four times that in dizygotic twins, multiple alternative gene determination is probable.

Examples.

In the survey of the familial incidence of mammary cancer mortality (*Penrose, MacKenzie* and *Karn* [1948]), for parents of affected propositi, the incidence was 25 against an expected value of 11.3 based upon the general population, leading to a K-value of 2.21.

For sibs, the incidence was 24 against 7.05, leading to a K-value of 3.38. We might reasonably suppose that the lower incidence in parents than in sibs indicates recessivity of a single causal gene and that such a gene could have a frequency of about 0.4. This corresponds to a predisposition frequency of 0.16. The high prevalence of the disposition agrees with the observation that the condition is responsible for some three per cent of the mortality of all females. In males, the manifestation is very low, only about one-hundredth of the female rate, although it could be assumed that the disposition is equally frequent in the two sexes. Such an explanation is not inconsistent with the view that the manifestation of the disease is also determined by environment or by very prevalent cytoplasmic agents transmitted through the parents, especially the mother. The main point to be emphasized, however, is not the possibility of single gene causation but the very great prevalence, which any recessive gene must have, to account for the observed familial incidence.

In the material of *Doll* and *Buch* [1950] on peptic ulcer the K-values are consistent with causation by a single dominant gene with frequency rather less than 10 per cent, or with an intermediate or multiple gene hypothesis with predisposition incidence of the same order of magnitude. Since the incidence in the adult control population was over 5 per cent in males and only 1 per cent in females, the gene may be manifested almost fully in males in some age groups though always incompletely in females. There is no hint of sex linkage in either this or the material on mammary cancer, so that it is reasonable to think in terms of autosomal genes as causal factors in both examples with very marked degrees of sex limitation. Further data were supplied by the same writers on the probable genetical distinction between gastric and duodenal sites. Since the K-value for ulcers at the duodenal site is higher than that for the gastric site, its predisposing genic background should be the less common of the two.

These analyses emphasize that the degree of familial incidence is not in itself an indication of degree of hereditary causation: it is an index of frequency of the hereditary predisposition in the population which is being studied. When the genetical basis of a trait is very common, however, as in the examples just given, the genetical causation is of less practical interest than environmental causes. In the example of acute rheumatic fever, the value of K being barely 1.5 is compatible with a frequency of at least one third for an intermediate gene; the susceptibility must, on any single gene hypothesis, be present in nearly half the population. There might be, on the other hand, many alternative additive genes all with about the same frequency but giving rise to a much less common susceptibility. A relatively high value of K in the monozygotic twin of an affected propositus, as compared with K for sibs, would point towards multiple genes rather than to a single gene pair. In fact, there is some evidence on this point compiled by *Stevenson* and *Cheeseman* [1953]. Numerous instances of monozygotic twin pairs, both affected, have been reported but few of such pairs with only one affected. The incidence in monozygotic twins is certain to be more than 50 per cent whereas the corrected incidence in sibs or in dizygotic twins of affected propositi is less than 10 per cent. Many alternative common genes are therefore indicated as probable part causes.

Summary

The principles of genetical analysis of very common traits by the method of comparing familial with population incidence are stated. The effects of imperfect correspondence between genotype and phenotype are discussed. Autosomal gene frequencies and predisposition frequencies, agreeing with various hypotheses, are tabulated and some examples of their application described.

Résumé.

L'auteur expose les principes d'une analyse génétique de caractères très fréquents en comparant leur presence dans les familles avec leur présence dans la population. Les effets d'un accord imparfait entre le génotype et le phénotype sont discutés. La fréquence des gènes autosomiques et la fréquence de prédisposition, s'accordant avec des hypothèses différentes, ont été groupées sous forme de tables, et leur application est illustrée par quelques exemples.

Zusammenfassung.

Der Verfasser legt die Prinzipien einer genetischen Analyse sehr gewöhnlicher Eigenschaften dar, indem er Vorkommen in Familien mit solchem innerhalb der Bevölkerung vergleicht. Die Auswirkungen unvollständiger Übereinstimmung zwischen Genotypus und Phänotypus wurden erörtert. Autosomatische Genfrequenzen und Prädispositionsfrequenz, die mit verschiedenen Hypothesen übereinstimmten, wurden in Tabellenform gebracht und einige Beispiele für deren Anwendung beschrieben.

REFERENCES

Doll, R. and J. *Buch:* Ann. Eugen., Lond. *15*, 135, 1950. – *Luxenburger, H.:* Zbl. ges. Neurol. Psychiat. *84*, 585, 1936. – *Pearson, K.:* Tables for statisticians and biometricians. Cambridge University Press, London 1914. – *Penrose, L. S., H. MacKenzie* and M. N. *Karn:* Ann. Eugen., Lond. *14*, 234, 1948. – *Roberts, L.:* Med. Offr. *86*, 159, 1951. – *Stecher, R. M.:* Amer. J. med. Sci. *201*, 801, 1941. – *Stevenson, A. C.* and E. A. *Cheeseman:* Ann. Eugen., Lond. *17*, 177, 1953.

Part III

MAN'S GENETIC VARIABILITY: EXISTENCE AND MAINTENANCE

Editors' Comments
on Papers 21 Through 26

Eighteenth- and nineteenth-century political philosophy repeatedly extolled the uniqueness of man. A commonplace assertion became then and remains now that all human beings have been uniquely endowed by their creator. The intent of this assertion was generally, of course, to justify equality of civil rights and liberties, but implicit in it is the assumption of broad genetic diversity. But how great, in fact, is this diversity? What means exist to estimate its magnitude? These are central issues in contemporary biology. Their resolution is made more urgent by the growing realization that man's exposure to mutagenic agents mounts rapidly as a consequence of changing lifestyles and the exuberance of the petrochemical and pharmaceutical industries. It is rightly feared that man may be doing irreparable damage to his species. But need this be so?

Historically, observations on the existence and extent of genetic variability in natural populations date, at least, from Charles Darwin's famous voyage aboard *The Beagle,* although knowledge of the mechanism of Mendelian inheritance was then still to come. After the rediscovery of Mendel's work, it soon became apparent that much of the variability in morphologic and quantitative characters that exists is genetic in origin (Fisher 1918). Real insights into variation at the gene level were not soon to come, however, for the exact relationship between many of man's characteristics and genes is so complicated. Nevertheless, answers to questions such as what causes this variation or how is it maintained have been sought since the days of Darwin. Studies of such variation have broadened in the interim into a spectrum—from observable, large morphologic differences to miniscule, unnoticeable chemical changes in the codon sequences of molecules that lead to little or no alteration whatsoever in the size or function of the molecule. The origin and growth of population genetics as a separate field of biological research owes much to this basic issue of variability, which remains a central biological problem.

Our aim, in this section, is not merely to bring the reader up to date in the literature on the existence and maintenance of man's genetic variability but to expose some of the pioneering attempts to understand its origin. We have intentionally avoided an extended account of the recent controversies between the selectionist and neutralist points of view. Even so, it is not easy to select a set of unequivocally accepted papers in the limited space available to us when thousands of journal pages have been devoted to this topic in recent decades. A more measured selection, one that should satisfy the specialists as well as the serious beginning student, will be found in Li's *Stochastic Models in Population Genetics* (1977).

Man's nature—that is, his genetic constitution—and his nurture, or environment, have provided the basis for a continuing, often acrimonious and unenlightening dialogue for a century or more. Francis Galton (1889) was, perhaps, the first to call attention to the fact that we are the result of the interaction of our native abilities and the circumstances under which we are reared. Numerous others have elaborated on this theme and described in exquisite and occasionally tiring detail the complexities of such interactions (see, e.g., Haldane 1936 and Paper 21; Hogben 1939). Paper 21 in particular is addressed to the logic that should be pursued in the evaluation of this problem. Haldane was concerned primarily with a simple classification of the possible types of interactions that can occur between two genetically different groups and two environments. He also sets forth the general expressions for determining the possible types for m, n, and k genotypes, environments, and environmental attributes, respectively. His language brooks no improvement. His analysis of the nature-nurture issue is, we believe, one of the best ways in which we can introduce this section on genetic diversity in man.

Genetic diversity is perhaps most readily described by a single measure: the degree of polymorphism observed in the species concerned. Simply put, a polymorphism merely implies the occurrence within a species of two or more different forms and as such could be construed to embrace both discontinuous and continuous variation. It has come to have a more restricted meaning, however. It is generally taken to imply the existence of discrete morphologic alternatives, and indeed, most of the classical theory of natural selection developed by Haldane (1924a,b; 1926, a,b), Fisher (1930), and Wright (1931) is addressed to the origin and maintenance of such variation.

Ford (1940) has stated: "Polymorphism may be defined as the occurrence together in the same habitat of two or more distinct forms of a species in such proportions that the rarest of them cannot be maintained by recurrent mutation," and this definition has enjoyed wide acceptance. In 1945 Ford (Paper 22) reviewed this definition by noting that a polymorphism may be either balanced or transient. Clearly, factors other than mutation are acknowledged to contribute to the maintenance of variation. His view is also deterministic in nature, although Wright and Fisher had by then called attention to the role of stochastic factors in the maintenance of variation in genetic traits. This later definition of Ford excludes seasonal and geographic variation and furthermore does not relate to the continuous variation that falls within the limits of a single unimodal distribution. He viewed those variations detectable by physiologic methods, such as color blindness and PTC-tasting, or immunologic methods, such as the blood groups ABO, MN, and Rh, as outstanding examples of his definition of polymorphism. Ford had difficulty, however, in citing polymorphisms that were demonstrable by morphologic means and were relatively frequent and at the same time of a clearly "discontinuous form." He took the occurrences of industrial melanism in some species of *Geometridae* and *Thyatridae* to be instances of transient polymorphism. In man, red hair and blue eyes tend to segregate as autosomal recessive characters. However, McKusick (1968) cites hair and eye color, attached or unattached ear lobes, clockwise or counterclockwise hair whorl, and so forth as having some features of continuous characters and thus are not susceptible to simple Mendelian interpretation.

In the context of current controversies on the mechanisms of the maintenance of genetic variability, it is important to note that proponents of either view tend to avoid any restriction associated with what causes such variation in their definitions of polymorphism. For example, a somewhat arbitrary but operationally convenient modern definition of a polymorphic genetic locus would be one in which the most common identifiable allele has a frequency no greater than 99 percent (Kimura 1971). Note that this definition does not preclude polymorphisms due to rare disease alleles constantly eliminated by natural selection and replaced by mutation, whereas Ford's definition does.

A somewhat different attitude toward the maintenance of polymorphisms in a species or alternatively how polymorphic variability is important in speciation is represented in the next paper by Haldane (Paper 23). Here Haldane suggests that the struggle against disease in general and infectious disease in particular may have had important evolutionary implications. General arguments are presented by Haldane that would attribute to genetic polymorphism an important role in the survival of populations under attack by a number of biochemically different races of pathogens. In some sense his hypothesis constitutes a mechanism that favors polymorphism because it gives a selective value to a genotype so long as it is rare. Along with several examples in plants and animals, Haldane cites Rh incompatibility in man. Interesting is the fact that Haldane did not suggest the association between genetic traits and malaria in this paper for which it is often cited (e.g., Cavalli-Sforza and Bodmer 1971). The correlation between the geographic distributions of the thalassemia gene and malaria, which he first observed, was reported in his address on the rate of mutation of human genes at the International Congress of Genetics in Stockholm (Haldane 1949). Our selection of the more general arguments of Haldane also stems from the obscurity of the journal in which it appeared.

The paper by Hirschfeld and Hirschfeld (Paper 24) represents the first attempt to use serologic differences within a species to solve anthropological questions regarding the origin of races and to use population frequencies to illustrate the mechanism of inheritance of such serological diversities. A clear-cut distinction between hetero-antibodies specific for a species and iso-antibodies specific for individuals within a species is illustrated through the ABO iso-agglutinins first studied by Landsteiner. Although Bernstein (1924) was subsequently to establish that the ABO blood group polymorphism is explicable in terms of Mendel's law of inheritance of a single locus with multiple alleles, Hirschfeld and Hirschfeld's demonstration that the extent of polymorphism varies from population to population had immense impact on the use of genetic arguments in anthropological research. In this sense the field of anthropological genetics really stems from this paper. Implicit in the paper is another notion, namely, the examination of the consequences of racial admixture on the frequencies of polymorphic genes in a population, which has since received considerable attention (see, e.g., Bernstein 1931; Reed 1969; Chakraborty 1975). It is also in this paper that the earliest evidence of clinal variation in gene frequencies in man is observed.

This idea led Brues (1954) to ask, "What causes the restrictions in the range of ABO blood group frequencies in the now existing races?'' She examined the effect of random genetic drift and soon came to the conclusion that the extreme variation in ABO gene frequencies seen among different population groups of Amerindians cannot be readily explained by genetic drift alone. Her suggestion that this variability might be maintained by overdominant selection was later studied more extensively using computer simulation methods (Brues 1963). She noted that the assumptions of appreciable incompatibility loss, with relative variabilities of 0.79 for 00, 0.74 for AA, 0.66 for BB, 0.89 for OA, 0.86 for OB, and 1.0 for AB, and an allowance for random variation due to finite size of populations gave rise to a distribution corresponding reasonably well with the observed worldwide distribution of ABO frequencies. The approach seems to be a convincing one, although in view of the absence of suitable immunological or biochemical techniques to distinguish the genotypes AO from AA, and BO from BB, there exists substantial room for uncertainty in judging the real significance of the derived estimates. Although Brues' views were later strengthened somewhat by the finding that similar polymorphisms exist not only in man but in other higher primates (chimpanzee, gorilla, and orangutan) and monkeys, current evidence of heterozygote advantage remains at best speculative in nature (see Reed 1975 for a recent review).

Presently, perhaps the best known example of balanced polymorphism in man is the sickle cell trait. To many, the insight into the dynamics of human population genetics afforded by the hemoglobinopathies is no less interesting than the information we have gained from them on gene action. Allison (Paper 25) was the first to support vigorously the notion that the high frequencies of the sickle cell gene in Africa were attributable to a balanced polymorphic system wherein the heterozygote enjoyed a selective advantage as a consequence of increased resistance to malaria. The evidence advanced by Allison in Paper 25 is circumstantial. In the main, it rests upon a correlation between the frequency of the sickle cell gene and the endemicity or hyperendemicity of malaria. Although these efforts by Allison and those of Beutler, Dern, and Flanagan (1955) to demonstrate directly that the heterozygote was more resistant to infection with *Plasmodium falciparum* led to somewhat conflicting results, the intervening years have seen the accumulation of additional data in support of Allison's general thesis.

Excellent reviews of the evidence and the problem of documenting the association of malaria with the sickle cell gene are to be found in Neel (1956), Livingstone (1958), and Cavalli-Sforza and Bodmer (1971, pp. 148–54). The precise basis for the relative immunity that must be involved is not known, although there are a number of clues, among which is Raper's (1956) data suggesting that sicklers do not suffer from cerebral malaria. There are other findings on the heterozygote, such as hyposthenuria, that have not been fitted into the puzzle as yet. With respect to hyposthenuria, there are reasons for believing that the lowered specific gravity of the urine of the heterozygote reflects a renal pathology. If this is so, we have the rather interesting situation where the heterozygote operates at an advantage in a malarious zone, but at a disadvantage in a nonmalarious area. Moreover, the advantage of the heterozygote in Africa, say, must be such as to replace not only genes lost through the death of persons with sickle cell anemia but possibly also some fraction of heterozygotes whose fitness is below average because of renal disease.

A variety of hypotheses have been advanced regarding the mechanism of increased fitness of the heterozygote. For example, Livingstone (1957) suggested that the increased fitness of the sickle cell heterozygote may be mediated to a large degree through the carrier female. He believes that *P. falciparum* infections of the placenta may not develop to the same extent in sickling as opposed to nonsickling mothers and that as a consequence the former would have a proportionately higher number of live-born children. Much of the available evidence is consistent with this hypothesis. Edington (1955), for example, finds that sickling mothers have a slightly lower stillbirth rate and that their infants have a higher mean birth weight and a higher survival rate. There is as yet, however, no direct evidence that placental infections are either less common or less severe in sickling mothers.

Up until the mid-fifties antigenic reactions and serotyping were the principal tools for decoding man's genetic diversity. Karl Landsteiner's four red blood cell phenotypes had grown to no less than 29,952 through essentially the same experimental technique, and the one series of reactions had become eight. A decade later, in 1965, the immunological distinctions that could be more or less routinely made resulted—in theory, in no less than 2,717,245,440 distinct phenotypes—and if the secretor phenomenon (which results in the production of water soluble forms of the ABH antigens) is included, this number doubles. These distinctions, we emphasize, do not include pathological characteristics of the red blood cell nor, more importantly, the increasing number of inherited serum proteins or red cell enzymes, nor the histocompatibility antigens, which are possibly the most variable of all.

A more complete answer to determine the extent of genetic variability came through the development of molecular biology. On the theoretical side, by taking into account the fact that at the molecular level almost an infinite number of alleles may be produced at a locus, Kimura and Crow (1964) showed that the number of alleles at a locus that can be maintained in a finite population is fairly large. (It should be stated that Wright (1948) indicates this possibility even though the structure of the basis of heredity, DNA, was not then known to Wright.) The development of starch gel electrophoresis (Smithies 1955) in combination with a simple staining technique for a specific enzyme activity (Hunter and Markert 1957) provided a valuable tool by which genetic heterogeneity in proteins and isozymes could easily be detected. By 1965, it was already known that natural populations contain a large amount of polymorphism with respect to proteins and enzymes. In a review, Shaw (1965) stated that "enzymes which vary within populations are rather the rule than the exception."

An important step in the study of genic variation in populations was made by Lewontin and Hubby (1966) and Harris (Paper 26). In our final selection for Part III, Harris describes the methodology presently in use to determine the extent of genetic diversity in man's nature at the "electrophoretic" level. The current theory that the primary structure and probably the rate of synthesis of different polypeptides are under genetic control suggests that the genetic variation in a natural population might be measured best by the enzymic diversity present. Clearly, one cannot presently hope to study all of the enzymes produced by the genome to determine the proportion of genetic loci that show variation. It is possible, however, at least in theory to study a representative sample of the enzymes of a species and thereby estimate the extent of its genetic variability. But an unbiased estimate will be obtained only if genetic variation is detectable at several loci and the loci are chosen randomly with respect to how variable they are. With the electrophoretic identification of isozymic variation, it has become possible to sample genetic loci (by selecting enzymes) in a manner which at least approaches randomness.

As might have been anticipated, even this approach to the sampling of genetic loci falls somewhat short of the ideal. First, only alterations in structural genes are efficiently detected. Detection of variation in the activity of enzymes is poor by electrophoresis and is biased in that we do not know how many regulator genes go undetected. Second, the proteins studied must be soluble, and third, a histochemical staining method for identification of the protein after electrophoresis must be available. These limitations notwithstanding an increasing body of evidence exists to support the notion that a substantial fraction of genetic loci exhibit variation in polymorphic proportions.

Harris (1970) has reported findings on twenty arbitrarily chosen enzymes in Europeans and Negroes. Some twenty-seven genetic loci are involved in the structural control of these enzymes. The average heterozygosity for these twenty enzymes, assuming twenty-seven loci, he finds to be 0.054 for Europeans and 0.052 in Negroes. (*Excursus:* Average heterozygosity is measured in the following manner. Let p_i be the frequency of the i-th allele at a locus. The expected proportion of individuals heterozygous for this gene will be $H = 1 - \Sigma p_i^2$, assuming the population is large, panmictic, and no selection obtains at the locus. The "average heterozygosity" is merely the individual heterozygosities, the H's, summed over all loci under consideration, and divided by the number of loci.) Harris estimates that the average heterozygosity per locus for alleles determining all structural enzyme and protein variants will be three times the values cited above. He argues that approximately one-third of the total variability will be recognizable electrophoretically. This factor, one-third, rests in part on the knowledge that almost two-thirds of all random amino acid substitutions in the hemoglobin molecule will produce no change in the net charge of the molecule and hence will be undetectable electrophoretically.

Lewontin (1967) has suggested another, albeit somewhat similar, approach to the estimation of the average heterozygosity in man. His argument, which utilizes serological data, is based upon the supposition that as more and more bloods are tested, the "detected" heterozygosity will converge on the "true" heterozygosity. Somewhat differently stated, data on blood groups will be less and less biased toward polymorphic loci as more and more bloods are studied.

As of 1962, Lewontin noted that some thirty-three blood groups were known; of these 36.4 percent were polymorphic and the average heterozygosity was 0.162. Of the next seventeen blood groups to be recognized, only one, Dombrock, was polymorphic. If these data, and the three blood groups that Lewontin failed to tabulate are considered, in 1968 among the fifty-three known systems, thirteen were polymorphic. The average heterozygosity was correspondingly lower. Nei and Roychoudhury (1974) have recently reexamined these data, but as they and others have noted, the loci associated with the blood groups may not be a random sample of the human genome. To this end, the table from Harris and Hopkinson (1976) provides an up-to-date summary of our knowledge of enzyme-protein variation in man.

Over the past three decades, population genetics has seen the emergence of an important concept, namely, the genetic load of a population. This concept has been defined as the extent to which the population is impaired by the fact that not all persons are of the optimum type. "Optimality" in this context implies optimal fitness in the Darwinian or reproductive sense. The usual metric for the latter is the expected number of children of an individual, either relatively or absolutely, counted at the same age as the individual.

Although the values these approaches give rise to are somewhat different, all of the previous procedures for estimating man's genetic diversity attest to a potential for the existence of tremendous variability. This potential is, of course, at the species level, and it might be much less within individual populations. The latter, however, seems not

279

Polymorphism and Average Heterozygosity as Obtained from Electrophoretically Detectable Enzyme Variants in European and U.S. White Populations

	Polymorphic loci (av. het. >0.02)	*Non- polymorphic loci (av. het. <0.02)*	*Number of loci screened*	*Av. het. per locus*
1. Oxidoreductases	4	20	24	0.042
2. Transferases	8	21	29	0.059
3. Hydrolases	11	27	38	0.087
4. Lyases	1	9	10	0.050
5. Isomerases	–	3	3	–
6. Ligases	–	–	–	–
Total	24	80	104	0.063

Adapted from Harris and Hopkinson 1976.

to be the case, for measurements of the genetic distance between villages within a primitive tribe as contrasted with intertribal differences suggest the former to be a very large part of the total variability (Lewontin 1972; Nei and Roychoudhury 1972, 1974). Thus, it would seem that even within small populations, interindividual variability is large. Genetic diversity is ubiquitous. But controversy is still the the air. Answers to such questions as whence does it come, how it is maintained, and is there an upper limit to variation have yet to emerge.

REFERENCES

Bernstein, F. 1924. Ergebnisse einer biostatischen zusammenfassenden betractung uber die erblichem blutstrukturen des menschen. *Clin. Wsch.* 3:1495–97.

Bernstein, F. 1931. *Die geographische verteilung der blutgruppen und ihre anthropologische bedeutung. Comitato Italiano per lo Studio dei Problemi della Populazaione.* Instituto Poligrafico della Stato, Roma, pp. 227–43.

Beutler, E., R. J. Dern, and C. L. Flanagan. 1955. Effect of sickle-cell trait on resistance to malaria. *Brit. Med. J.* 1:1189–91.

Brues, A. M. 1954. Selection and polymorphism in the ABO blood groups. *Am. J. Phys. Antrhop.* 12.559–97.

Brues, A. M. 1963. Stochastic tests of selection in the ABO blood groups. *Am. J. Phys. Anthrop.* 21:287–99.

Cavalli-Sforza, L. L., and W. F. Bodmer. 1971. *The Genetics of Human Populations.* Freeman and Co., San Francisco.

Chakraborty, R. 1975. Estimation of race admixture—A new method. *Am. J. Phys. Anthrop.* 42:507–12.

Edington, G. M. 1955. The pathology of sickle-cell disease in West Africa. *Trans. Royal Soc. Trop. Med. Hyg.* **49**:253–67.

Fisher, R. A. 1918. The correlation between relatives on the supposition of Mendelian inheritance. *Trans. Royal Soc. Edinburgh* **52**:399–433.

Fisher, R. A. 1930. *The Genetical Theory of Natural Selection*. Clarendon Press, Oxford.

Ford, E. B. 1940. Polymorphism and taxonomy. In J. S. Huxley (ed.), *New Systematics*. Clarendon Press, Oxford, pp. 493–513.

Galton, F. 1889. *Natural Inheritance*. Macmillan, London.

Haldane, J. B. S. 1924a. The mathematical theory of natural and artificial selection. Part I. *Trans. Camb. Phil. Soc.* **23**:19–41.

Haldane, J. B. S. 1924b. The mathematical theory of natural and artificial selection. Part II. *Trans. Camb. Phil. Soc., Biol. Sci.* **1**:158–63.

Haldane, J. B. S. 1926a. The mathematical theory of natural and artificial selection. Part III. *Trans. Camb. Phil. Soc.* **34**:363–72.

Haldane, J. B. S. 1926b. The mathematical theory of natural and artificial selection. Part IV. *Trans. Camb. Phil. Soc.* **23**:607–15.

Haldane, J. B. S. 1936. Some principles of causal analysis in genetics. *Erkenntnis* **6**:346–61.

Haldane, J. B. S. 1946. The interaction of nature and nurture. *Ann. Eug.* **13**:197–205.

Haldane, J. B. S. 1949. The rate of mutation of human genes. *Proc. 5th Int. Cong. Genet.*, Stockholm, pp. 267–73.

Harris, H. 1970. *The Principles of Human Biochemical Genetics*, 1st ed. North Holland Publishing, Amsterdam.

Harris, H. 1976. Enzyme variants in human populations. *The Johns Hopkins Med. J.* **128**:245–52.

Harris, H., and D. A. Hopkinson. 1976. *Handbook of Enzyme Electrophoresis in Human Genetics*. North Holland-American Elsevier, New York.

Hogben, L. 1939. *Nature and Nurture*, rev. ed. W. W. Norton and Co., New York.

Hunter, R. L., and C. L. Markert. 1957. Histochemical demonstration of enzymes separated by zone electrophoresis in starch gels. *Science* **125**:1294–95.

Kimura, M. 1960. A measurement of the rate of accumulation of genetic information in adaptive evolution. *Bull. Int. Stat. Inst.* **28**:239–48.

Kimura, M. 1971. Theoretical foundation of population genetics at the molecular level. *Theor. Pop. Biol.* **2**:174–208.

Kimura, M., and J. F. Crow. 1964. The number of alleles that can be maintained in a finite population. *Genetics* **49**:725–38.

Lewontin, R. C. 1967. An estimate of average heterozygosity in man. *Am. J. Hum. Genet.* **19**:681–85.

Lewontin, R. C. 1972. The apportionment of human diversity. *Evol. Biol.* **6**:381–98.

Lewontin, R. C., and J. L. Hubby. 1966. A molecular approach to the study of genic heterozygosity in natural populations. II: Amount of variation and degree of heterozygosity in natural populations of *Drosophila pseudoobscura*. *Genetics* **54**:595–609.

Li, W. H. 1977. *Stochastic Models in Population Genetics*. Dowden, Hutchinson & Ross, Stroudsburg, Pa.

Livingstone, F. B. 1957. Sickling and malaria. *Brit. Med. J.* **1**:762–63.

Livingstone, F. B. 1958. Anthropological implications of sickle-cell gene distribution in West Africa. *Am. Anthrop.* **60**:553–62.

McKusick, V. A. 1968. *Human Genetics.* Prentice-Hall, Englewood Cliffs, N.J.

Neel, J. V. 1956. The genetics of human haemoglobin differences: Problems and prospectives. *Ann. Hum. Genet. Lond.* **21**:1–30.

Nei, M., and A. K. Roychoudhury. 1972. Gene differences between Caucasian, Negro, and Japanese populations. *Science* **177**:434–36.

Nei, M., and A. K. Roychoudhury. 1974. Genetic variation within and between the three major races of man, Caucasoids, Negroids, and Mongoloids. *Am. J. Hum. Genet.* **26**:421–43.

Raper, A. B. 1956. Sickling in relation to morbidity from malaria and other diseases. *Brit. Med. J.* **1**:965.

Reed, T. E. 1969. Caucasian genes in American Negroes. *Science* **165**:762–70.

Reed, T. E. 1975. Selection and the blood group polymorphisms. In F. M. Salzano (ed.), *The Role of Natural Selection in Human Evolution.* North Holland-American Elsevier, New York, pp. 231–46.

Shaw, C. R. 1965. Electrophoretic variation in enzymes. *Science* **149**:936–43.

Smithies, O. 1955. Zone electrophoresis in starch gels: Group variations in the serum proteins of normal human adults. *Biochem. J.* **61**:629–641.

Wright, S. 1931. Evolution in Mendelian Populations. *Genetics* **16**:97–159.

Wright, S. 1948. Genetics of populations. *Encyclopedia Britannica* **10**:111, 111A–D, 112.

21

Reprinted from *Ann. Eug.* **13**:197–205 (1946)

THE INTERACTION OF NATURE AND NURTURE

BY J. B. S. HALDANE

The interaction of nature and nurture is one of the central problems of genetics. We can only determine the differences between two different genotypes by putting each of them into a number of different environments. We compare two pure lines of mice not only as regards colour, hair form, and other characters which are little affected by nurture, but for such characters as resistance to different bacterial and virus infections, each of which must be tested by appropriate changes of environment. We may require a wheat to be resistant to each of a number of un-favourable conditions, such as drought, frost, and attacks by several different varieties of rust. The line must be exposed to each of these separately.

The problem is, of course, exceedingly complex, but certain facts about it are so simple that they are apparently never stated. As they are matters of logic rather than experiment, I first put them forward (Haldane, 1936) at a congress mainly composed of philosophers, though in a rudimentary form which is here expanded.

Suppose that we have two genetically different populations A and B, and two different environments X and Y. The results will be clearest if A and B are clones or pure lines, and X and Y constant in all measurable respects. But this need not be the case. For example, A and B may be subspecies or even species, and X and Y geographical localities. Suppose also that the individuals show a character which can be measured, such as length of tail or life, number of eggs laid in two years, or mean butter-fat percentage; or a character which can be graded, such as colour intensity, dominance in combats, or intelligence as estimated by a battery of tests. The differences between all, some, or none of the combinations may be statistically significant.

First suppose that all four experiments give results which differ significantly, so that they can be placed in an order, as in Fig. 1. Arrangement 1a means that race A in environment X did

		X	Y				X	Y
Type 1a					Type 1b			
	A	1	2			A	1	3
	B	3	4			B	2	4
Type 2		X	Y					
	A	1	4					
	B	2	3					
Type 3		X	Y					
	A	1	2					
	B	4	3.					
Type 4a		X	Y		Type 4b		X	Y
	A	1	3			A	1	4
	B	4	2			B	3	2

Fig. 1.

'best', i.e. had the highest mean weight, milk yield, colour intensity, etc., while race A in environment Y did 'second best' and so on. Of course, with a different criterion the order might be different. Now there are just twenty-four ways of arranging the numbers 1, 2, 3, 4 in a square; and if we choose A and X so that AX is the best combination, there are just six ways, which are shown in Fig. 1.

If we reverse the order, e.g. grade the lightest colour instead of the darkest as number 1, the six types are unchanged. For example, type 1 a becomes

$$
\begin{array}{ccc}
 & X & Y \\
A & 4 & 3 \\
B & 2 & 1
\end{array}
\quad \text{or} \quad
\begin{array}{ccc}
 & Y & X \\
B & 1 & 2 \\
A & 3 & 4
\end{array}
$$

as before.

Each represents a type of interaction. In interactions of type 1, A does better than B in each environment, and X is superior to Y for each race. Types 1 a and 1 b do not differ essentially, for if the difference between X and Y is diminished sufficiently, 1 b passes over into 1 a. Eugenists claim that interaction of type 1 a is the most important for our own species. Environmentalists stress the importance of type 1 b, even when they differ as to whether the 'best' environment X is socialism, free-for-all competition, or life in accordance with their own religious beliefs. But both are apt to forget the existence of the four other types of interaction. Numerous examples of both types 1 a and 1 b are to be found in the literature. For example, if A are white recruits, B negro recruits to the U.S. army in 1917–18, and the criterion the median score in the army Alpha test, then if X is New York and Y Tennessee we find an interaction of type 1 a; if X is New York and Y Arkansas, we find one of type 1 b, according to the data of Yerkes (1921). The median values were;

	New York	Tennessee	Arkansas
Whites	58·3	44·0	35·6
Coloured	38·6	29·7	16·1

It is clear that, whichever pair of states is compared, the interaction is of the same kind. The fact that the Tennessee whites did a little better, and the Arkansas whites a little worse than the New York coloured people, does not alter its character. Of course, on the above facts it is quite impossible to decide whether if the whites and negroes had been placed in identical environments from birth (or better, as a result of ovarian transplantation, from conception) there would have been any significant differences between the white and coloured people. The possession of a pigmented skin was a handicap to a person born in the U.S.A. about 1900. Whether a part of this handicap was due to congenital intellectual inferiority is quite unknown. It is only certain that much of it was due to other causes.

The second type of interaction is highly characteristic of domesticated plants and animals. A specialized race A gives a higher yield than a race B which is nearer to the wild type, in a favourable but highly artificial environment. But in more primitive conditions it does worse. Consider two of the Scottish breeds of beef cattle, A the Aberdeen Angus, and B the Galloway. Under optimal conditions X, A produces more beef. The mean weight of show steers with an average age of 1 yr. 10 months at Smithfield from 1902 to 1911 was 1737 lb. for A and 1224 lb. for B. The Galloways, to quote *British Breeds of Livestock* (1920), 'will thrive and put on flesh on very poor grazing, and are admirably suited to cold, wet, climatic conditions'. But they 'bear heavy feeding very well'. Thus if Y represents a poor pasture and cold wet climate, the Galloway B does not do badly, but it does better on good pasture X. The hill farmers who breed Galloways would breed Aberdeen Angus if it paid them. But in environment Y, the breed A is not so good an economic proposition as B. It would probably have a high death-rate, and the survivors would not thrive. Unfortunately, I can find no figures comparing the performance of these breeds in poor environmental conditions. Cases are, however, available in plants, where, for example, wheat A does better than B when sparsely sown, and worse when densely sown. In fact it would be more accurate to say that our domestic animals and plants have been selected for variable response to their environment than for a consistently favourable response. It has been suggested that the highest human types only excel in a favourable environment, and fail completely in an adverse one which had a less disastrous effect on the average individual. I doubt if this is generally true, but it may well be true for some genotypes.

With interactions of types 3 and 4 the change of environment from X to Y affects the two genotypes in opposite directions. This situation is very common. Bridges & Brehme (1944) and others list responses of mutant types in *Drosophila melanogaster* to temperature which are summarized in Table 1. It will be seen that abnormalities of many different types are enhanced by heat, and a slightly but not significantly greater number enhanced by cold.

Table 1

Eye colour	Eye size	Eye surface	Venation	Wing size	Wing shape	Bristles	Mis llaneous
doughnut +	Bar +	morula +	Abruptex +	Beadex^r –	balloon –	bobbed –	bithorax +
light +	Deformed -	pebbled +	bifid +	Crimp +	bent +	condensed +	cloven thorax –
maroon-like –	eyeless +	sparkling –	Cell +	Clipped +	blistered –	Dichaete +	cut (head effect) –
mottled of	Lobe +	...	cubitus	dumpy +	Curly +	hook +	pentagon (body colour) –
In (2LR) 40d –	scutenick –	...	interruptus –	fringed +	fluted –	morula +	reduplicated (legs) –
mottled-28 –	welt +	...	heavy vein –	jagged +	jaunty +	Notopleural –	sable (body colour) –
white-buff –	...	...	knot –	narrow –	pleated –	polychaetoid –	scutenick (scutellum) +
white-blood +	...	...	Plexate –	roughish –	Pufdi +	polychaetous +	staroid (pleiotropic) +
			suppressor of	vestigial –	pupal –	scabrous +	short-wing (pleiotropic) +
			veinlet –		Stubblerec^2 –	...	tetraltera –
			thick veins –	...	vesiculated –	...	...

+ Abnormality increased by heat.
– Abnormality increased by cold.

Suppose A is fringed^2, B vestigial, X is 19° C., Y 30° C., and the flies are graded by wing area. AX is slightly scalloped, AY more deeply so, BY a narrow strap, and BX a mere scale. Another example of type 3 interaction may be taken from our own species. Let A be a group of normal children, B a group of mild mental defectives. Let X be an ordinary school, Y a special school for defectives. A will beat B on educational tests in either environment, but A will be better adjusted in the ordinary school, B in the special school.

In type 4 interaction each genotype does best in an environment to which it is adapted. One cannot say that either race or either environment is superior. The adaptation may be to climate or other physical conditions, or to parasites. Thus Europeans have a longer expectation of life than negroes in European cities, partly owing to their comparative immunity to tuberculosis. But in many parts of West Africa the negroes probably have the advantage, largely through their greater resistance to yellow fever.

Quite analogous 'pre-adaptations' are found among the *D. melanogaster* mutants. Thus polychaetous and polychaetoid are autosomal recessives in chromosomes 2 and 3 respectively, both producing extra bristles. Polychaetous overlaps the wild type at 19° C., but gives a good crop of extra bristles at 28–30° C. Polychaetoid overlaps the wild type at 25° C. and higher temperatures, but is clearly distinguishable at 19° C.

Many of the published results on photoperiodicity give excellent quantitative examples of type 4 interaction. For example, Hawkes (1943) obtained the following results for mean weight of tubers from two clones of *Solanum andigenum* in long and short days respectively over three years. The short days were 8½ hr., the long varied from 13 to 16 hr.

	Short	Long
Clone 1108	134	37
Clone 1068	91	131

In selection of stocks for agricultural purposes in different districts one is only concerned with interactions of types 2 and 4, in which each race is superior in a particular environment. Where the demand is heavy and transport is an important item in the costs, a crop will be grown or a stock raised in comparatively unfavourable environments. Interaction of type 2 may then be important. Where demand is less urgent or transport costs small, interaction of type 4 is more likely. Thus in Britain we find a place for tough but slow-growing breeds of sheep and cattle in our mountainous districts, but in peacetime we do not grow wheat unless a pretty high yield per acre is obtainable. So for sheep and cattle interactions of type 2 are fairly common, whereas type 4 is probably commoner with wheat so far as Britain is concerned. We require somewhat different types of wheat for the dry chalk uplands and the fens. For the planet as a whole this is probably not so. The spring wheats grown near the northern limit of cultivation give a poor yield per acre, but winter wheats would give less. The interaction is thus of type 2.

To obtain a further insight into the relation between the different types of interaction, let us suppose that the difference between the two environments X and Y is made very small. This is always possible when the difference is one of a single factor such as temperature, humidity, or length of day. It is not necessarily possible for differences of locality, since all the above, besides differences in soil, biological environment, and so on, may be important, and they may interact in a complex manner. If X and Y are very similar, then AX and AY can differ little. Similarly, BX and BY can differ little. Thus the interaction is of type $1a$ if the change of environment affects the two genotypes in the same way, and of type 3 if it affects them in opposite ways.

Next consider what occurs if X and Y differ considerably, but A and B are very similar. This can, of course, be achieved if for B we substitute a mixture of 99 % A and 1 % B, the organisms being kept separate (e.g. mice in individual cages) to avoid biological interactions such as infection or crossing. But we cannot, in general, find a single genotype intermediate between two others in all respects, any more than we can find a locality. In particular, alleles in a multiple series which are intermediate in one respect are not necessarily so in another. If, however, A and B are very similar the interaction is of type $1b$ if the genetic change has the same effect in both environments, of type 2 if it has the opposite effect. For example, if A is superior to B in both environments we get type $1b$ interaction. But if, for example, B is slightly tougher but less specialized than A we may expect an interaction of type 2 if X is a more favourable environment than Y.

It follows that if A is very similar to B, and X to Y, we can only get an interaction of type 1. Type 1 may, in fact, be regarded as the limiting type when differences are small. And type 4 is only likely to come into operation when both A and B and X and Y differ markedly. But in almost all species the range of environments and genotypes is so large that interactions of this type must occur. Professor Penrose has pointed out to me that it is important that the performance measured or graded in all four tests should be exactly the same. Thus at first sight it might seem legitimate to say that a group A of children proved more intelligent than group B when confronted with a verbal test X, and less so when confronted with a maze test Y. But in such a case we cannot really compare the performance of A in respect to X and Y. Suppose that A are English and B foreign children, it is probable that the test scores will have been standardized so that a representative group of English children scores about 100. If so the only possible types of results are as follows (X and Y being interchanged if necessary):

	X	Y				X	Y	
A	102	97	($1a$)	or	A	102	97	(3)
B	90	81			B	81	90	

if the English children are consistently superior. If they are consistently inferior we get the same two types. If, however, the English children do better on one test, but worse on the other, we get:

	X	Y				X	Y	
A	102	97	(2)	or	A	97	102	($4a$)
B	117	84			B	117	84	

Interactions of types $1b$ and $4b$ with significant differences cannot be obtained. This is, however, possible if the different types of performance can be compared by some objective standard, for example, the economic value of milk or beef produced. If so it is of course clear that the order may alter with a change in prices or transport conditions, and it is clearly very hard to make such comparisons in the case of men.

In many practical cases the differences between two or more of the four measured characters are not significant. If this is so it may or may not be possible to specify the type of interaction. There are two types of indetermination. First of all two or three of the characters may be indistinguishable, though significantly different from the other or others. Thus if environment X is sterile, and Y infected, then AX and BX will probably give no deaths, and AY will not do so if A is immune to the disease in question, whereas BY will do so if B is susceptible. Secondly, two characters may differ significantly. But another may lie between them without differing significantly from either. Thus if each measurement were based on a fairly large sample, we might get such a result as AX 9.0 ± 0.5, BX 8.2 ± 0.6, AY 7.5 ± 0.4. Here AX and AY differ by 2·34 times the standard error of their difference. But AX and BX only differ by 1·02 times this quantity, BX and AY by 0·97 times.

Type $1a$ or $1b$		X	Y			X	Y			X	Y			X	Y
	A	1	2
	B	2	4
Type $1a$ or 3		X	Y			X	Y			X	Y			X	Y
	A	1	2	A	1	1	A	1	1	A	1	1
	B	4	4	B	3	4	B	4	3	B	3	3
Type $1b$ or 2		X	Y			X	Y			X	Y			X	Y
	A	1	3	A	1	4	A	1	3	A	1	4
	B	1	4	B	1	3	B	1	3	B	2	4
Type 2 or $4b$		X	Y
	A	1	4
	B	2	2
Type 3 or $4a$		X	Y
	A	1	2
	B	4	2
Type $4a$ or $4b$		X	Y			X	Y			X	Y			X	Y
	A	1	3	A	1	4	A	1	4	A	1	3
	B	4	1	B	3	1	B	4	2	B	3	1

Fig. 2.

Fig. 2 deals with the first case. The top left-hand square means that AX had the highest score, AY and BX equal second, and BY fourth. The other possible results give no information about the type of interaction. For example, if A was an immune variety, B susceptible to a disease, X a sterile environment, Y an infected one, we might get

	X	Y
A	1	1
B	1	4

if all individuals survived the experiment except some of the infected B population. If, however, the experiment lasted longer it could turn out that, in environment X, A was longer-lived or shorter-lived than B. It could also turn out that the infection slightly shortened A's expectation of life. But it might however lengthen it. Some bacteria are useful symbionts, for example, those which produce vitamin K in the human gut. But when their ancestors first invaded the guts of our ancestors, they may well have killed off some genotypes. Thus all six types of interaction are possible.

Where one score lies between two others without differing significantly from either, things are more complicated. Thus consider the situation

	X	Y
A	1	(2)
B	3	4

in which AX and BY differ significantly, but AY does not differ significantly from either. If the experiment were repeated on a larger scale we might get a result:

	X	Y	Type 1a			X	Y	Type 3	or		X	Y	Type 1b
A	1	2			A	2	1				A	1	3	
B	3	4			B	3	4				B	2	4	

In any given case it is easy to work out the possibilities.

In practice we generally have to consider a large number of genotypes and environments, and the number of possible types of interaction is very great. Thus if we have m genotypes and n environments, but can rearrange the order of the genotypes and environments after the experiment is completed, we have $\dfrac{(mn)!}{m!\,n!}$ possible types of interaction; for example, if $m = n = 10$, there are $7 \cdot 09 \times 10^{144}$ types. Even for the simplest case but one, of two genotypes in three environments or three genotypes in two environments, there are sixty types of interaction. Similarly, if we classify our environments in several different ways, e.g. hot and cold, dry and moist, there are, in the simplest case of a $2 \times 2 \times 2$-fold table, 5040 types of interaction; for a (2^n)-fold table $(2^n - 1)!$ types. If, on the other hand, we choose several different criteria of performance, e.g. milk yield, beef yield, and capacity for ploughing, there are seventy-two different types of interaction in the simplest case of two genotypes, two environments, and two criteria, and in general for m genotypes, n environments, and k criteria, $\dfrac{(mn!)^k}{m!\,n!\,k!}$ types of interaction. Clearly there would be no advantage in attempting to generalize our classification. One point is, however, worth making.

Of the $\dfrac{(mn)!}{m!\,n!}$ types of interaction of m genotypes and n environments, only $\dfrac{(mn)!}{(m!)^n n!}$ are such that the order of merit of the m genotypes is the same in every environment, and only $\dfrac{(mn)!}{m!\,(n!)^m}$

are such that the order of merit of the n environments is the same for every genotype. This can readily be seen by considering the number of ways in which an order of this rare type can be deranged. For example, three races in three environments have 10,080 types of interaction. In only 280 of these is the order of the three races the same in each environment. It is much harder to calculate the number of types of interaction in which the order of merit of the genotypes is the same in every environment, and the order of merit of environments the same for every genotype. This number is 1 (type 1a) for $m = n = 2$, and 5 for $m = 3$, $n = 2$, or $m = 2$, $n = 3$.

A few words may be said about the application of the ideas here developed to agriculture and to eugenics. The agricultural problem is enormously simpler because there is comparatively little crossing between stocks of the same species, and these stocks are homozygous for many of the genes responsible for the differences between them. But our own species is not so divided up, and even if extreme eugenical principles were accepted by a people or practised by a dynasty of tyrants, many centuries would elapse before races or classes attained the degree of genetical homogeneity of the average breed of cattle. Still more important, each domestic plant or animal breed is judged in the main by a single economic criterion. No doubt the criterion is biologically complicated. Thus besides milk yield and quality, fertility, rate of maturing, resistance to disease, and meat production by male castrates, are all of some importance in a breed of dairy cattle. And within a whole domestic species, only a few types of performance are required. Thus sheep are bred with a view to meat production, and also to that of several types of wool, of fleeces for coat linings and the like, and of several types of ornamental fur. In a few countries they are used for milk production and transport. But the division of labour in a modern society, quite apart from any question of cultural or spiritual excellence, demands many different human abilities, each of which may give a different ranking order.

At first sight it might be regarded as a sound agricultural policy to select the genotypes which give the best performances in each of a fairly large series of typical environments. Unfortunately in current practice far too much attention is paid to performance in highly favourable environments, and there is a tendency to lose sight of type 2 interactions. In consequence a breed of cattle may be 'graded up' so as to give a high milk production under optimal conditions. But it may actually give a lower one under poorer conditions, particularly when allowance is made for deaths. It may also turn out that the highest ranking genotype is a heterozygote, perhaps a multiple heterozygote. This is certainly true for some fancy breeds, and for economically valuable breeds such as Dexter cattle. If so one or both of the homozygous forms must be kept as breeders, though their actual performance may be relatively low. In sheep-breeding it is a common practice to cross certain breeds, the hybrids not being used for further breeding. Thus Scottish Blackfaced ewes are commonly crossed with Border Leicester rams. This practice is sometimes carried so far that a breed is mainly valued for its performance as a parent. Thus to quote *British Breeds of Livestock*, 'now practically all the purebred flocks of Cotswolds are maintained to provide rams for crossing purposes'. In the United States maize for agricultural purposes is largely grown from crosses of homozygous lines of comparatively low yield. Finally, first crosses between poultry of different breeds are reared on a large scale not mainly on account of hybrid vigour but because their sex can be determined on hatching.

In our own species we know little as to the good or bad results of interracial crosses, since race mixture is always complicated by culture mixture. But it is a curious commentary on human racial theories that if we could argue from animal breeding to human breeding, the first cross between different races would probably often be more vigorous than either pure race, whatever might happen in later generations. So a dictator who combined the ruthlessness of one lately deceased with a greater knowledge of biology might have encouraged racial crossing, whilst preventing the further breeding of the hybrids; and a far-sighted human biologist, while not recommending widespread interracial crossing in the present state of our ignorance, might favour the preservation of pure stocks of even the least progressive races, not merely for their possible future performance, but for their possible value in hybridization at a later date.

It might, however, be thought a sound agricultural policy to discard all stocks other than those which give the best performance in one or other of a well-chosen series of environments, and those which may be needed for breeding heterozygous types. This view is based on the false assumption that the environment will not change greatly. The following are among the obvious changes which may occur:

(1) Emergence of new diseases.

(2) Changes in agricultural technique, e.g. introduction of machinery, the use of new types of fertilizer, the use of hormones, vernalization, and so forth.

(3) Changes in demand, causing, e.g. a great increase in the area under a given crop.

(4) Changes in environment due to export.

The first of these is a fundamental reason for preserving considerable genetical diversity. It is particularly important with clonally grown plants, where a very high degree of homogeneity is obtained. A striking example occurred in the banana. Large areas of Central America and the West Indies were grown with the triploid clone called Gros Michel. This is particularly susceptible to the fungus *Fusarium cubense*, causing Panama disease, and the spread of this disease has therefore had serious effects on the banana industry, and greatly stimulated genetical work intended to produce other sterile clones. Similar dangers are to be expected if a single pure line of wheat is grown over wide areas. Since the evolutionary steps by which a parasite may adapt itself to a new host cannot be foreseen, much less prevented, such events can only be guarded against by the maintenance of a high level of genetical diversity. All that we can be sure of is that the biological environment will change.

Similarly, the possibility of changes in agricultural technique call for a large reserve of geno-types. Vernalization permits the use of autumn wheats in cold climates where only spring wheats could formerly be grown. But it is unlikely that the varieties of autumn wheat which give the best results when vernalized are those which give the best results when sown in autumn in a warmer climate. Similarly the variety of sheep which will give the highest yield of wool in a given area when their fertility is increased by hormone injections is not necessarily that which will give the highest yield when not so treated.

In fact the type of classification here adopted is not really adequate for the practical problems of agriculture because it does not take time, and particularly environmental change, into consideration. If the environment X is liable to change into environment Y we need, besides our A stock, a B stock in reserve which will give an interaction of type 2 or type 4. As we do not know the nature of Y beforehand this means that we should have a considerable number of stocks in reserve, including even definitely inferior stocks which, though they are not themselves likely to be of value, may carry valuable genes.

An increased area is likely to call for interactions of type 2, since the organism in question will have to be grown in less favourable environments than before. Export, for example of British cattle to tropical countries, calls for interactions of type 4, though probably type 2 is more likely to be achieved in practice.

The same arguments apply to eugenic policy. We are not justified in condemning a genotype absolutely unless we are sure that some other genotype exists which would excel it by all possible criteria in all possible environments. We can only be reasonably sure of this in the case of the grosser types of congenital mental and physical defect. A moderate degree of mental dullness may be a desideratum for certain types of monotonous but at present necessary work, even if in most or all existing nations there may turn out to be far too many people so qualified.

In a society which was perfect from the eugenical point of view there would be no interactions of types 1 or 3; genotypes of consistently inferior performance would have been eliminated. Similarly, in a society which was optimal from the point of view of environment there would be no interactions of types 1 or 2. People would not be placed in environments where they could not do their best, at least in some respect. It follows that interactions of type 4 are the ideal at which we should aim. That is to say in the ideal society there would be a diversity of social functions and of human endowments, but no social function could be dispensed with without damage to the society, and no individual could exercise a different function without performing it less efficiently and happily than those who were actually doing so. This ideal is obviously remote and probably never fully attainable, but it is perhaps worth stating. Meanwhile our efforts should be mainly concentrated on the elimination of interactions of type 1, that is to say the elimination of environments which are unfavourable to all genotypes, and of genotypes which are inferior in all environments.

SUMMARY

A simple classification of the possible types of interaction between two stocks and two environments is given. This classification is applied to a number of concrete cases arising in genetics, agriculture and eugenics.

I should like to acknowledge my debt to Dr J. M. Rendel for many of the ideas here put forward concerning breeds of livestock, while acquitting him of any erroneous views of which I may have been guilty.

REFERENCES

C. B. BRIDGES & K. S. BREHME (1944). The mutants of Drosophila melanogaster. Publ. Carneg. Instn, no. 552.
British Breeds of Livestock (1920). H.M. Stationery Office. London.
J. B. S. HALDANE (1936). Some principles of causal analysis in genetics. Erkenntnis, 6, 346.
G. HAWKES (1943). Photoperiodism in the Potato. Cambridge: Imp. Bur. Plant Breeding and Genetics.
R. M. YERKES (1921). Psychological examining in the U.S. army. Mem. Nat. Acad. Sci. 15, 689.

22

Reprinted from *Biol. Rev.* **20**:73–88 (1945)

POLYMORPHISM

By E. B. FORD, Reader in Genetics in the University of Oxford

(*Received* 9 *December* 1944)

I. INTRODUCTION

A considerable amount of information on polymorphism has accumulated since I last summarized one of its aspects, that bearing upon taxonomy (Ford, 1940*b*); for, in addition to the results of recent research, Huxley (1942) has brought together a great variety of data upon the subject which have never before become accessible in a single work. Moreover, it is only in the last few years that the bearing of polymorphism upon human genetics has been recognized, a consideration which now invests it with added interest.

I have defined polymorphism as the occurrence together in the same habitat of two or more distinct forms of a species in such proportions that the rarest of them cannot be maintained by recurrent mutation (Ford, 1940*b*). It is at the outset necessary briefly to explain and analyse this definition. In the first place, it excludes seasonal and geographical variation, for it concerns solely those forms which occur 'together in the same habitat'. Secondly, it relates not to 'continuous variation', such as that falling within a single normal distribution, but to relatively sharply contrasted differences which either do not overlap or else give rise to a bi-modal (or multi-modal) curve. The phenomenon is, however, one in which the frequencies of the different phases play an essential part, and in order to make this clear it is convenient to classify genetic variation under three headings from the point of view of its selective effect, while that due to the environment must be considered separately.

Genetic variability must be due to genes which are either (1) harmful, (2) neutral, or (3) beneficial, in their influence upon the organism. Now the fundamental attribute of mendelian inheritance as an evolutionary agent is the extreme permanence of the hereditary factors, so that those which arose far apart in time or in space can be brought together and combined in new, and possibly advantageous, ways. This is ensured partly by the fact that the genes do not contaminate one another when present in the same individual. But such protection from the possibility of genic blending would be unavailing unless they were themselves intrinsically stable. Thus we are driven to the conclusion on the general grounds of evolutionary theory that the mutation rate must inevitably be very low, and so in practice we find it: indeed, in normal circumstances, its upper limit seems to be approached by genes mutating

in one out of 80,000 individuals (Ford, 1945*a*). Yet it must be remembered that the rarity of mutation is not demonstrated merely by observations upon its frequency in experimental conditions, but (as originally pointed out by Fisher, 1930*a*) upon the existence and production of pure lines of organisms; for these would be impossible if mutation were common. Consequently, any given gene subject to adverse selection must always be infrequent in the population; because, being constantly eliminated, it is dependent for its existence upon mutation.

The second group of genes, those which are neutral from the point of view of survival value, contribute but little to the variation of organisms in nature, and this for two reasons. First, as Fisher (1930*b*) has shown, they are very rare, since effective neutrality of this kind requires that the balance of selective advantages between the members of an allelomorphic pair must be remarkably exact. Secondly, the spread of such a gene is exceedingly slow, for the number of individuals which possess it cannot greatly exceed the number of generations since its occurrence, if it be derived from a single mutation (Fisher, 1930*a*); and the mutation rate is too low materially to hasten the process.

From what has just been said, it will be clear that if a genetically controlled variety occupies even a few per cent. of a population, we can be fairly sure that it possesses some advantage. If multi-factorial, such variation will be of the 'continuous' type; but if clear cut, depending upon a genetic or environmental switch mechanism, it must fall within the definition of polymorphism. Yet this includes two distinct conditions: (1) that in which an advantageous gene is in process of spreading through the population, and (2) that in which it is maintained at some fixed level by a balance of selective agencies. These I have called *transient* and *balanced polymorphism* respectively (Ford, 1940*b*).

It was first shown by Fisher (1927) that, as will presently be explained, a permanent equilibrium between genetically controlled forms can only be established by means of what is now called balanced polymorphism. Indeed, I only included the transient type as a form of polymorphism on the grounds of practical necessity, owing to the great difficulty, often encountered, of deciding in what instances stability is in fact attained. It will, however, be realized that both situations involve the spread of an advantageous variation, which in the one advances through the population unopposed and in the other is checked by selection at a given level. It will be convenient to begin the detailed study of the subject by considering these two types.

II. TRANSIENT POLYMORPHISM

Transient polymorphism is maintained only during the time that a gene is spreading through a population and is in process of replacing its allelomorph. When it has reduced this to the status of a rarity, though it was formerly the normal one, virtual uniformity, of a new type, has been reached and polymorphism is ended. If organisms lived in constant environments, it would hardly be possible for a mutant gene to increase in frequency save in special genetic conditions. For mutation is a recurrent phenomenon, so that any advantageous gene should long ago have been incorporated into the gene complex. In such circumstances, further evolution would depend upon the selection of gene combinations which interacted to produce favourable effects. But environments are not stable, and it may well be that a mutant gene constantly eliminated in the past may be a success to-day. The great changes brought about by man, often creating environments never before known or approached, will evidently accentuate this tendency. For not only in industrial regions has human activity produced situations of a quite unprecedented kind, but in many parts of the world what we are inclined to regard as 'unspoiled country' is in reality far removed from conditions which have ever arisen, or indeed could arise, by natural means. We may expect, therefore, that the phenomenon of transient polymorphism, due to the spread of genes with fairly clear-cut effects, should not only be occasionally observable but that, at the present time, there should be exceptional opportunities for observing it. These general conclusions appear to be realized, and it will be useful briefly to discuss them with the help of examples. These must be illustrative only, for it is not my intention to bring together from the literature a wide selection of instances demonstrating the spread of genes in natural conditions, as this subject is but a side issue of the present study. Moreover, this process has repeatedly been assumed on inconclusive evidence, since the possibility is often not excluded that stability has long ago been reached. Such situations may fall within the general definition of polymorphism without being assignable to one or other of its two types, and a selection of them, demonstrating certain general principles, will be discussed on pp. 80–82.

Transient polymorphism has, however, been fully established in the hamster, *Cricetus cricetus* L., which is a typical rodent of the dry zone of the Russian steppes, and a blackish form, inherited as a simple recessive, is known as a rarity throughout its range. In the latter part of the eighteenth century, the geographer Lepekhin found that these melanics occupied a high proportion of the population in a small area in the north-eastern corner of its habitat, not far from the foothills of the Ural mountains. Since that time they have spread steadily along the northern limit of the species across Central Russia, until to-day they are common even as far west as the Dnieper; a process which it has been possible to follow from the statistics of the Russian fur trade. The line of their advance represents the moister wood-steppe subzone which checks the spread of the hamster northwards, and here the black variety appears to be at some advantage which it does not possess in the dry country which is the true home of this animal (Timoféeff-Ressovsky, 1940). This example is typical of a large number of other instances which we lack the data to interpret. Apart from the possible effects of trapping, human interference can hardly play a part in the change which has taken place. Yet the boundary between the dry and the moister steppe must have altered constantly over a long period of time since the last ice age, and it is conceivable that it now coincides with geographical features which make it possible for the black form to colonize it; but this of course is pure conjecture. Since, however, these melanics occur widely though rarely in the normal population, this must contain a not inconsiderable proportion of heterozygotes. It is not clear, therefore, why the variety did not spread at a much earlier date, unless we may invoke the onset by natural means of conditions favourable to it. We have, however, no ground for supposing that the selection for or against this form depends solely upon its colour, for there is evidence that many melanics, as well as albinos, differ from the normal in their physiology.

Another example which may usefully be compared with this one is provided by the relative increase of the black form of the opossum, *Trichosurus vulpecula* Kerr, in Tasmania, described by Pearson (1938) and further discussed by Huxley (1942). In the first place, the evidence against the establishment of a stable balance between the two forms is indirect only, though it appears to be fairly conclusive. This variety is very rare in Australia but common in Tasmania, where it occupies approximately the whole population of the species in the north-western corner of the island. From here its frequency declines until at the present time only about 25 % of the animals on the east coast are black. Moreover, the extremely narrow Tasman Peninsula seems to be acting as a partial barrier to its advance, for here it is proportionately less frequent than on the adjoining district of the main island.

The genetics of this variety have not been studied, but it is probably under uni-factorial control, for intermediates are very rare or absent; while, since its frequency is not correlated with climatic gradients, it is not likely to be environmentally produced: without such evidence, this would be at least a possibility, since genes are known which produce pale pigment above a critical temperature (which in mammals must be near the constant body temperature) and dark pigment below it (e.g. the Himalayan rabbit, Schultz, 1920; other instances are discussed by Ford, 1938). The absence of melanics, save as aberrations, in Australia suggests that they began to spread after Tasmania had been severed from the continent; a view strengthened by their absence from the opossum population on an island representing a fragment of the former land-bridge to Victoria. The success of this gene may be due to changes which have taken place in Tasmania since that date, nor is it certain that those resulting from human influence can wholly be excluded.

Both the instances of transient polymorphism so far mentioned relate to the increase of black forms in wild populations, and there is one aspect of this occurrence of such outstanding importance and interest that it demands somewhat detailed examination. This is the establishment of industrial melanism, and some related phenomena, in moths.

abruptaria Thnbg. is not infrequent in London and its suburbs. The spread of var. *albigensis* Warn. of *Tethea or* Fb. was followed with some care in the Hamburg district. It was very rare in 1904, while by 1910–11 over 90 % of the larvae collected in that area produced it.

A few other species may be mentioned in which melanism is probably, but not quite certainly, industrial (see pp. 77, 78 for non-industrial melanism). These include the following:

Lymantria monacha L. (Lymantriidae). Dark forms of this moth, leading up to the coal-black *atra* Linst., have greatly increased on the European continent during the last forty years. Their genetics have been studied by Goldschmidt (1921), who showed that the condition is due to the combined action of three genes. (1) A sex-linked dominant (C) produces black hind wings and a suffusion of black on the fore wings. (2) An autosomal 'dominant' (B) increases the black pigment in the central area of the fore wings, but its effects are more extreme in

Table 1. *Examples of industrial melanism. All the melanic forms are simple dominants*

Family	Species	Melanic form		
		Name	Distribution	Genetic status
Geometridae	*Gonodontis bidentata*	*nigra* Prout	N. England	Bowater, 1914
,,	*Phigalia pedaria*	*monacharia* Stgr.	,,	Buckley, see Ford, 1937
,,	*Biston betularia*	*carbonaria* Jordan	,,	Lemche, 1931
,,	˙*Cleora repandata*	*nigricata* Fuchs.	,,	Walther, 1927
,,	*Hemerophila abruptaria*	*fuscata* Tutt	London	Onslow, 1921 c
,,	*Boarmia punctinalis*	*humperti* Hum.	Germany	Hasebroek, 1934
,,	*Ectropis extersaria*	*cornelseni* P. Hoff.	,,	,,
Thyatiridae	*Tethea or*	*albigensis* Warn.	,,	Gerschler, 1915

In certain industrial areas almost the whole population of moths, belonging to several species, has become black during the last eighty years, resulting in the most considerable evolutionary change which has ever actually been witnessed. The phenomenon has been particularly marked in England, but it is also well known in continental Europe and elsewhere. The species in which it was first encountered is *Biston betularia* L. (Geometridae), a specimen of the black variety *carbonaria* Jordan having been taken near Manchester in 1850. Since that date it has largely replaced the normal form, which is white speckled with black and, when sitting on a tree trunk, resembles a patch of lichen. To-day the melanics are predominant not only in the Manchester district but throughout the 'black country' of the north. Numerous other species have followed the same course in various manufacturing areas and in the neighbourhood of large towns. Among them may especially be noticed those listed in Table 1, in which the phenomenon has been established by careful observation. The melanic forms of the first four of these are now common in the manufacturing areas of northern England, and that of *Hemerophila*

homozygotes than in heterozygotes and in females than in males. (3) Another autosomal dominant (A) intensifies the pattern and increases the effect of the other two genes. However, the effect of C is so much greater than of B or A that a distinct polymorphism is produced when the melanic form becomes frequent.

A black variety, *fuscana* Sheldon, of *Acalla comariana* Zell. (Tortricidae) is now not uncommon in Lancashire. It is produced by a gene whose effect is intermediate in the heterozygote (Fryer, 1931).

Orthosia populeti F. has become black in the neighbourhood of Meissen and Wilsdruff in Germany owing to the spread of var. *nigra* Tutt, which is a simple dominant (Walther, 1927). The district is one in which large porcelain works are situated, but it is perhaps doubtful if the change in this insect can be included as 'industrial melanism' in the strict sense.

A large number of species could be cited in which the spread of melanism in manufacturing areas appears to be in its early stages. We are very deficient in exact statistics demonstrating the rate of this change, and these are greatly needed. Anyone obtaining such data would be performing a useful service. It is important that this information should in each year be related to estimates of the absolute numbers of the population in the area studied. There are reasons for thinking that the rate at which a gene will spread may vary with total abundance (p. 82), and indeed Walther (1927) was led to suspect this for the increase of melanic *Orthosia populeti* in the Wilsdruff neighbourhood; for he says that this variety appeared to be relatively commoner in the years of greatest plenty. (A method for obtaining the absolute numbers of a population of Lepidoptera is given by Dowdeswell, Fisher & Ford, 1940.)

Though some of the species involved in industrial melanism are closely allied, others are but distantly related. The feature which they have in common is, in fact, not affinity but their manner of concealment: for they appear to be insects which do not hide in crevices, but depend for protection upon cryptic colouring, which allows them to rest unobserved even in exposed situations such as tree trunks, walls and fences. Consequently, it has been suggested that the phenomenon is due to simple selection, favouring dark forms in industrial regions merely because such insects match the blackened vegetation better than would those of normal colouring, while the reverse is true in unpolluted country. Yet this explanation entirely fails to account for an additional fact which is one of the most remarkable ever met with in the study of wild populations: that is to say, the melanics appear to be hardier than the typical forms, although they have entirely failed to supplant them in ordinary conditions.

On the other hand, Prof. J. W. Heslop Harrison has attempted to show experimentally that such melanism is due to the induction of mutation produced by salts of lead and manganese, which are present in the soot which may be found upon leaves in industrial districts (Harrison & Garrett, 1926; Harrison, 1928). Space does not allow a detailed discussion of this work, except briefly to point out that Harrison does not appear to have substantiated his claim. He reared larvae on food impregnated with lead and manganese salts and obtained melanics after several generations. If their occurrence were really due to mutation induced by these substances, Fisher (1933) has drawn attention to its astonishingly high frequency, which is much greater than that ever produced in *Drosophila melanogaster* by X-rays! Perhaps a more serious difficulty is introduced by Harrison's choice of material, for he worked principally with *Tephrosia bistortata* Goeze and *Selenia bilunaria* Esp., species in which industrial melanism is unknown. In both these insects the melanic varieties are recessive, so that insects heterozygous for the condition might be introduced into the experimental material, and Fisher (1933) shows that Harrison's controls did not exclude this possibility. Moreover, both Hughes (1932) and Thomsen & Lemche (1933) have repeated Harrison's work, using

very large numbers of *Selenia bilunaria*, and they entirely failed to confirm his results. This Harrison (1935) attributes to larval mortality acting differentially upon the melanics, which he admits to be the less hardy in this species. Here he introduces what appears to be a serious criticism of his own experimental design. That is to say, he attempted to demonstrate a phenomenon which has never occurred in industrial, or in any other conditions: the spread of *recessive* melanism in the Lepidoptera. Now recessive melanic varieties are known not only in the two species with which Harrison worked but in many others. Examples of them are provided by *Ephestia kühniella* Zell. (Kuhn & Henke, 1929–36); *Abraxas grossulariata* L., vars. *varleyata* Porritt and *hazeleighensis* Raynor (Onslow, 1921 a); and *Zygaena trifolii* Esp. (Grosvenor, 1926–7). As far as the evidence goes it appears always to reduce viability, and it is unknown except as a rare sporadic occurrence. Here we have an entirely different situation from that under discussion, for all the melanic forms of Lepidoptera which have become established in nature are either dominants or semi-dominants (see also p. 78) and, as already indicated, it is constantly reported of them that they are hardier than the typical insects.

Thus not only are the theories so far mentioned unsubstantiated, but at the outset they fail even as explanations; for they give no indication why the relatively more viable melanics have not long ago displaced the non-melanic forms. Accordingly, I suggested in *Biological Reviews* (1937) a new interpretation of industrial melanism, one important aspect of which I have since been able to verify.

Briefly, the view which I put forward is this. All genetic factors which improve the viability of the organism must have spread through the population, displacing their allelomorphs, and become incorporated in the gene complex of the species, except those which in addition carry with them some disadvantage sufficient to outweigh the superior hardiness which they confer. Among these latter we may reasonably include any that are responsible for such an excess of melanin production as to destroy the cryptic pattern of the insect, upon which its safety depends. However, black colouring will not be such a handicap, perhaps even an asset, in industrial districts, and this for two reasons. (1) It will tend to match the smoke-grimed tree-trunks and fences. (2) There may well be fewer predators in such places, so that the selection pressure in favour of the correct colouring will be less intense. Here then the species may at last be able to avail itself of those genes which confer greater viability, even though they may involve an excessive formation of melanin.

It will be noticed that, on this theory, melanics might be successful even if the disadvantage due to their colour is merely reduced though still present, and that appearance is not the quality for which they are selected. Also that they are not due to the induction of mutation; for I assume that these species have constantly had experience of the genes responsible for industrial melanism, but hitherto they have not been able to use them. Of this, indeed, we have good evidence, and it should have been apparent long ago that industrial conditions are not necessary for the *production* but for the *spread* of the black varieties. For example, the black form (*cornelseni* P.Hoff.) of *Ectropis extersaria* Hb. is, in fact, known in England, but only as a rare variety; it has never established itself here since the distribution of the species does not include manufacturing areas; but it does so in Germany, where *cornelseni* has proved a success and become an industrial melanic.

One of the essential features of this interpretation is the greater viability of the melanics, which has been reported by several competent breeders who have studied species in which melanism has become established either in manufacturing districts or in small non-industrial areas (p. 78). The excess of melanics which they frequently obtained is, however, not a significant one in any instance, except for one series of back-crosses involving the dark form of *Hemerophila abruptaria*. The larvae were subject to heavy elimination owing to severe winter frosts (Onslow, 1921c). They produced fifty-six insects in all, distributed as thirty-seven melanics: nineteen typical (for which $\chi^2 = 5.8$). However, Brett (1935–6) attributes this departure from equality to complications arising from the presence of a second melanic form which Onslow had not distinguished, though I have given reasons for thinking that it is at least in part due to differential viability (Ford, 1940a).

Clearly the matter could not be allowed to remain at this stage. I therefore undertook an investigation of the comparative viability of melanics and normals in *Cleora* ('= *Boarmia*) *repandata* L., and obtained back-cross results from three families bred under normal conditions. These produced a total of 192 insects, consisting of 101 melanics and 91 typical specimens. This is the type of result often obtained in the past by those who have studied the genetics of melanism in the Lepidoptera; one in which there is a slight, but obviously not significant, excess of the black form. It might well become significant if very large numbers could be used, but it seemed to me more practicable to test its reality in another way, by increasing the severity of selection. Accordingly, I reared seven further back-cross broods which I deliberately starved—leaving the larvae without food on alternate days and providing only a restricted supply when feeding was permitted. This treatment proved too rigorous for two large broods, which produced no imagines. The remaining five gave, in all, a total of eighty-three moths. The two forms were no longer in equality among them, for there were fifty-two melanics and thirty-one typical specimens. The departure from expectation is now a significant one: $\chi^2 = 5.8$, giving $P < 0.02$ with one degree of freedom (Ford, 1940a).

By this means the superior viability of the melanic form was established, and we may reasonably assume that the same thing could be done in the other industrial melanics in which a slight excess of the dark form has been reported. Indeed, I think we must concede that expert breeders were able to detect a difference in viability which, though real, did not reach formal significance. Thus the transient polymorphism due to the spread of industrial melanism appears to be promoted by selection favouring the more viable form which, in normal circumstances, is probably held in check by counter selection against individuals in which the cryptic colouring is obliterated by excess melanin formation.

Industrial melanism in the Lepidoptera is not limited to the imagines, for in a number of species the larvae have also become black in manufacturing districts (an excellent survey of this phenomenon is given by Harrison, 1932). Larval genetics have been studied very little, but it has been established that black forms give rise to normal moths in several species in which melanism of the imago is known (e.g. *Meganephria oxyacanthae* L.). Indeed the characters of larvae and adults are very generally under independent genetic control. As I have already pointed out (Ford, 1937), this is not surprising, for variation even on similar lines is likely to have very different adaptive values in these instars, whose interrelations with the environment are so dissimilar.

A circumstance of considerable interest is the recent spread of melanic forms of certain moths in restricted areas outside manufacturing districts. Some of these have established themselves as industrial melanics elsewhere within the range of the species, while others have not. The occurrence is of importance, since it has given rise to misconceptions in regard to the selective basis of transient polymorphism. The black forms of *Boarmia punctinalis* Scop. (var. *humperti*) and of *Ectropis extersaria* (var. *cornelseni*) have become common in one or more woods in Kent. They occur elsewhere within the English range of these insects only as rare aberrations though, as already mentioned, both have become industrial melanics in Germany. The deep greyish variety *delamerensis* B. White of *Ectropis crepuscularia* Hb. has spread in several manufacturing districts abroad, but it has also become common in Delamere Forest, Cheshire, which is not affected by industrialism.

On the other hand, the black variety *nigra* Banks of *Ectropis consonaria* Hb., has also established itself in a wood in Kent, though it has never become an industrial melanic; but I am in doubt if the moth anywhere enters manufacturing country. The same is true of *Cleora ribeata* Cl., for var. *sericearia* Curtis has much increased, and is now well established in several woods in Surrey and in the New Forest; also of *Polia nebulosa* Hufn., the dark varieties of which are now common in Delamere Forest.

Perhaps the most remarkable of these examples is provided by *Cleora rhomboidaria* Bkh., for it is an abundant and widespread species certainly entering industrial areas. Yet the black variety *rebeli* Ainger has gained a footing nowhere but in one or two woods in Kent.

In all these instances, the melanic forms are complete, or almost complete, dominants,* except in *P. nebulosa*. Here the heterozygotes, *robsoni* Collins, are intermediate between the two homozygotes, *thompsoni* Arkle and the normal pale grey *nebulosa* (Bowater, 1914). This species belongs to the family Agrotidae, while all the others are Geometridae.

Several of those who have studied the genetics of these insects have reported that the black forms are hardier than the normals: Onslow for *E. consonaria*, *C. ribeata*, and *C. punctinalis*, and Williams for *C. rhomboidaria*. Thus the situation seems to be similar to that found in true industrial melanism, and it is especially to be noticed that the explanation of the latter phenomenon which I have adopted is entirely consistent with the establishment of melanic forms here and there in unpolluted country. On the other hand, such non-industrial melanism is completely at variance with the view that the black forms are produced by mutation due to lead and manganese salts contaminating the larval food in manufacturing areas, or that they are spread by simple selection favouring dark coloration in districts where the vegetation is begrimed with soot. On the contrary, it appears probable that the selective balance between advantage (that of greater viability) and disadvantage (the loss of cryptic colouring) is in normal circumstances but slightly weighted against the black forms, so that these are able to establish themselves in certain rural areas where the ecological adjustment has been upset in such a way as to favour them (Ford, 1945b). This is quite likely to happen here and there, owing to the great changes brought about by civilization; for, as already mentioned, most of the English woodlands and their fauna are now vastly different from their primeval condition.

Thus the increase of melanic varieties in moths, which is so remarkable a feature of evolution at the present time, appears to be quite consistent with the general interpretation of transient polymorphism; that it is due to selection favouring the spread of genes beyond the frequency at which they are maintained by recurrent mutation. This has already progressed so far in certain areas as to affect nearly the whole population; nor have we any reason to suspect that it will reach a stable equilibrium except in those regions immediately around a centre of melanism where the black forms, which spread from it into rural districts, are being eliminated by counter selection.

* *E. crepuscularia* (Harrison & Garrett, 1926), *E. consonaria* (Onslow, 1919, 1920a), *C. ribeata* (Onslow, 1920b) *C. rhomboidaria* (Williams, 1932–3).

III. BALANCED POLYMORPHISM

The instances so far discussed are those in which an advantageous character spreads unhindered through a population, leading first to a transient polymorphism and finally back to uniformity. True balanced polymorphism, however, is that in which this spread is checked at some level during its course, so as to produce a stable equilibrium. Now the character initially favoured may be either environmentally or genetically controlled, while the barrier imposed upon its advance may also be either environmental or genetic. These various situations require a brief initial analysis.

A balanced polymorphism in which the variation involved is environmental is generally the product of rather exceptional conditions. It is well illustrated by certain castes in insects. In the honey-bee the two types of female, queens and workers, are produced by differential feeding. The one is necessary for the survival of the community and the other for its social organization. A balance must be struck between these needs, and any tendency to produce one type only would be checked by selection.

However, the main interest of balanced polymorphism centres round the spread of genes and the types of check imposed upon it. We will first consider a few illustrative situations in which diversity is maintained by the ecological balance of the species. A well-known instance of this is provided by Batesian mimicry. Here an advantage is conferred upon a palatable species owing to its resemblance to one that is protected either by a nauseous flavour, the possession of a sting, or some other device. Evidently the mimic must not become too common relative to the model, otherwise the warning, and therefore conspicuous, colours which it copies will become a danger instead of an asset. This will happen when, on the average, they come to be associated by predators with edibility rather than inedibility. Clearly, therefore, the numbers of a mimetic form are dictated by those of its model. Indeed, it is rather surprising that Batesian mimics should exist which are without the elasticity conferred by polymorphism, as, for example, the Aegeriidae (Clearwing moths). These resemble various Hymenoptera, but they are all monomorphic. Yet it will evidently be a great advantage to mimic several protected species, or even to possess, in addition, a non-mimetic phase; for however dissimilar are the advantages which the forms receive, they will necessarily be adjusted by selection to such proportions that all obtain equal protection. At that level they will reach stability, and a balanced polymorphism will be produced.

It is not my intention to review mimetic polymorphism, as this has often been done before (Carpenter & Ford, 1933; Eltringham, 1910; Ford, 1937), but only to draw attention to a few of its features which are relevant to this discussion.

Clearly, it must always be governed by a switch mechanism which can produce clear-cut alternative forms. In the simplest situation, that of dimorphism, the segregation of a single pair of allelomorphs is entirely appropriate for this purpose. This is, however, complicated by the circumstance that mimicry usually involves the selective adjustment of a number of distinct characters, all of which must be brought within the unifactorial control, by selection operating on the gene complex and modifying the effect of the pair of allelomorphs which determine the alternative phases. Thus, on the west coast of Africa are to be found two forms of the nymphaline butterfly *Hypolimnas dubius* Beauv. One of them, *dubius*, copies *Amauris psyttalea damoclides* Staud., and the other, *anthedon* Dbl., copies *Amauris niavius niavius* L. The distinction between them is uni-factorial, *dubius* being a simple dominant (Lamborn, 1912), but they differ widely both in their pattern and habits in order to produce a convincing resemblance to their models. When three or more forms are involved, the genetic control takes the form of an interacting system of allelomorphs. For example, in the Oriental *Papilio polytes* L., the non-mimetic form *cyrus* Fabr. is a simple recessive to *polytes*, which mimics *Papilio aristolochiae* Fabr. This result is, however, obtained only if the individual is homozygous for a second pair of genes **rr**; for **R** has a dominant effect converting *polytes*, but not *cyrus*, into a third form *romulus* Cram., mimicking *Papilio hector* L. Two pairs of allelomorphs are thus responsible for a trimorphism, but it is one which operates in the female only: the males are monomorphic and *cyrus*-like. This work was conducted by Fryer (1913) who encountered numerous sterile unions in his matings, and Fisher (1927) has pointed out that these are probably due to a reduced fertility of the homozygous dominants. I have also detected a nearly significant departure from 3 : 1 towards 2 : 1 in the segregation of the mimetic female forms *misippus* and *inaria* Cram. of *Hypolimnas misippus* L. The possible existence in mimetic species of a differential viability (capable itself of leading to polymorphism) is a remarkable circumstance which will be discussed on pp. 83, 84.

In general terms it may be said that polymorphism is adjusted to promote diversity, and this appears to be carried much further in the characters associated with courtship in certain birds. Perhaps the most remarkable of these is the extreme polymorphism of the Ruff, *Philomachus pugnax* L. This involves the crests and the ruff round the neck. Its genetics are unknown, but it appears to be maintained because the *diversity* of display going on around the females on the common courting ground is necessary to provide the stimulus required for copulation (Ford, 1940*b*). For Fraser Darling (1938) has shown that in some species of birds this is obtained not from the courtship of a single male, which is insufficient, but from that of a group.

The main effect of selection in the foregoing instances is to promote variability, but a balanced polymorphism will also arise when a genetic barrier checks the spread of an advantageous gene. The simplest situation of this kind is that in which the heterozygote is at a greater advantage than either homozygote. This will evidently lead to permanent diversity, and it is one which can be studied in experimental conditions. For example, heterozygotes for the mutant gene *ebony* in *Drosophila melanogaster* have a higher viability than normal flies, but the viability of the homozygotes is inferior to that of the normals. Consequently, if a few wild-type *Drosophila* are introduced into a laboratory culture of *ebony*, the population of dark-bodied flies at first rapidly declines, but later becomes stabilized at a low level when a balanced polymorphism has been attained (L'Héritier & Teissier, 1937).

Very similar conditions are frequently encountered in nature. Many species of the butterfly genus *Colias* are remarkable for their female dimorphism. In all of them the males are yellow and the females either yellow or white. The latter form is invariably the rarer, ranging from about 2 to 20 % of the females, but actual data on their proportion seem only to have been obtained for two of the species. The results of Gerould (1923, 1941) on *Colias philodice* Godart are shown in Table 2. It will be observed that there is considerable variation in the frequency

Table 2. *The proportion of the pale female form of* Colias philodice *in New Hampshire*

Locality and date	Yellow ♀	Pale ♀	Total
Hanover, 1911	52	10 (16·1 %)	62
„ 1923	37	4 (9·8 %)	41
„ 1940	135	26 (16·1 %)	161
Goffstown, 1917	50	2 (3·8 %)	52

of the pale females, but that at Hanover it was the same in 1911 and 1940, suggesting that it is probably fluctuating round a stable equilibrium. The proportions of the pale form of the European *Colias croceus* Fourc. have been studied in some detail by Ford (1942), and a similar distribution to that of *C. philodice* was obtained in 1941, and by previous observers (Clogg, 1865–6; Nash, 1929), as shown in Table 3. The genetics of this situation have been

Table 3. *The proportion of the pale female form of* Colias croceus *in England*

Locality and date	Yellow ♀	Pale ♀	Total
Cornwall, 1865	182	8 (4·2 %)	190
Suffolk, 1928	133	17 (11·3 %)	150
Cornwall, 1941	55	9 (15·8 %)	64

studied in several species, including *C. croceus*, sufficiently to demonstrate that the pale form is determined as a simple sex-controlled dominant in each of them, but it has been subjected to detailed analysis only in *C. philodice*, by Gerould (1923). He found that the gene responsible for the dominant pale variety *alba* Stgr. is linked with a recessive lethal. Consequently, only the heterozygotes normally survive. However, crossing-over freed *alba* from the lethal in his stocks; yet matings with wild specimens always brought in the lethal again, so that it is apparently widespread in the population.

A corresponding condition exists in the nymphaline butterfly *Argynnis paphia* L., in which the males are also monomorphic and the females dimorphic—one form being brownish, like the male, and the other (*valezina* Esp.) blackish-green. It is noteworthy that the rarer of the two is again the one differing from the male, and that it is a simple dominant. Goldschmidt & Fischer (1922) report that it also is linked with a recessive lethal, so that the analogy appears to be carried yet further.

Certain elements in this situation deserve brief attention. In the first place, it is apparent that the 'recessive lethal', which reacts fatally with the pale forms of *Colias* and the *valezina* form of *Argynnis paphia* must, in other circumstances, be of some advantage to the species, otherwise it could not be widespread in them (it is not of course suggested that the lethal is the same gene in all). Similarly, these rarer dominant forms must themselves be at some advantage in the heterozygous phase, otherwise they would not have spread until they occupy even as much as 2 % of the population (p. 73). We have here a curious balance of selective advantages of a type frequently encountered in polymorphism. It would at first seem remarkable that the viability of the homozygous dominants when interacting with the 'recessive lethal' has not been improved by selection so as to permit the uninterrupted spread of the rarer but advantageous polymorphic form. It is not possible to explain this in the present state of our knowledge. It may, however, be conjectured that to modify favourably the response of the organism to the rare homozygote might endanger the advantage which the recessive lethal must undoubtedly confer (on account of its frequency in the population). Indeed, we may assume that all advantageous genes which can be spread even as homozygotes, without interfering seriously with the existing genetic adjustments, have in fact been incorporated into the normal gene complex; those which cannot, remaining to produce a form of polymorphism.

An essential feature of balanced, as compared with transient, polymorphism is its stability, which can be proved, or inferred, in a variety of instances. Thus it is evident that a polymorphism must be perpetuated whose advantage lies in the actual diversity which it involves, as in the examples discussed on p. 79, or those related to sex which will be considered later (pp. 83–85). Moreover, the may

proportions of the two types of *Colias* females strongly suggest a dimorphism fluctuating somewhat widely round an average frequency distribution, on account of their essential similarity in two species, one American and one European (Tables 2 and 3), and the wide distribution of this dimorphism throughout the genus. We have, however, one remarkable observation indicating the great permanence of a polymorphism. Diver (1929) showed that the proportions of land snails with banded or unbanded shells, a distinction controlled by a single factor pair, is in certain colonies the same to-day as it was in Pleistocene times. This is in marked contrast to the distribution of rare genes maintained, ultimately, by mutation pressure. Their occurrence in wild populations of *Drosophila melanogaster* in the Caucasus has been studied by Dubenin *et al.* (1934) in an exceptionally thorough way. They showed that such genes are widespread in these populations, and that they are subject to extreme fluctuations in frequency in different years.

There are, however, great numbers of instances in which the stability of a polymorphism may reasonably be presumed, though it cannot be proved, and others in which it is doubtful whether or not it is attained. Some of these provide valuable data on an aspect of the subject which has been greatly developed in recent years: its bearing upon geographical distribution.

Any species may vary geographically in a number of distinct ways. It may be divisible into discontinuous groups, each with its own special features, and these constitute a set of subspecies. The sharp differentiation between them is due to the fact that they normally evolve in isolation from one another. Subsequently, however, these groups may extend their ranges and meet. Yet when they do so they generally retain their identity, merging into one another only along a narrow belt where they come into contact. For each will have built up a harmoniously balanced gene complex suited to its needs; but the respective balances cannot be preserved in the hybrid population, so that this will be at a disadvantage compared with either of the pure subspecies. Such intermediates will therefore be eliminated by selection as constantly as they are produced, and the integrity of the two subspecies themselves preserved. However, if a species has a wide but continuous range, the need to adjust itself to a variety of habitats will produce one or more character gradients in it, to which Huxley (1939) gave the name *clines*. The more abruptly that conditions change, the sharper will be the gradient in proportion between two forms at that place. This may lead merely to differences in the frequency of various characters at opposite ends of the cline. Alternatively, one character may be wholly replaced by another, the cline thus connecting two forms which, were they isolated, would often unhesitatingly be judged distinct subspecies. Indeed, they be called Dependent Subspecies (Huxley, 1942), in distinction to the Independent Subspecies which are not united by an intergrading series.

From what has just been said it will be evident that clines will often lead to polymorphism, and a few examples of this may usefully be quoted. Zimmermann (1935) has studied a dental abnormality in the vole *Microtus arvensis* Schinz. This consists in the absence of the last ridge of enamel on the molar teeth—the 'simplex' condition. It is found in 80–90 % of the population in Schleswig-Holstein and Mecklenberg. From that centre it occurs with decreasing frequency to the east, south, and west, giving rise to a 'dimorph cline' which finally leads to populations in which the character is unknown, as in the upland country of central Germany which appears to have proved a complete barrier to its spread. It seems probable that this is a transient, not a balanced, polymorphism, but further data are needed in order to establish the point.

On the other hand, stable gene ratios must almost certainly be involved in the cline of mimetic female forms of the African swallow-tail butterfly *Papilio dardanus* Brown, for these are related to the numbers of their models. The frequencies of four of them, in three of the geographical races of this insect, are shown in Table 4. These appear to be dependent

Table 4. *The percentage frequencies of four of the female forms of* Papilio dardanus *in three of its subspecies. The estimates of* planemoides *and* trophonius *are approximate only. (Based on Ford, 1936.)*

Distribution	West coast and west central	East central	East coast and south
Subspecies	*dardanus*	'transitional'	*cenea*
♀ form:			
hippocoonides	98	60	9
planemoides	1	21·5	—
cenea	—	7	85
trophonius	1	4·5	2

subspecies, and there is a gradual south to north cline, in which the abundance of *hippocoonides* Haase increases and that of *cenea* Stoll decreases, as we pass up the east coast (subsp. *cenea*) and east to west through Tanganyika and Uganda (the 'transitional' subspecies). Between these two is interposed another subspecies, *tibullus* Kirby, in which the proportions are not known with sufficient accuracy, but they are in general those of the transitional rather than the *cenea* race. Indeed there is a region in which the cline changes rapidly both in the neighbourhood of Delagoa Bay, where the *tibullus* and *cenea* races meet, and again west of Lake Victoria, where the transitional and *dardanus* races intermingle rather more gradually. It will be noticed that one form,

planemoides, is practically confined to the transitional race, and that *trophonius* Dbl. does not take part in the cline, for it occurs at a low, and approximately similar frequency in all three races (as it does also in *tibullus*). Indeed, it is important to observe that clines often involve only individual characters of a species.

These mimetic forms are controlled by a simple genetic mechanism in which *cenea* is determined by a single gene and is dominant both to *hippocoonides* Haase and *trophonius*. Another gene converts *hippocoonides* into *trophonius*, which is dominant to it.* Yet *hippocoonides* and *trophonius* are not alike throughout their range, for on the west coast, where they are called *hippocoon* Fab. and *trophonissa* Auriv. respectively, they are subject to minor adjustments to accord with the geographical variation of their models (*Amauris niavius* L. and *Danaus chrysippus* L.). This circumstance draws attention to an important principle, that of the adaptation of the effects of a gene by selection operating on the gene complex. It may be necessary to modify the expression of one or more polymorphic forms in different parts of their range, though they may be under uni-factorial control. Moreover, the spread of any gene will require an adjustment in the gene complex which will have to be modified to work harmoniously in the conditions imposed by its establishment in the population. Consequently, as Timoféeff-Ressovsky (1940) shows, the success of one gene will be accompanied by the establishment of other minor heritable differences independently controlled. He illustrates the point by means of the lady-bird *Epilachna chrysomelina* F., in which the spread of a form *elaterii*, giving rise to a conspicuous change in the colour pattern, has produced a distinct geographical race in Corsica, north Sardinia, and north-west Italy, which is a small part only of the range of the species. *Elaterii* is inherited as a simple semi-dominant, but it is accompanied also by additional small genetic differences due to other and independent genes. Dobzhansky (1937) has investigated the polymorphism of the Asiatic lady-bird *Harmonia axyridis* Pall., which involves five main forms controlled by simple genetic means. The proportions of several of them form clines, and some are absent from a considerable part of the total range of the species.

The red fox, *Vulpes vulpes* L., exists in two sub-species, *V. v. vulpes* in the Old World and *V. v. fulva* in the New. Each can be subdivided into a number of geographical races† in all of which polymorphism exists. This is chiefly due to the recombinations of three pairs of allelomorphs (Huxley, 1942, p. 185),

---

* The genetics of *planemoides* are not fully analysed, but it is certainly determined by a simple switch mechanism.

† By some taxonomists these are regarded as sub-species, and *v. vulpes* and *v. fulva* as closely related but distinct species, but the point is immaterial for the present purpose.

though the true silver-fox is produced by a gene absent from the Old World subspecies. The frequencies of the various forms involved in this polymorphism—white, blue, red, 'cross', and silver—vary from one extreme situation to another where the species is monomorphic. The abundance of the Canadian fox is subject to a regular four-year cycle (Elton, 1942) in some parts of its range and a ten-year cycle in others (Hewitt, 1921), during which these frequencies may alter greatly, but there seems to be evidence that they follow the cycle in absolute numbers, a given distribution of polymorphic forms being characteristic of the species at any one place for any given numerical density. Such a situation is indeed to be anticipated on theoretical grounds, for the ecological balance of a species is by no means the same when common as when rare, and we should expect that the selection pressure determining the distribution of polymorphic forms should be correlated with the total numbers. Very few data actually demonstrating this effect exist, for it is one to which little attention has been paid. It would, however, well repay investigation, and in my own studies on the moth *Panaxia dominula* L., shortly to be published, the spread of a gene in nature has been strictly related to the absolute numbers of the population. However, cyclical changes in the proportions of polymorphic forms very similar to the type just described have been demonstrated in the lady-beetle *Adalia bipunctata* L. in the neighbourhood of Berlin. Here the species is polymorphic, black and red forms, determined by a single pair of allelomorphs, being found together. The black occupy about 55–70 % of the population before hibernation and about 30–45 % of it afterwards, these proportions being repeated in successive years (Timoféeff-Ressovsky, 1940).

It is not infrequently found that a monomorphic species may become dimorphic (or polymorphic) in one part only of its range, a cline leading up to the region where the more restricted form becomes commonest. The squirrel, *Sciurus vulgaris* L., is dimorphic, black or red, in most parts of the European continent but monomorphic, red only, in the British subspecies. This must represent a balanced polymorphism, in which the black phase is not favoured by the ecological situation in Britain or, more probably, by the gene complex of the subspecies here. Also the percentage of the bridled variety of the guillemot, *Uria aalge* Pont., studied by Southern & Reeve (1941) increases from south to north and, to a lesser extent, from east to west. On the Iberian peninsula and along the west coast of France, it seems to be maintained only as a rare variety, occupying well under 1 % of the population. In Britain it increases from rather under 1 % in southern England to over 20 % in the Shetlands. This cline is not even, but the frequency rises rather sharply at Kintyre and between Fair Isle and the Shetlands. In Iceland its direction is reversed, for the bridled

form decreases from over 50 % in the south to about 8 % in the north. It is not yet certain whether this dimorphism is transient or balanced, though its coincidence with an increase in humidity northwards and westwards over a considerable part of its range suggests the latter alternative.

A condition to be contrasted with that just discussed is seen in the moth *Diaphora mendica* Clerck. The sexes of this insect are very different in appearance, the ground colour of the males being deep chocolate-brown and of the females white. In a few restricted localities in Europe and throughout Ireland, the males are white like the females, though this variety, *rustica* Hb., is almost unknown in the normal population. It is due to a single factor pair acting without dominance, the heterozygotes (var. *standfussi*) being of a sandy colour, but very variable: indeed they may overlap *rustica* and approach, but not overlap, *mendica* (Adkin, 1927; Onslow, 1921 b). It is possible, as Huxley (1942) points out, that the situation encountered in Ireland is the final stage in a transient polymorphism.

Many dimorph clines connect regions where the alternative forms are found alone. For example, a cline in the proportions of the black and white forms of the hawk *Accipiter novae-hollandiae* Gm. passes across Australia. In several islands in the extreme north only the black birds are found, and in Tasmania only the white, the cline passing gradually from one to the other of these extremes (Stresemann, 1923–6). If this represents a balanced polymorphism, as seems probable, we have good reason to think that the dimorphic situation itself has some advantage over the greater part of the species' range. Such a situation could not be produced merely by contamination due to genes spreading from the places where one or the other type alone is at an advantage: as already explained (p. 80), that would lead to a rather narrow belt of dimorphism, not to a long cline. Indeed, there are instances known in which the dimorphism between alternative monomorphic colonies shows no cline; as in the lady-beetle, *Sospita vigintiguttata* L., studied by Dobzhansky (1937) in the Kiev neighbourhood. Forms with black or yellow elytra generally occur together in about equality, though he found one colony containing only the black and another only the yellow type.

Finally a special type of plant polymorphism associated with cline formation must be mentioned, that in which certain species are highly polymorphic, the various forms being adapted to different ecological situations (ecotypes). Different proportions of these are then maintained by selection along any environmental gradient, so producing a cline, though a very plastic one, changing according to the ecological needs of the species. An excellent example of this type of polymorphism is provided by *Plantago maritima* L. (Gregor, 1938, 1939), which is maintained by the need for adaptation to an extensive range of closely interwoven habitats. Were these habitats more widely spaced, they would lead to the selection of a particular ecotype within each, to the exclusion of the others except as rare varieties: this would not be polymorphism but ecological, or geographical, variation, and it is exemplified by a form of *Hieracium umbellatum* L. which has become highly adapted to life among shifting sands. Turesson (1922) shows that it cannot maintain itself at points where woods approach the shore, owing to cross-pollination upsetting the balanced gene complex of the dune-living plants, for these seem to be maintaining themselves in a difficult habitat by strict selection.

Certain kinds of balanced polymorphism máy now briefly be discussed which depend upon rather special genetic mechanisms for their maintenance. One of these is that due to close linkage, in which the various forms include a single 'universal recessive'. Though involving quite an exceptional situation, it is to be found somewhat widely: in Insecta (the grouse-locusts *Apotettix* and *Paratettix*) Mollusca (various land-snails of the genus *Cepaea*), and in Vertebrata (the fish *Lebistes*). Haldane (1930) has pointed out that the association of these two phenomena may be attributed to the occurrence of sectional translocations, associated with duplications of the translocated segments.

The universal recessive is always the commonest of the forms, and the rarer ones, though fully dominant to it, produce intermediate effects when they are themselves brought together. This is due to the fact that such combinations will very seldom arise in nature, so that there have not been sufficient opportunities for dominance modification to take place in respect of these rare types, as there have been between each of them and the abundant universal recessive. It has been suggested by Fisher (1930 c) that the advantage which maintains the rarer dominants in the population is probably balanced by a reduction in viability associated with the homozygous duplication, and he has been able to demonstrate a significant deficiency of homozygous dominants in Nabour's data on the genetics of grouse-locusts.

It seems that the close linkage of the genes responsible for these various forms is the chief feature underlying such polymorphism, for it reduces the number of combinations upon which selection can operate to improve them. However, a duplication, even though lethal when homozygous, may provide a tract of chromatin in which modifiers affecting the heterozygotes can be selected. At the same time, these genes are sheltered in the duplication from outside competition, a danger to which they would otherwise be exposed when the rest of the gene complex is being adjusted to improve the universal recessive. Thus the rarer dominants can be maintained as polymorphic forms, while they retain their genetic flexibility.

A highly specialized and instructive type of poly-morphism is that leading to the production of plants bearing different types of flowers adjusted to produce outbreeding: this is the heterostyled condition so well exemplified by the genus *Primula*, the effects of which have been analysed with much success by Mather & de Winton (1941). The system is one in which a pair of allelomorphs controls the development of two forms of flowers borne on different plants, thrum (dominant) and pin (recessive), whose structure encourages outbreeding. But this is one only of the effects of these genes, for in addition they are always responsible for a physiological outbreeding mechanism, that of illegitimacy. Thus in distylic flowers, as those of the genera *Hottonia* and *Primula* among the Primulaceae, thrum × thrum and pin × pin matings are less likely on morphological grounds than thrum × pin (or its reciprocal), but. even if attained, few or no seeds are set.

Mather & de Winton point out with evident justice that the real significance of the morphological differences involving pistils and stamens lies in the physiological differences which follow them, and that the alternative possibility, that these are independent outbreeding mechanisms developed side-by-side to provide 'double assurance', is contrary to all the evidence; for the two systems are always found together, never apart. On the contrary, the morphological effects are presumably selected for the physiological differences which they produce.

The essential feature in the heterostyly of *Primula* is the fact that the homozygous dominants (thrums) are less viable than either of the other two genotypes. But dominance is achieved by selection tending to make the heterozygote resemble the more favourable homozygote (Fisher, 1931). Consequently, as Mather & de Winton point out, the reduced viability of thrum has been imposed since its dominance was achieved. We may presume, therefore, that pin arose as a deleterious mutant which was pressed into recessiveness in the normal manner. For the reduction in viability of thrum follows as a natural consequence of the combined outbreeding mechanism constituted by the illegitimacy and the morphological flower difference. In a heterostyled population, pollination is normally always taking the form of a back-cross between heterozygous thrum (Ss) and recessive pin (ss). Accordingly, homozygous thrums (SS) are not produced, and during the evolution of the process they would become progressively rare. In consequence of this, the section of the chromosome carrying S will be sheltered from selection in the immediate neighbourhood of that gene, so that deleterious recessives can accumulate there without reducing the viability of thrum plants, since these are always heterozygotes. But, when by artificial or other means, homozygous thrums arise, the full effect of the harmful recessives accumulated near the S locus becomes manifest.

At the start, the physiological and morphological differences produced by the genes Ss and ss respectively were presumably slight, sufficient only to ensure that crossing between them was relatively more frequent than selfing (or matings between like types). The physiological and morphological differences between thrum and pin as now seen will have been built up gradually by selection of the gene complex. So long as the advantage of outbreeding was outweighed by the disadvantageous effects of **ss**, pin would remain merely a rare variety, but as soon as any change in the environment placed a sufficient premium upon outbreeding, pins would begin to spread and heterostyly to evolve.

Indeed the advantage or otherwise of outbreeding depends strictly upon the evolutionary status of the plants relative to their environment. In some circumstances, it may be highly desirable for the species to be very plastic, such as the great variability conferred by outbreeding allows, while in others it may be best for a well-adapted form to be maintained, adjusted to its optimum conditions with as little variation as possible, and here selection would favour the homostyle condition which leads to inbreeding. Mather & de Winton point out that once a heterostyle mechanism has been evolved, a shift across from one to the other of these situations is rather easy, so allowing the species to modify its reproductive mechanism as may be needed. For the genes at the **s** locus determine (1) the height of the anthers, (2) the height of the stigma, and (3) the illegitimacy behaviour, and other allelomorphs at the same locus can give different combinations of these three effects. It is quite possible, therefore, for another allelomorph to arise, and in appropriate conditions to be favoured, which superimposes homostyly upon a previous heterostyly by failing to alter the anther and stigma length reciprocally and readjusting the illegitimacy balance. The homostyles so produced may evidently belong to either of two types, short in stigma and anthers or long in stigma and anthers, and both will be characterized by ability to produce progeny by self-pollination.

There is at the present time an area in south-western England in which selection is favouring the production of homostyles, at least up to certain frequencies, in the common primrose, *Primula vulgaris* Huds., and the situation there is under investigation. Moreover, the evolution of one of the British primroses has proceeded along these lines, for all the species are normally heterostyled except *P. scotica* Hook. This is closely related to *P. farinosa* L. and is sometimes treated as a subspecies of it. The two have, however, adapted themselves to different environments in Britain, and it has evidently become advantageous for them to adjust their reproductive mechanisms in different directions, *P. farinosa* being a heterostyled form from northern England and southern Scotland, while *P. scotica*, which is smaller with flowers of a deeper pink, is homostylic and an inhabitant of the north of Scotland. We may be sure

that *P. scotica* has evolved from *P. farinosa* and not the reverse, for homostyly is a carefully controlled inbreeding mechanism superimposed upon heterostyly and is not a primitive condition.

The fundamental features of heterostyly are present whether the situation is the distylic one just discussed or the more complex tristyly of *Lythrum salicaria* L., in which are found plants bearing respectively three forms of flower, short-, mid- and long-styled, and possessing two sets of anthers at two different levels. This arrangement is clearly one which favours cross-pollination, and it is associated with an illegitimacy mechanism of such a type that pollination fails unless the pollen is brought from anthers growing at the same height as the stigma and, therefore, from another plant. Thus pollen from the long stamens of a short-styled plant will fertilize successfully a long-styled plant, while pollen from the mid-stamens of the same long-styled plant will fail to do so. The length of the stamens is controlled genetically, but the behaviour of the pollen is determined developmentally. The difference between long- and short-styled *Lythrum salicaria* is analogous to that of pin and thrum, 'longs' being homozygous recessives (carrying homozygous **s**) and 'shorts' heterozygous dominants (with **S**). The genetic control of mid-style, however, defeated analysis until Fisher (1941) showed that it is determined by another gene **M**, dominant in effect, but complicated by autopolyploid inheritance, a condition apparently very rare in wild plants though common in cultivated ones. The longs do not possess the gene for mid-, but carry its recessive allelomorph **m** in the homozygous state, while all long- and mid-genotypes have counterparts among the short plants, their activity being suppressed by **S**. It is not yet certain whether *L. salicaria* is a tetraploid or a hexaploid, and chromosome counts have so far supplied no clear evidence on the point. However, Fisher has genetic tests in hand which should decide the question.

Attention has already been drawn to instances in which there is some evidence that the polymorphism of animals involves a lower viability of the homozygous dominants than of the other genotypes, in some of which, as in mimicry, the ecological situation itself provides all that is needed to maintain the diversity. The explanation of this additional genetic complication should now be plain, for any system which leads to the production of an undue excess of heterozygous dominants carries with it the possibility of reducing the viability of the dominant homozygotes.

Not only does the system of reproductive mechanisms which has just been described fall within the definition of polymorphism but so does sex itself, and it must be of the balanced type. In any one species the two sexes will clearly be maintained at optimum proportions, which are generally near equality, and a tendency for one of them to spread at the expense of the other would be checked by selection. The switch mechanism involved is generally of the simple genetic type which has been so thoroughly studied, but instances are also known in which it is environmental, as in two genera, *Bonellia* and *Hamingia*, of the Echiuroidea, a highly aberrant group of marine Annelida. Here the embryo always develops into a female unless it becomes attached to the proboscis of an adult specimen. In that event its growth is arrested and it becomes a male, living as a minute parasite on or within the body of the female.

Whatever type of control is employed, the numerous adaptations of the two sexes must be capable of selection and adjustment, although the whole sexual outfit segregates sharply into the male and female conditions. Indeed the possibility of genetic variation within a rigid framework producing alternative forms, itself genetic or environmental, is an essential feature of all polymorphism.

IV. THE HUMAN POLYMORPHISMS

It is a matter of much interest and considerable importance that certain potentially advantageous qualities have not established themselves in an entire human population but affect a proportion of it only, giving rise to polymorphism. Some aspects of this situation must briefly be reviewed.

The ability to taste phenyl-thio-urea in aqueous solutions of 50 parts to the million is a dominant character controlled by a single factor pair, while the inability to do so, except at much higher concentrations (400 parts to the million), is recessive. This slight failure in sensory perception is, from one point of view, quite different from the great majority of human inherited defects, for it is found in 25–30 % of the population according to the race (Boyd & Boyd, 1937), instead of being merely a rare abnormality maintained, ultimately, by mutation. We can be sure that the allelomorph which arose the more recently must confer some advantage on its possessor, or it could not have become so common (p. 73). On general grounds, it seemed more probable that a selective balance must be involved here than that we are witnessing the spread of a gene which has lately become advantageous owing to a change in environment or in the genetic constitution of mankind (due, for example, to racial crossings). However, evidence for the stable nature of the gene ratio involved has been provided by a study of anthropoid apes; for it has been found not only that chimpanzees can also be scored as 'tasters' or 'non-tasters' of phenyl-thio-urea, but that the proportions of the two classes are approximately the same in apes as in man (Fisher, Ford & Huxley, 1939). Without the conditions of a stable equilibrium it is hardly possible that the gene ratios involved could have persisted unaltered, within the

limits of the test, for the million or more generations since the separation of the anthropoid and hominid stocks. Therefore this observation provides evidence that the polymorphism involved is a balanced one of long standing. The effects of this gene upon which selection acts must clearly be other than the taste test by which we recognize its presence, but no data are available to indicate their nature, nor is it known what opposes the advance of the more favourable allelomorph: perhaps the heterozygotes are more viable than either homozygote, but this is mere conjecture.

All other instances of polymorphism so far detected in man relate, directly or indirectly, to the blood groups. This is far too large a subject to summarize, however briefly, in the present article,[*] and it is possible only to draw attention to certain relevant aspects of it.

These blood-group polymorphisms are due to genes at four loci which are, as far as is known, situated in different chromosomes. A series of multiple allelomorphs exists at two of them. One of these is responsible for the *O*, *A*, *B*, *AB* set of blood groups, with its two subdivisions A_1 and A_2. This is controlled by a series of four allelomorphs, and all of them must certainly be maintained on a polymorphic basis; for in England the two rarest groups are A_2 and *B*, but both of these occupy over 7 % of the population (Taylor & Prior, 1938, have shown that in this country 17·9 % of group *A* belongs to A_2), and elsewhere they may be still more frequent. Other members of the series (A_3, A_4, A_5) have been reported, but their status seems to be an entirely different one, for they are of great rarity, and there is no reason to regard them as polymorphic at all. They are probably eliminated by selection without a counterbalancing advantage, and are in the position of great numbers of other genes known in man which are maintained merely by recurrent mutation.

It is reasonable to conclude, from what we know of polymorphism, that individuals belonging to the different blood groups are not equally viable (Ford, 1940*b*), and we may expect elimination to fall upon the *AB* class—a subject now under investigation. A valuable line of enquiry which does not yet seem to have been pursued in any detail would be to study the blood-group distributions in patients suffering from a wide variety of diseases. It is possible that in some conditions, infectious or otherwise, they would depart from their normal frequencies, indicating that persons of a particular blood group are unduly susceptible to the disease in question. It is highly unlikely that the effect would be sufficiently large to suggest methods of treatment, though this is at least conceivable. For

[*] A knowledge of the serology, genetics, and distribution of the human blood groups is presumed in the discussion which follows: a short and fully up-to-date account of the subject may be obtained from Taylor, 1944.

example, if the *AB* group were found, by reason of its undue excess, considerably to predispose towards any given illness, it might be worth while to transfuse an *AB* patient suffering from it with group *O* blood. It must, however, be remembered that the effect of the genes concerned may not be limited to their control of antibody and antigen formation, so that even in such circumstances the type of blood might not be a factor in the case.

It should be noticed that the frequencies of the *O, A, B, AB* series alter greatly with the race. In western Europe *B* is always below 10 % (in non-Jewish populations). However, 37 % of Hindus belong to this group, and approximately this same level is found among the gipsies of western Europe who are derived from them (see Ford, 1945 *b*, Table B). This racial variation does not necessarily militate against the assumption that these blood groups are maintained as a balanced polymorphism, for their optimum frequencies within one race are unlikely to be the same as in another. We have, moreover, evidence to show that the characteristic racial differences of the blood groups are preserved among populations which have migrated to other parts of the world, as in the instance of the gipsies just quoted.

Two pairs of allelomorphs associated with the human blood groups are known in no more than alternative forms. One of these determines the *M, N, MN* blood-group series, in which the three classes are in England distributed as *M* 29·7, *MN* 49·0, *N* 21·2 % (Taylor & Ikin, 1939), but the values vary greatly according to the race (Wiener, 1939). These three classes also represent the distribution of the genotypes, for *MN* are heterozygotes. Some of the problems raised by this polymorphism are clearly of the same type as those encountered in that of the *O, A, B* series, but here the number of genotypes is smaller and the *M* and *N* antibodies are absent except as a rarity.

The second pair of allelomorphs known only in alternative forms controls the secretion of the *A* and *B* antigens into the saliva and the semen, in which they are either present in much greater concentrations than in blood (the dominant type) or totally absent. This latter, which is the recessive class, occupies about 18–28 % of the population in England. It is not unlikely that it is ultimately concerned in the complex balance which is probably attained in the distribution of the *O, A, B* types, and it may play a part, which deserves investigation, in the relationship between mothers and embryos who belong to different blood groups of that series.

The second set of multiple allelomorphs associated with the blood groups is that controlling the rhesus factor (*Rh*). Seven members of this series are now known, six being responsible for forms of the dominant rhesus-positive condition and one for the rhesus-negative. However, only three of them, two of the dominants and the recessive (Rh_1, Rh_2, *rh*)

are common; the other four are all scarce or very rare, and they may not represent true polymorphisms at all. 85 % of the population in England, and of white Americans, are rhesus-positive, but the values differ considerably in other races.

These features are, in certain respects, evidently parallel to those described for the *O, A, B* series, but the selective value of the rhesus factor is clearly established and not a matter of inference only. For the condition is of much importance when an individual has to receive more than one transfusion, and when it is required to transfuse a mother after the birth of a child. This is due to the formation of antibodies for rhesus, which do not normally exist in the human plasma; but they are produced when a rhesus-negative person is transfused with rhesus-positive blood, and they may also be formed in a rhesus-negative mother in response to antigens in the foetus. In addition, the passage through the placenta of such maternally formed antibodies is responsible for haemolytic diseases in infants.

Accordingly Haldane (1942) has suggested that the polymorphism of rhesus is of the transient kind, owing to the elimination of heterozygotes ensured by such haemolytic disease of the newborn. If unopposed this would tend rather rapidly to reduce the *rh* gene (rhesus-negative) to its mutation frequency, and Haldane considers that its present status is due to crossings between races which were respectively rhesus-positive and rhesus-negative to a predominant extent. The whole subject has been well reviewed by Race (1944). His article, which describes a suggestion due to R. A. Fisher, shows that the rhesus genes may yet be balanced in spite of haemolytic disease. Fisher points out that deaths in infancy are usually compensated for by additional births. This will partly make up for the loss even when the fathers are homozygous, for not all affected infants die. With heterozygous fathers, on the other hand, half the births are normal (for the condition arises only when the mother is rhesus-negative, and therefore recessive), and these will have the higher survival rate; consequently, even a partial replacement of lost children may maintain the *rh* gene.

The human blood groups give rise to many problems, and it seems that some of them may usefully be considered from the aspect of polymorphism. This unifies them in certain respects and opens up lines of enquiry from which the subject may ultimately benefit.

SUMMARY

(1) Polymorphism is the occurrence together in the same habitat of two or more distinct forms of a species in such proportions that the rarest of them cannot be maintained by recurrent mutation. (2) If a genetically controlled form occurs in even a few per cent. of a population it must have been favoured by selection. (3) Poly-

morphism may either be *transient*, in which a gene is in process of spreading through a population unopposed, or *balanced*, in which it is maintained at a fixed level by a balance of selective agencies. (4) Owing to the recurrent nature of mutation, transient polymorphism is generally due to changes in the environment, which make the effects of a previously disadvantageous gene beneficial. (5) A notable instance of transient polymorphism is provided by industrial melanism. (6) The explanation of this phenomenon, either on a basis of simple selection for the black forms in industrial country or of induced mutation, produced by mineral salts in the food, cannot be supported. (7) On the contrary, industrial melanism is probably due to the spread of genes improving the viability of the species, but those responsible for excess melanin production could not be used in normal circumstances, for this destroys cryptic colouring. It is now known that the genes giving rise to industrial melanism occur as rarities in the ordinary population, and that they confer a greater viability than the normal forms which they have been unable to replace. Such genes can, however, be used in industrial areas, where black colouring is a less serious handicap. (8) The variability involved in balanced polymorphism

may either be environmental or genetic. (9) A balanced polymorphism generally involves a high degree of permanence in the ratios of the respective forms. (10) Variations in the relative frequencies of polymorphic forms may, however, follow cycles in absolute numbers. (11) Polymorphic forms may be distributed as clines, which may or may not lead from groups which are monomorphic. (12) Though balanced polymorphism necessitates the existence of a switch mechanism determining alternative types, these are themselves capable of genetic adjustment within the framework of the polymorphism. (13) A special kind of balanced polymorphism is associated with close linkage dependent upon sectional translocations of the chromosomes. (14) The polymorphism of heterostyled plants is due rather to the physiological than the morphological differences between the forms. (15) Sex itself is a true balanced polymorphism, environmentally or genetically controlled. (16) The human polymorphisms include the ability to taste phenyl-thio-urea in low concentrations and the blood groups (and conditions related to them). The relative claims of the transient and the balanced systems of polymorphisms are considered for each of them.

V. REFERENCES

ADKIN, R. (1927): *Proc. ent. Soc. Lond.* **2**, 15–16, 66.

BOWATER, W. (1914): *J. Genet.* **3**, 299–315.

BOYD, W. C. & BOYD, L. G. (1937): *Ann. Eugen., Lond.*, **8**, 46–51.

BRETT, G. A. (1935–6): *Trans. S. Lond. Ent. Nat. Hist. Soc.* pp. 84–92.

CARPENTER, G. D. HALE, & FORD, E. B. (1933): *Mimicry.* London.

CLOGG, S. (1865, 1866): *Entomologist*, **2**, 338–9; **3**, 1.

DARLING, F. FRASER (1938): *Bird Flocks and the Breeding Cycle.* Cambridge.

DIVER, C. (1929): *Nature, Lond.*, **124**, 183.

DOBZHANSKY, TH. (1937): *Genetics and the Origin of Species.* New York.

DOWDESWELL, W. H., FISHER, R. A. & FORD, E. B. (1940): *Ann. Eugen., Lond.*, **10**, 123–36.

DUBININ, N. P. *et al.* (1934): *Biol. Zh., Mosk.*, **3**, 166–216.

ELTON, C. (1942): *Voles, Mice, and Lemmings.* Oxford.

ELTRINGHAM, H. (1910): *African Mimetic Butterflies.* Oxford.

FISHER, R. A. (1927): *Trans. R. Ent. Soc. Lond.* **75**, 269–78. — (1930a): *The Genetical Theory of Natural Selection.* Oxford. — (1930b): *Proc. Roy. Soc. Edinb.* **50**, 204–19. — (1930c): *Amer. Nat.* **64**, 385–406. — (1931): *Biol. Rev.* **6**, 345–68. — (1933): *Proc. Roy. Soc. B*, **225**, 197–226. — (1941). *Ann. Eugen., Lond.*, **11**, 31–8.

FISHER, R. A., FORD, E. B. & HUXLEY, J. S. (1939). *Nature, Lond.*, **144**, 750.

FORD, E. B. (1936): *Trans. R. Ent. Soc. Lond.* **85**, 435–66. — (1937): *Biol. Rev.* **12**, 461–503. — (1938): *The Study of Heredity.* Oxford. — (1940a): *Ann. Eugen., Lond.*, **10**, 227–52. — (1940b): *The New Systematics*, pp. 493–513. Oxford. — (1942): *Entomologist*, **75**, 1–6. — (1945a): *Genetics for Medical Students* (2nd ed.). London. — (1945b): *Mendelism and Evolution* (4th ed.). London.

FRYER, J. C. F. (1913): *Philos. Trans. B*, **204**, 227–54. — (1931): *J. Genet.* **24**, 195–202.

GEROULD, J. H. (1923): *Genetics*, **8**, 495–551. — (1941): **26**, 152.

GERSCHLER, M. W. (1915): *Z. indukt. Abstamm.- u. VererbLehre*, **13**, 58–87.

GOLDSCHMIDT, R. (1921): *Z. indukt. Abstamm.- u. VererbLehre*, **25**, 89–163.

GOLDSCHMIDT, R. & FISCHER, E. (1922): *Genetica*, **4**, 247–78.

GREGOR, J. W. (1938): *New Phytol.* **37**, 15–49. — (1939): **38**, 293–322.

GROSVENOR, T. H. L. (1926–7): *Proc. S. Lond. Ent. Nat. Hist. Soc.* pp. 95–6.

HALDANE, J. B. S. (1930): *Amer. Nat.* **64**, 87–90. — (1942): *Ann. Eugen., Lond.*, **11**, 333–40.

HARRISON, J. W. H. (1928): *Proc. Roy. Soc. B*, **102**, 338–47. — (1932): **111**, 188–200. — (1935): **117**, 78–92.

HARRISON, J. W. H. & GARRETT, F. C. (1926): *Proc. Roy. Soc. B*, **99**, 241–63.

HASEBROEK, K. (1934): *Zool. Jb. (Abt. Zool. Phys.)*, **53**, 411–60.

HEWITT, C. G. (1921): *The Conservation of the Wild Life of Canada.* New York.

HUGHES, A. W. McK. (1932): *Proc. Roy. Soc. B*, **110**, 378–402.

HUXLEY, J. S. (1939): *Bijdr. Dierk.* **27**, 491. — (1942): *Evolution: The Modern Synthesis.* London.

KUHN, A. & HENKE, K. (1929–36): *Abh. ges. Wiss. Göttingen*, N.F. **15**, 1–272.

LAMBORN, W. A. (1912): *Proc. Ent. Soc. Lond.* p. iv.

LEMCHE, H. (1931): *J. Genet.* **24**, 235–41.

L'HÉRITIER, P. & TEISSIER, G. (1937): *C.R. Soc. Biol., Paris*, **124**, 882–4.

MATHER, K. & DE WINTON, D. (1941): *Ann. Bot., Lond.*, **5**, N.S. pp. 297–311.

NASH, W. G. (1929): *Entomologist*, **62**, 111–12.

ONSLOW, H. (1919): *J. Genet.* **9**, 53–60. — (1920*a*): **9**, 339–46. — (1920*b*): **10**, 135–40. — (1921*a*): **11**, 123–39. — (1921*b*): **11**, 277–92. — (1921*c*): **11**, 293–8.

PEARSON, J. (1938): *Pap. Proc. Roy. Soc. Tasmania,* for 1937, p. 21.

RACE, R. R. (1944): *Brit. Med. Bull.* **2**, 164–5.

SCHULTZ, W. (1920): *Arch. EntwMech. Org.* **47**, 43–75.

SOUTHERN, H. N. & REEVE, E. C. R. (1941): *Proc. Zool. Soc. Lond.* A, **111**, 255–76.

STRESEMANN, E. (1923–6): 'Mutationstudien, I–XXV', in *Orn. Mber.* **31–4**; *J. Ornith.* **71–4**.

TAYLOR, G. L. (1944): *Brit. Med. Bull.* **2**, 160–4.

TAYLOR, G. L. & IKIN, E. W. (1939): *Brit. Med. J.* **1**, 1027–31.

TAYLOR, G. L. & PRIOR, A. M. (1938): *Ann. Eugen., Lond,* **8**, 343–61.

THOMSEN, M. & LEMCHE, H. (1933): *Biol. Zbl.* **53**, 541–60.

TIMOFÉEFF-RESSOVSKY, N. (1940): *The New Systematics,* pp. 73–136. Oxford.

TURESSON, G. (1922). *Hereditas, Lund,* **3**, 211–350.

WALTHER, H. (1927): *Iris,* **41**, 32–49.

WIENER, A. S. (1939): *Blood Groups and Blood Transfusion* (2nd ed.). London.

WILLIAMS, H. B. (1932–3): *Proc. S. Lond. Ent. Nat. Hist. Soc.* pp. 1–10.

ZIMMERMANN, K. (1935): *Arch. Naturgesch.* N.F. **4**, 258–73.

23

Reprinted from *Ricerca Sci. (Suppl.)* **19**:3–10 (1949)

DISEASE AND EVOLUTION

J. B. S. Haldane

[Editor's Note: Summary has been omitted at this point.]

It is generally believed by biologists that natural selection has played an important part in evolution. When however an attempt is made to show how natural selection acts, the structure or function considered is almost always one concerned either with protection against natural "forces" such as cold or against predators, or one which helps the organism to obtain food or mates. I want to suggest that the struggle against disease, and particularly infectious disease, has been a very important evolutionary agent, and that some of its results have been rather unlike those of the struggle against natural forces, hunger, and predators, or with members of the same species.

Under the heading infectious disease I shall include, when considering animals, all attacks by smaller organisms, including bacteria, viruses, fungi, protozoa, and metazoan parasites. In the case of plants it is not so clear whether we should regard aphids or caterpillars as a disease. Similarly there is every gradation between diseases due to a deficiency of some particular food constituent and general starvation.

The first question which we should ask is this. How important is disease as a killing agent in nature? On the one hand what fraction of members of a species die of disease before reaching maturity? On the other, how far does disease reduce the fertility of those members which reach maturity? Clearly the answer will be very different in different cases. A marine species producing millions of small eggs with planktonic larvae will mainly be eaten by predators. One which is protected against predators will lose a larger proportion from disease.

There is however a general fact which shows how important infectious disease must be. In every species at least one of the factors which kills it or lowers its fertility must increase in efficiency as the species becomes denser. Otherwise the species, if it increased at all, would increase without limit. A predator cannot in general be such a factor, since predators are usually larger than their prey, and breed more slowly. Thus if the numbers of mice increase, those of their large enemies, such as owls, will increase more slowly. Of course the density-dependent check may be lack of food or space. Lack of space is certainly effective on dominant species such as forest trees or animals like *Mytilus*. Competition for food by the same species is a limiting factor in a few phytophagous animals such as defoliating caterpillars, and in very stenophagous animals such as many parasitoids. I believe however that the density-dependent limiting factor is more often a parasite whose incidence is disproportionately raised by overcrowding.

As an example of the kind of analysis which we need, I take Varley's (1947) remarkable study on *Urophora jaceana*, which forms galls on the composite *Centaurea nigra*. In the year considered 0.5% of the eggs survived to produce a mature female. How were the numbers reduced to 1/200 of the initial value?

If we put $200 = e^k$, we can compare the different killing powers of various environmental agents, writing $K = k_1 + k_2 + k_3 + \ldots$, where k_p is a measure of the killing power of each of them. The result is given in Table 1. Surprisingly, the main killers appear to be mice and voles (*Mus*, *Microtus*, etc.) which eat the fallen galls and account for at least 22%, and perhaps 43% of k. Parasitoids account for 31% of the total kill, and the effect of *Eurytoma curta* was shown to be strongly dependent on host density, and probably to be the main factor in controlling the numbers of the species, since the food plants were never fully occupied.

When we have similar tables for a dozen species we shall know something about the intensity of possible selective agencies. Of course in the case of *Urophora jaceana* analysis is greatly simplified by the fact that the imaginal period is about 2% of the whole life cycle, so that mortality during it is unimportant.

A disease may be an advantage or a disadvantage to a species in competition with others. It is obvious that it can be a disadvantage. Let us consider an ecological niche which has recently been opened, that of laboratories where the genetics of small insects are studied. A number of species of *Drosophila* are well adapted for this situation. Stalker attempted to breed the related genus *Scaptomyza* under similar conditions, and found that his cultures died of bacterial disease. Clearly the immunity of *Drosophila* to such diseases must be of value to it in nature also.

Now let us take an example where disease is an advantage. Most, if not all, of the South African artiodactyls are infested by trypanosomes such as *T. rhodesiense* which are transmitted by species of *Glossina* to other mammals and, sometimes at least, to men. It is impossible to introduce a species such as *Bos taurus* into an area where this infection is prevalent. Clearly these ungulates have a very powerful defence against invaders. The latter may ultimately acquire immunity by natural selection, but this is a very slow process, as is shown by the fact that the races of cattle belonging to the native African peoples have not yet acquired it after some centuries of sporadic exposure to the infection. Probably some of the wild ungulates die of, or have their health lowered by the trypanosomes, but this is a small price to pay for protection from other species.

A non-specific parasite to which partial immunity has been acquired, is a powerful competitive weapon. Europeans have used their genetic resistance to such viruses as that of measles (rubeola) as a weapon against primitive peoples as effective as fire-arms. The latter have responded with a variety of diseases to which they are resistant. It is entirely possible that great and, if I may say so, tragic episodes in evolutionary history such as the extinction of the Noto-ungulata and Litopterna may have been due to infectious diseases carried by invaders such as the ungulates, rather than to superior skeletal or visceral developments of the latter.

A suitable helminth parasite may also prove a more efficient protection against predators than horns or cryptic coloration, though until much more is known as to the power of helminths in killing vertebrates or reducing their fertility, this must remain speculative.

However it may be said that the capacity for harbouring a non-specific parasite without grave disadvantage will often aid a species in the struggle for existence. An ungulate species which is not completely immune to *Trypanosoma rhodesiense* has probably (or had until men discovered the life history of this parasite) a greater chance of survival than one which does not harbour it, even though it causes some mortality directly or indirectly.

TABLE 1.

Month	Density per square metre	Cause of death	*k*
July	203.0	—	—
July	184.7	Infertile eggs.	0.095
July	147.6	Failure to form gall.	0.224
July	144.6	? Disease.	0.021
July	78.8	*Eurytoma curta*	0.607
Aug., Sept.		Other parasitoids.	0.222
Aug., Sept.	50.2	Caterpillars.	0.234
Winter	19.2	Disappearance, probably mice.	0.957
Winter	7.0	Mice.	1.009
Winter	5.2	Unknown.	0.297
May–June		Birds.	0.090
May–June	3.6	Parasitoids	0.270
July	2.03	Floods.	0.581
			4.606

Mortality of *Urophora jaceana* in 1935-1946, after Varley.

The winter disappearance was probably due to galls carried off by mice. The mortality attributed to mice is based on counts of galls bitten open. The *k* due to *Eurytoma curta* in the preceding year was 0.069, in the subsequent year 0.137. This cause of death depends very strongly on host and parasitoid densities. The caterpillars killed the larvae by eating the galls.

I now pass to the probably much larger group of cases where the presence of a disease is disadvantageous to the host. And here a very elementary fact must be stressed. In all species investigated the genetical diversity as regards resistance to disease is vastly greater than that as regards resistance to predators.

Within a species of plant we can generally find individuals resistant to any particular race of rust (Uredineae) or any particular bacterial disease. Quite often this resistance is determined by a single pair, or a very few pairs, of genes. In the same way there are large differences between different breeds of mice and poultry in resistance to a variety of bacterial and virus diseases. To put the matter rather figuratively, it is much easier for a mouse to get a set of genes which enable it to resist *Bacillus typhi murium* than a set which enable it to resist cats. The genes commonly segregating in plants have much more effect on their resistance to small animals which may be regarded as parasites, than to larger ones. Thus a semiglabrous mutant of *Primula sinesis* was constantly infested by aphids which however are never found on the normal plant. I suppose thornless mutants of *Rubus* are less resistant to browsing mammals than the normal type, but such variations are rarer.

Anyone with any experience of plant diseases will of course point out that the resistance of which I have spoken is rarely very general. When a variety of wheat has been selected which is immune to all the strains of *Puccinia graminis* in its neighbourhood, a new strain to which it is susceptible usually appears within a few years, whether by mutation, gene recombination, or migration. Doubtless the same is true for bacterial and virus diseases. The microscopic and sub-microscopic parasites can evolve so much more rapidly than their hosts that the latter have little chance of evolving complete immunity to them. It is very remarkable that *Drosophila* is as generally immune as it is. I venture to fear that some bacillus or virus may yet find a suitable niche in the highly overcrowded *Drosophila* populations of our laboratories, and that if so this genus will lose its proud position as a laboratory animal. The most that the average species can achieve is to dodge its minute enemies by constantly producing new genotypes, as the agronomists are constantly producing new rust-resistant wheat varieties.

Probably a very small biochemical change will give a host species a substantial degree of resistance to a highly adapted microorganism. This has an important evolutionary effect. It means that it is an advantage to the individual to possess a rare biochemical phenotype. For just because of its rarity it will be resistant to diseases which attack the majority of its fellows. And it means that it is an advantage to a species to be biochemically diverse, and even to be mutable as regards genes concerned in disease resistance. For the biochemically diverse species will contain at least some members capable of resisting any particular pestilence. And the biochemically mutable species will not remain in a condition where it is resistant to all the diseases so far encountered, but an easy prey to the next one. A beautiful example of the danger of homogeneity is the case of the cultivated banana clone "Gros Michel" which is well adapted for export and has been widely planted in the West Indies. However it is susceptible to a root infection by the fungus *Fusarium cubense* to which many varieties are immune, and its exclusive cultivation in many areas has therefore had serious economic effects.

Now every species of mammal and bird so far investigated has shown a quite surprising biochemical diversity revealed by serological tests. The antigens concerned seem to be proteins to which polysaccharide groups are attached. We do not know their functions in the organism, though some of them seem to be part of the structure of cell membranes. I wish to suggest that they may play a part in disease resistance, a particular race of bacteria or virus being adapted to individuals of a certain range of biochemical constitution, while those of other constitutions are relatively resistant. I am quite aware that attempts to show that persons of a particular blood group are specially susceptible to a particular disease have so far failed. This is perhaps to be expected, as a disease such as diphtheria or tuberculosis is caused by a number of biochemically different races of pathogens. The kind of investigation needed is this. In a particular epidemic, say of diphtheria, are those who are infected (or perhaps those who are worst affected) predominantly drawn from one serological type (for example *AB*, *MM*, or *BMM*)? In a different epidemic a different serological type would be affected.

In addition, if my hypothesis is correct, it would be advantageous for a species if the genes for such biochemical diversity were particularly mutable, provided that this could be achieved without increasing the mutability of other genes whose mutation would give lethal or sublethal genotypes. Dr. P. A. Gorer informs me that there is reason to think that genes of this type are particularly mutable in mice. Many pure lines of mice have split up into sub-

311

lines which differ in their resistance to tumour implantation. This can only be due to mutation. The number of loci concerned is comparable, it would seem, with the number concerned with coat colour. But if so their mutation frequency must be markedly greater.

We have here, then, a mechanism which favours polymorphism, because it gives a selective value to a genotype so long as it is rare. Such mechanisms are not very common. Among others which do so are a system of self-sterility genes of the *Nicotiana* type. Here a new and rare gene will always be favoured because pollen tubes carrying it will be able to grow in the styles of all plants in which it is absent, while common genes will more frequently meet their like. However this selection will only act on genes at one locus, or more rarely at two or three. A more generally important mechanism is that where a heterozygote is fitter than either homozygote, as in *Paralettix texanus* (Fisher 1939) and *Drosophila pseudoobscura* (Dobzhansky 1947). This does not, however, give an advantage to rarity as such. It need hardly be pointed out that, in the majority of cases where it has been studied, natural selection reduces variance.

I wish to suggest that the selection of rare biochemical genotypes has been an important agent not only in keeping species variable, but also in speciation. We know, from the example of the *Rh* locus in man, that biochemical differentiation of this type may lower the effective fertility of matings between different genotypes in mammals. Wherever a father can induce immunity reactions in a mother the same is likely to be the case. If I am right, under the pressure of disease, every species will pursue a more or less random path of biochemical evolution. Antigens originally universal will disappear because a pathogen had become adapted to hosts carrying them, and be replaced by a new set, not intrinsically more valuable, but favouring resistance to that particular pathogen. Once a pair of races is geographically separated they will be exposed to different pathogens. Such races will tend to diverge antigenically, and some of this divergence may lower the fertility of crosses. It is very striking that Irwin (1947) finds that related, and still crossable, species of *Columbia*, *Streptopelia*, and allied genera differ in respect of large numbers of antigens. I am quite aware that random mutation would ultimately have the same effect. But once we have a mechanism which gives a mutant gene as such an advantage, even if it be only an advantage of one per thousand, the process will be enormously accelerated, particularly in large populations.

There is still another way in which parasitism may favour speciation. Consider an insect in which a parasitoid, say an ichneumon fly, lays its eggs. And let us suppose both host and parasite to have an annual cycle, the parasite being specific in that particular host. To simplify matters still further, we shall suppose that the parasite is the only density-dependent factor limiting the growth of the host population, whilst the density-dependent factor limiting the growth of the parasite population is the difficulty of finding hosts. It is further assumed that the parasitoid only lays one egg in each host, or that only one develops if several are laid. Varley showed that all these assumptions are roughly true for *Urophora jaceana* and *Erytoma curta*.

Let e^k be the effective fertility of the host, that is to say let k be the mean value of the natural logarithm of the number of female-producing eggs laid per female. Let k_1 be the killing power of agents killing during the part of the life cycle before the host is not infested. Let k_2 be the killing power of agents other than the parasite killing during that part of the life cycle when it is infested, and therefore killing the same fraction $1 - e^{-k_2}$ of parasites as of hosts. Let k_3 be the killing power of agents killing after the parasites have emerged as imagines.

Let x be the equilibrium density of adult hosts, y that of parasitoids, and a the mean area of search of the parasitoid.

Then the fraction of hosts which are not parasitized must be e^{-k_4}, where $k_1 + k_2 + k_3 + k_4 = K$. Nicholson and Bailey (1933) showed that $k_4 = ay$. But the host density available for parasitism is e^{K-k_1}. Of these $(1 - e^{-k_4})e^{K-k_1}$ are parasitized. And $e^{K-k_1-k_2}(1 - e^{-k_4})$ of the parasites live to emerge. This is diminished by a factor e^{-k_1}. The equilibrium densities are given by

$$x = \frac{e^{k_1-k_3}\,y}{e^{K-k_1-k_2-k_3} - 1}\ ,$$

$$y = \frac{K^{k_1-k_2-k_3}}{a}\ ,$$

though there is perpetual oscillation round this equilibrium. Now consider the effect of changes in these parameters. Any gene in the host which increases $K - k_1 - k_2 - k_3$ will give its carriers a selective advantage over their fellows, and will therefore spread through the population. This will cause an increase in the density of parasitoids. If it acts before or during parasitisation, thus diminishing k_1 or k_2, it will diminish the equilibrium value of x. If it acts after parasitisation is over, thus diminishing k_3, it will increase the equilibrium value of x, though not very much. But since we have supposed that the parasitoids only emerge shortly before the end of the hosts life cycle, every increase in its adaptation to environmental factors other than the specific parasite will diminish the numbers of adult hosts, though it may increase the number of their larvae at an early stage. This is a striking example of the way in which the survival of the fittest can make a species less fit.

A concrete case would be a gene which, by improving cryptic or aposematic coloration of the larvae, enabled more of them to escape predators, and therefore more parasites to do so. Since the host population is denser they will parasitise a larger fraction of the hosts and thus reduce their number. Since a larger fraction of parasites escape, equilibrium will be reached with a lower host density. Fewer of the caterpillars "nati a far l'angelica farfalla" will achieve this end. More of them will give rise to ichneumons or chalcids.

Natural selection will also favour genes which enable the host to resist the parasitoid, but the latter will also increase its efficiency by natural selection. As Nicholson and Bailey showed, every increase in the area searched by it will diminish the density of both hosts and parasites.

The best that can be said for this tendency, from the host's point of view, is that it makes it less likely to become extinct as the result of other agencies. For the parasitoid being dependent on the density of the host population, will allow its numbers to increase rapidly after any temporary fall.

The host can hardly hope to throw off the parasite permanently by changing its life cycle, developing immunity (if this is possible) or otherwise. But it can reduce its numbers by speciating. For suppose that the pair of species is replaced by two host-parasitoid pairs, the population will be doubled, in so far as the parasitoid is the regulator. It is unlikely that a species can divide sympatrically, but the reduction in numbers caused by parasitoidism will leave food available for other immigrant species of similar habits, even if they are equally parasitised.

Thus certain types of parasitism will tend to encourage speciation, as others encourage polymorphism. This will specially be the case where the parasite is very highly adapted to its host, the most striking cases of adaptation being probably those of the parasitoid insects.

We see then that in certain circumstances, parasitism will be a factor promoting polymorphism and the formation of new species. And this evolution will in a sense be random. Thus any sufficiently large difference in the times of emergence or oviposition of two similar insect species will make it very difficult for the same parasitoid to attack both of them efficiently. So will any sufficiently large difference in their odours. We may have here a cause for some of the apparently unadaptive differences between related species.

Besides these random effects, disease will of course have others. It is clear that natural selection will favour the development of all kinds of mechanisms of resistance, including tough cuticles, phagocytes, the production of immune bodies, and so on. It will have other less obvious effects. It will be on the whole an antisocial agency. Disease will be less of a menace to animals living singly or in family groups than to those which live in large communities. Thus it is doubtful if all birds could survive amid the faecal contamination which characterises the colonies of many sea birds. A factor favouring dispersion will favour the development of methods of sexual recognition at great distances such as are found in some Lepidoptera.

Again, disease will set a premium on the finding of radically new habitats. When our ancestors left the water, they must have left many of their parasites behind them. A predator which ceases to feed on a particular prey, either through migration or changed habits, may shake off a cestode which depends on this feeding habit. When cerebral development has gone far enough to make this possible, it will favour a negative reaction to faecal odours and an objection to cannibalism, and will so far be of social value. A vast variety of apparently irrelevant habits and instincts may prove to have selective value as a means of avoiding disease.

A few words may be said on non-infectious diseases. These include congenital diseases due to lethal and sub-lethal genes. Since mutation seems to be non-specific as between harmful and neutral or beneficial genes, and mutation rate is to some extent inherited, it follows that natural selection will tend to lower the mutation rate, and this tendency may perhaps go so far as to slow down evolution. It will also tend to select other genes which neutralise the effect of mutants, and thus to make them recessive or even ineffective, as Fisher has pointed out. Whether the advantage thus given to polyploids is ever important, we do not know. But the evolution of dominance must tend to make the normal genes act more intensely and thus probably earlier in ontogeny, so that a character originally appearing late in the life cycle will tend to develop earlier as time goes on.

Deaths from old age are due to the breakdown of one organ or another, in fact to disease, and the study of the mouse has shown that senile diseases such as cancer and nephrosis are often congenital. In animals with a limited reproductive period senile disease does not lower the fitness of the individual, and increases that of the species. A small human community where every woman died of cancer at 55, would be more prosperous and fertile than one where this did not occur. Senile disease may be an advantage wherever the reproductive period is limited; and even where it is not, a genotype which lead to disease in the 10% or so of individuals which live longest may be selected if it confers vigour on the majority. As Simpson (1944) has emphasized, some of the alleged cases of hypertely can be explained in this way.

Deficiency diseases are due to the lack of a particular food constituent which an organism or its symbionts cannot synthesize or make from larger molecules. They must act as a selective agent against the loss of synthetic capacity which is a very common type of mutation in simple organisms at least, and in favour of genotypes with a varied symbiotic flora. They might thus have speeded up the evolution of the ruminants, whose symbionts probably make vitamins as well as simple nutrients like acetic acid. To be precise it might be an advantage to have a small rumen where symbionts made B vitamins before it got large enough to add appreciably to the available calories.

On the other hand Rudkin and Schulz (1948) have shown that deficiency diseases can select mutants which utilize the nutrilite in question less than does the normal type. In particular the vermillion mutant of *Drosophila melanogaster* does not form the brown eye pigment ommatin from tryptophane. Thus deficiency diseases may cause a regressive type of evolution characterized by the loss of capacity to utilise rare nutrilites for synthesis.

In this brief communication I have no more than attempted to suggest some lines of thought. Many or all of them may prove to be sterile. Few of them can be followed profitably except on the basis of much field work.

REFERENCES

DOBZHANSKY, T. 1917. Genetics of natural populations XIV. A response of certain gene arrangements in the third chromosome of *Drosophila subobscura* to natural selection. *Genetics*, 32:142.

FISHER, R. A. 1939. Selective forces in wild populations of *Paratellix texanus*. *Ann. Eugen.*, 9:109.

IRWIN, M. R. 1949. Immunogenetics. *Advances in Genetics*, 1:133.

NICHOLSON, I. J. and BAILEY, V. A. 1935. The balance of animal populations Pt. 1. *Proc. Zool. Soc. London*:551-598.

RUDKIN, G. T. and SCHULZ, J. 1948. A comparison of the Tryptophane requirements of mutant and wild type *Drosophila melanogaster*. *Proc. Int. Cong. Genetics, Stockholm*. July 1948.

SIMPSON, G. G. 1944. *Tempo and mode in Evolution*. New York.

VARLEY, G. C. 1947. Natural control of population balance in the knapweed gallfly (*Urophora jaceana*). *Journ. Anim. Ecol.*, 16:139.

[Editors' Note: Discussion has been omitted at this point.]

Reprinted from *Lancet* II:675–679 (1919)

SEROLOGICAL DIFFERENCES BETWEEN THE BLOOD OF DIFFERENT RACES

The Result of Researches on the Macedonian Front

Ludwik Hirschfeld and Hanka Hirschfeld

Race Problems and Researches in Immunisation.

IT is a well-known fact that it is possible to produce antibodies by injecting an animal of one species with the red blood corpuscles of an animal of a different species. These antibodies, which we call hetero-antibodies, are capable of reacting with the erythrocytes of all representatives of the species used for immunising. A rabbit immunised with the blood of a man of any race will produce agglutinins or hæmolysins which can influence to a greater or lesser degree the blood corpuscles of men of any race. The hetero-antibodies are thus specific for a species and cannot bring us nearer to the solution of the race problem.

But, as Ehrlich showed in goats and von Dungern and Hirschfeld[1] in dogs, we do possess a means of finding serological differences within a species. This is effected by immunisation *in the species*. The reason for this can be explained in a few words. Those antigen properties which are common to the giver and receiver of blood cannot give rise to any antibodies, since they are not felt as foreign by the immunised animal. The

* Paper read before the Salonika Medical Society, June 5th, 1918, and submitted to us for publication in September 1919.

[1] Von Dungern: Münchener medizinische Wochenschrift, 1911. Von Dungern and Hirschfeld: Zeitschrift für Immunitätsforschungen, 1911, Comm. I., II., III.

antibodies produced within the species which we call iso-antibodies do not, therefore, act against the whole of the antigen properties of the species, but only against the differences between the blood of the animal which provides the blood for injection and that of the recipient. The iso-antibodies thus do not influence all representatives of the species, but only the blood used for injection and other kinds of blood similar to it. If we inject into dogs the blood of other dogs it is in many cases possible to produce antibodies. By means of these antibodies we have been able to show that there are in dogs two antigen types. These antigen types, which we recognise by means of the iso-antibodies, we may designate biochemical races.

It was, therefore, clear to us from the beginning that we could only attack the human race problem on serological lines if we could succeed in making use of antibodies of this kind, for the iso-antibodies alone are capable of selecting from the whole of the biochemical elements which serologically characterise human blood as such those elements that are characteristic of the blood of only a part of the human species.[2]

Such a differentiation of the human species is now possible by means of the iso-agglutinins first analysed by Landsteiner. If the serum and the blood corpuscles of different pairs of human beings are brought together agglutination sometimes occurs. Accurate analysis of the agglutinable properties of the blood and of the agglutinating properties of the serum showed that this phenomenon has nothing to do with disease. It depends on the following physiological facts. There are present in human blood two agglutinable properties, which, however, are not equally marked in all individuals. In the serum there is never present an agglutinin reacting with its own blood, but always an agglutinin which reacts with that property which is absent in that particular blood. As has often been pointed out in the literature of the subject, these properties are of great importance in blood transfusion, for a blood must never be injected which can be agglutinated by the recipient. Either of the two agglutinable properties may be present or both together, or both may be absent, and we can therefore distinguish four different combinations in man. In the literature of the subject we therefore most often find the statement that there are in the human species four different groups. But since we have, indeed, four groups but only two agglutinable properties, von Dungern and Hirschfeld[1] introduced another definition which is shown in the accompanying table. (Table I.) They

Group O. In accordance with Landsteiner's rule that in the blood there are always agglutinins against the qualities lacking in the corpuscles we have here both the shaded anti-A and the black anti-B arrows. Finally, in the fourth square are represented the red corpuscles, which have both A and B properties and accordingly no agglutinins. Below our diagram we have given the classification used by English writers on the subject. The English Group I. corresponds to our A B. Group II. is our A, and Group III. our B. As against Mosse, we wish to point out that Group I. does not represent any special individuality, but merely, as can easily be proved by absorption experiments, the combination of the properties A and B. For the differentiation of the groups, therefore, we require not three, but only two sera, A and B.

Results of Experiments.

What biological significance, then, has this peculiar differentiation of the blood within the human species, and how are these marks of the biochemical race inherited? Von Dungern and Hirschfeld first undertook experiments on the inheritance of these characteristics in dogs, and found that the biochemical properties of the blood are sometimes inherited and sometimes disappear in the offspring. They established, further, that the anatomical and biochemical characteristics are inherited independently of one another. Young dogs which had the general structure and colour of the mother showed the agglutinable properties of the father and vice versa.

Researches on the inheritance of the bio-chemical group properties in man as differentiated by means of Landsteiner's iso-antibodies permitted von Dungern and Hirschfeld to come to an important conclusion as to the nature of these properties. We succeeded in showing that the A and B properties are generally inherited, but sometimes may disappear in the offspring. When the parents had A or B we found sometimes the Group O occurring in the children. On the other hand, we never found either A or B property in a child when it was absent in the parents. This observation will permit under certain circumstances of medico-legal decisions being made in order to find the real father of a child. If we find in a child either A or B property when it is absent in the mother it must be present in the real father.

The analysis of our numerical results proved that we can apply Mendel's law to the inheritance of the biochemical properties A and B. Since the A and B properties may disappear but never appear spontaneously, we can regard them according to Mendel's law as those properties which once present in the germ-plasm must also be outwardly visible. Mendel, as we know, named such a property or quality, which always gives the species its outward appearance, the dominant, while the absence of the property (which property may appear in children although absent in the parents) is considered as latent or recessive, and is set in contra-distinction to the dominant. If, for example, the dominant is *red* the Mendelian quality or property antagonistic to it—e.g., *white*—is described as *non-red*. If we introduce the Mendelian terminology we can in our cases speak of non-A

TABLE I.—*Landsteiner's Law of Iso-agglutinins.*

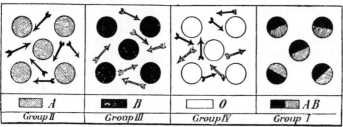

▨ A	▰ B	▭ O	◼ A B
Group II	Group III	Group IV	Group I

called the agglutinable property, which is common in Central Europe, A, the other, which is rare in Central Europe, B. In the figure the blood corpuscles possessing the A property are shown shaded, those with the B property are black. Landsteiner's rule lays down that there are always present in the serum agglutinins against the agglutinable property which is absent in the blood corpuscles of the same blood. If the individual has in the blood corpuscles the A property, he has in his serum agglutinin anti-B, and vice versa. These agglutinins are represented diagrammatically by arrows corresponding with the agglutinable property, the anti-A being shaded, the anti-B black. We see in the first square of the figure the shaded red blood corpuscles A surrounded by the black arrows anti-B, while in the second square are the black blood corpuscles B with the shaded anti-A agglutinins. In the third square we see the non-agglutinable red blood corpuscles which possess neither A nor B property, called by von Dungern and Hirschfeld the

and non-B. With certain premises, which are discussed in the second part of von Dungern and Hirschfeld's paper, the figures showing the frequency of occurrence of A and B in central Europe can be brought into harmony with Mendel's law, the properties A and B being recognised as dominant, the properties non-A and non-B (the combination of which gives Group O) as recessive. A and non-A thus may be regarded as Mendelian pairs and similarly B and non-B.

What are the laws, then, governing the relationships of A and B with each other. The calculation showed that they simply fit in with the calculation of probability according to which A and B can come together when they do not influence each other. When A, for instance, occurs in half of all cases and B in one-tenth, A B will be found together in about one-twentieth—i.e., in 5 per cent. Experience has shown that the occurrence of Group I. (our Group A B) approximates to this figure. If, therefore, Landsteiner's rule is regarded from the broad biological point of view it can be stated as follows: There are within the human species four properties of blood, A, non-A, B, and non-B.

A and non-A, B and non-B behave to each other according to Mendel's law, while A and B, non-A and non-B do not influence one another. Experience has shown that the inheritance of the biochemical blood properties in man is not influenced by sex and does not correspond with the inheritance of anatomical qualities (family resemblance, &c.), so that we have to do with an independent heredity. The agglutinable properties appear already in embryonic life ; we have observed them in a six-months' foetus. We have succeeded in finding these properties in the placental blood even when they were absent in the mother. The iso-agglutinins, however, do not appear until the second year of life. We can confirm Landsteiner's observation that these group-properties have nothing to do with disease. They appear also not to alter with time. We ourselves possess the same groups and agglutinins which we found in our blood eight years ago in spite of the fact that one of us has had typhoid and now suffers from chronic malaria. The experiments in heredity, also, point to the constancy of the biochemical properties of the blood. There exist, indeed, weakly agglutinable blood corpuscles and weakly agglutinating sera. An accidental coming together of these might simulate Group O where in reality A or B was present.

We have often examined hundreds of specimens of blood with 10 different sera and found mostly only quantitative differences between the different A and B sera for the agglutinin is seldom absent. Rarely there are to be found certain anti-A (Group B) sera which agglutinate the corpuscles of rather more individuals than other sera. The corresponding groups are distinguished by von Dungern and Hirschfeld as Large-A and Small-a. We have therefore always used several sera for the examination of a people or a race, and, on the other hand, always used the same set of sera for the examination of different races. We have never observed that malaria has any influence on the agglutinations. The clumping in the blood of anæmic patients has nothing to do with the agglutinations, as one of us has succeeded in proving.[3] The red blood corpuscles always sink in their own plasma as soon as their number is so small that they cannot support each other. The rate of sinking thus runs parallel with the anæmia. It seemed, therefore, that it would be of interest to make use of the properties of blood, as defined by Landsteiner's rule, to form an anthropological criterion for the discovery of hitherto unknown and anatomically invisible relationships between different races. Through the accident of the war we happened to come to a part of the globe where more than elsewhere various races and peoples are brought together, so that the problems we are discussing, which otherwise would have necessitated long years of travel, could be brought in a relatively short time nearer to solution.

Technique.

The technique is as follows. We add a few drops of blood from the ball of the finger to a mixture of sodium chloride and sodium citrate solution (normal saline 9 parts, 2.5 per cent., Na citrate 1 part). A drop of the blood mixture is brought in contact in a small test-tube with a drop each of the A and B sera, of which the activity and specificity have always been established by a control experiment. The result was never read off in less than half an hour—a point of great importance, as particularly Group B is but slowly agglutinated. We used only Serbians as providers of serum. For each race several sera are used, but the same set of sera for all races. Our material consists for the most part of soldiers from various districts. This is to be regarded as a very favourable circumstance, since they are mostly unrelated to each other, so that the possibility that the frequency of a group might depend on family relationship is excluded. The refugees on our statistics for the Jewish race particularly show, give evidence of great individual differences in the first, second, and other hundreds examined, so that the results must be confirmed by tests on more extensive material. The exact statistics with regard to the different tribes, provinces, anatomical structure, &c., will appear in an anthropological journal. We examined 500-1000 persons of each race. For the Germans we quote from memory the results of von Dungern and Hirschfeld as the statistical table for these is unfortunately not to hand. For the Austrians we use the Vienna statistics of L. Landsteiner. For the Jews we used the refugees from

Monastir belonging to a people which came from Spain about 400 years ago. For the Greeks we examined 300 soldiers from Old Greece and the Islands and 200 refugees from Asia Minor and Thrace. For the Turks we used Macedonian Mahommedans. These last must certainly contain a large admixture of Slav blood, and the statistics should be confirmed in Turkey. We wish to emphasise particularly that all these people, except the Indians, who are as a nation for the most part vegetarians, are receiving exactly the same food and are exposed to the same hardships and diseases, so that it would not be correct to refer the great differences we have found in the frequency of A and B Groups to special climatological or pathological conditions.

Statistical Analyses.

We wish to point out as the first important fact that we found the Groups A and B present in all races examined. Table II. shows the percentage of A and B. The figures in the right-hand column of Table II. represent the total numbers of each race examined. A glance at the table shows that we find marked differences in the incidence of A and B in the different races. From the English onward we see that A always diminishes, whereas B increases.[4] In order to give an exact analysis we will first look at our data from the point of view developed by von Dungern and Hirschfeld in the second part of their paper.[5] The group A B was regarded by them as the accidental coming together of A and B. If we analyse our statistics from this point of view we reach the following figures : If A is present in 43.4 per cent. and B in 7.2 per cent. For the calculation of probability we must multiply 43.4/100 by 7.2/100, which gives 301/10,000 = 3.1 per cent. In reality we found 3.1 per cent.

—	A in per cent.	B in per cent.	A B in per cent.	O in per cent.	Total No. examined.	A B. Really found.	A B. Calculation of probability.	O. Really found.	O. Calculation of probability.
						%	%	%	%
English ...	43.4	7.2	3.0	46.4	500	3.1	3.1	46.4	47.9
French ...	42.6	11.2	3.0	43.2	500	3.0	4.7	43.2	46.6
Italians ...	38.0	11.0	3.8	47.2	500	3.8	4.1	47.2	49.5
Germans...	43.0	12.0	5.0	40.0	ca.500	5.0	5.1	40.0	43.1
Austrians	40.0	10.0	8.0	42.0	?	8.0	4.0	42.0	42.6
Serbs... ...	41.8	15.6	4.6	38.0	500	4.6	6.5	38.0	42.7
Greeks ...	41.6	16.2	4.0	38.2	500	4.0	6.7	38.2	43.4
Bulgarians	40.6	14.2	6.2	39.0	500	6.2	5.7	39.0	42.3
Arabs ...	32.4	19.0	5.0	43.6	500	5.0	6.1	43.6	47.5
Turks ...	38.0	18.6	6.6	36.8	500	6.6	7.0	36.8	39.9
Russians...	31.2	21.8	6.3	40.7	1000	6.3	6.8	40.7	44.9
Jews... ...	33.0	23.2	5.0	38.8	500	5.0	7.6	38.8	44.5
Malagasies	26.2	23.7	4.5	45.5	400	4.5	6.2	45.5	49.5
Negroes (Senegal)	22.6	29.2	5.0	43.2	500	-5.0	6.5	43.2	47.6
Annamese	22.4	28.4	7.2	42.0	500	7.2	6.4	42.0	45.3
Indians ...	19.0	41.2	8.5	31.3	1000	8.5	7.8	31.3	43.7

TABLE II.   TABLE III.

Since A occurs in 43.4 per cent. and A B in 3.1 per cent., we have the property A altogether in 46.5 per cent., and correspondingly the non-A in 53.5 per cent. Similar calculations give for non-B 89.7 per cent. The calculation of probabilities thus gives for Group O : $\frac{53.5}{100} \times \frac{89.7}{100} = \frac{4798}{10,000} = 47.9$ per cent.

We actually found 46.4 per cent.

We made the same calculations for each people and race, and have shown the results in Table III. The table shows only slight differences between the calculation and the facts, the calculation giving somewhat higher results.

Basing our opinion on the whole of our material embracing about 8000 cases we look on Group O (the English Group IV.)

[4] In the English literature of the subject we find that the frequency of occurrence of B is higher and corresponds to the figures given by Landsteiner for Vienna. We cannot corroborate this. We examined our English material several times with seven different anti-B sera without finding a greater frequency than that here given.   [5] Loc. cit.

[3] L. Hirschfeld : Korrespondenzblatt für Schweizer Aerzte, 1917.

as the accidental conjunction of the two groups non-A and non-B, while Group I., our A B, is to be regarded as the accidental conjunction of the groups A and B. If, then, we wish to find exactly the frequency of A and B, we must add the English Group I. (our A B) to the Groups II. and III. Thus in the English subjects instead of 43·4 per cent. A, 7·2 per cent. B, and 3 per cent. A. B. we read 46·4 per cent. A and 10·2 per cent. B.

In Table IV. the figures arrived at are shown diagrammatically, Group A being shaded and Group B black. The result is remarkable.

The prevalence of the Group A is characteristic of the European peoples. Most European peoples have not less than 45 per cent. A; in the Italians alone we found 41 per cent. This frequency of A only applies to Europe. In Africa and in Asia we find far fewer cases of A; the

Russians and Jews 28 per cent. The Russians arranged according to districts show the following relationships:—

TABLE V.

—	A %	B %	Total.
Central Russia	37·6	25·2	400
Siberia	36·5	29·0	321
Ukraine	35·1	33·3	111
Perm, Vologda, &c.	36·8	34·5	84

Central Russia and Siberia show the intermediate type. Little Russia and the Volga District have the Asiatic type.

We see thus that A and B are present in different proportions in different races. The serological formula for a particular race is in no way affected by the anthropological

TABLE IV.—*Showing the Percentage Frequency of A and B Serological Reactions in Various National Types.*

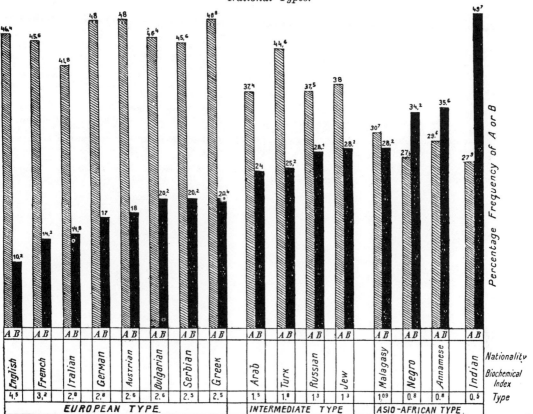

Malagasies have 30 per cent., Negroes 27 per cent., Annamese 29 per cent., and Indians 27 per cent. of A. The countries lying between Asia and Central Europe, with the exception of the Macedonian Turks who have the European A frequency, show the intermediate type; the Arabs 37 per cent., the Russians 37 per cent., the Jews 38 per cent. Thus we see that going southwards and eastwards the frequency of A constantly diminishes. If we now consider Group B we find the exact opposite: the English who are farthest west have only 10 per cent., the French and Italians 14 per cent., the German Austrians 17 to 18 per cent. Of the Balkan peoples the Serbians and Greeks have the same frequency of 20·2 per cent. the Bulgars 20·4. The contrary is found in Africa and Asia. Group B, which is numerically low in Europe, reaches 28 per cent. in the Malagasies, in Negroes 34 per cent., in Annamese 35 per cent., and finally in Indians a maximum of 49 per cent. The peoples of the Mediterranean basin and the Russians in this also show the intermediate type: the Arabs have 24 per cent., Turks 25 per cent.,

characteristics. The Indians, who are looked on as anthropologically nearest to Europeans, show the greatest difference from them in the blood properties. The Russians and the Jews, who differ so much from each other in anatomical characteristics, mode of life, occupation, and temperament, have exactly the same proportion of A and B. On the other hand, it is clear that the distribution of A and B corresponds with surprising accuracy to geographical situation. The nearer to Western and Central Europe the more A and the less B, the nearer to Africa and Asia, especially to India, the less A and the more B. The peoples lying between Central and Western Europe on one side, and Africa and Asia on the other—that is to say, the peoples of the Mediterranean basin—show the intermediate type. In order to designate these relationships by a number we will call the proportion of A to B the biochemical race-index. We see that the race-index in the European peoples varies between 4·5 and 2·5, in the Asio-African peoples it is 1 or less, while the intermediate type is characterised by the race-index 1-2. (Table V.)

This remarkable fact, that A and B are represented in very different proportions in different races may be discussed from two standpoints. One can imagine that when man appeared on the earth A and B were present in the same proportions in different races. In this case the differences which are now present in different races would depend on the assumption that for unknown reasons A is more suitable for increased resistance of the organism to disease in a temperate climate, while B is more suitable in a hot climate. The hypothesis is improbable.

We see that the Russians in Siberia have the same proportion of B as the natives of Madagascar. The Jews who have lived for centuries in Monastir show a type of blood different from that of the other Balkan peoples. It is also a priori improbable that the climatic conditions should influence the frequency of A and B. If B could not be acclimatised in a temperate climate it should have died out in those climates long ago in the extended periods of time which we can take into consideration. It is very difficult to imagine one single place of origin for the human race in view of our statistics, since it would then be inexplicable why A diminishes from west to east and south, while B increases. The figures arrived at by us are most easily explained on the assumption that A and B had different points of origin and that there are two different biochemical races which arose in different places. In this case the mutual infiltration of these races is the cause of the varying proportion of A and B. Since the greatest frequency of B is found in India,[6] we should then look for the moment on India as the cradle of one part of humanity—namely, of the biochemical race B. Both to the east (Indo-China) and to the west, towards Europe and Africa, a broad stream of Indians poured out, ever lessening in its flow, which finally, although continually diminishing, penetrated to Western Europe. A similar single place of origin for A cannot be

[6] Of the greatest importance are the researches from the Central Asia plateau.

indicated, since all European peoples show a greater or less preponderance of A, which only seems to diminish somewhat in the Italians. Since, however, the Asiatic and African peoples are poor in A, we must assume hypothetically that A arose in North or Central Europe and spread out thence southwards and eastwards. A detailed examination of the northern races may lead us to unexpected and important conclusions on this point.

Conclusion.

We see then that the analysis of the figures we obtained has led us to remarkable results and that the highly interesting problem of a possible double origin of the human race has become a question which can be studied by means of serological methods. A problem of such dimensions may be propounded, but can hardly be resolved by the experimental results of a science, the introduction of which into these problems is here attempted for the first time. Still, we believe that we have shown in this paper that experiments in immunisation deserve to be made use of for the solving of anthropological questions. A close coöperation would be necessary between anthropologists and serologists, and the researches should be conducted on an international basis. The present war has brought so many races and peoples together that the main problems should certainly be capable of solution in a short time and without great labour. A series of important special researches, such as the examination of various stocks, primitive races, and anthropoid apes should be begun without delay.

In our experiments we received much help from many medical and other officers of the Allied Armies in Macedonia. We owe our gratitude to the Directors of the Medical Services of the several armies for permission to examine soldiers and for other assistance. Major Stefanovitch and Captain W. L. Murphy, the liaison officers of the Serbian and British Army Medical Services, have undertaken the translation of the paper into English, for which we offer them our best thanks.

320

Copyright © 1954 by the British Medical Journal

Reprinted from *Brit. Med. J.* 1:290–294 (1954)

PROTECTION AFFORDED BY SICKLE-CELL TRAIT AGAINST SUBTERTIAN MALARIAL INFECTION

BY

A. C. ALLISON, D.Phil., B.M.*

(From the Clinical Pathology Laboratory, the Radcliffe Infirmary, Oxford)

The aetiology of sickle-cell anaemia presents an outstanding problem common to both genetics and medicine. It is now universally accepted that the sickle-cell anomaly is caused by a single mutant gene which is responsible for the production of a type of haemoglobin differing in several important respects from normal adult haemoglobin (Pauling *et al.*, 1949 ; Perutz and Mitchison, 1950). Carriers of the sickle-cell trait who are heterozygous for the sickle-cell gene have a mixture of this relatively insoluble haemoglobin and normal haemoglobin ; hence their erythrocytes do not sickle *in vivo*, whereas some at least of the homozygotes, who have a much greater proportion of sickle-cell haemoglobin, have sickle cells in the circulating blood, with inevitable haemolysis and a severe, often fatal, haemolytic anaemia. There is also a much

smaller group of sickle-cell anaemia patients who are heterozygous for the sickle-cell gene as well as for some other hereditary abnormality of haemoglobin synthesis (Neel, 1952).

It is thus possible to approach the problem from the clinical or the genetical side. From the clinical point of view it is important to distinguish between carriers of the sickle-cell trait who show no other haematological abnormalities and patients with sickle-cell anaemia, who have a haemolytic disease which can reasonably be attributed to sickling of the erythrocytes. From the genetical point of view the main distinction is to be drawn between those who are homozygous and those who are heterozygous for the sickle-cell gene. In the great majority of instances two classifications coincide—that is, most individuals with the sickle-cell trait are heterozygous and most cases of sickle-cell anaemia, in Africa at least, are homozygous for the sickle-cell gene.

*Staines Medical Research Fellow of Exeter College, in receipt of a grant from the Medical Research Council.

The sickle-cell trait is remarkably common in some parts of the world. Among many African negro tribes 20% or more of the total population have the trait, and frequencies of 40% have been found in several African tribes (Lehmann and Raper, 1949 ; Allison, 1954). In parts of Greece frequencies of 17% have been described (Choremis et al., 1953), and as many as 30% of the population in Indian aboriginal groups are affected (Lehmann and Cutbush, 1952).

Wherever the sickle-cell trait is known to occur sickle-cell anaemia will also be found. For a time it was thought by some workers that sickle-cell anaemia was rare among African negroes, but so many cases have been described during the past few years that this view is no longer tenable (Lambotte-Legrand and Lambotte-Legrand, 1951 ; Foy et al., 1951 ; Edington, 1953 ; Vandepitte and Louis, 1953).

The main problem can be stated briefly: how can the sickle-cell gene be maintained at such a high frequency among so many peoples in spite of the constant elimination of these genes through deaths from the anaemia ? Since most sickle-cell anaemia subjects are homozygotes, the failure of each one to reproduce usually means the loss of two sickle-cell genes in every generation. It can be estimated that for the lost genes to be replaced by recurrent mutation so as to leave a balanced state, assuming that the sickle-cell trait—that is, the heterozygous condition—is neutral from the point of view of natural selection, it would be necessary to have a mutation rate of the order of 10^{-1}. This is about 3,000 times greater than naturally occurring mutation rates calculated for man and, with rare exceptions, in many other animals—3.2×10^{-5} in the case of haemophilia (Haldane, 1947). A mutation rate of this order of magnitude can reasonably be excluded as an explanation of the remarkably high frequencies of the sickle-cell trait observed in Africa and elsewhere.

Possibility of Selective Advantage

Of the other explanations which can be advanced to meet the situation, one has received little attention : the possibility that individuals with the sickle-cell trait might under certain conditions have a selective advantage over those without the trait. It was stated for many years that the sickle-cell trait was in itself a cause of morbidity, but this belief seems to have been based upon unsatisfactory criteria for distinguishing the trait from sickle-cell anaemia. The current view is that the sickle-cell trait is devoid of selective value. Henderson and Thornell (1946) found that in American negro air cadets who had passed a searching physical examination the incidence of the sickle-cell trait was the same as in the general negro population of the United States. Lehmann and Milne (1949) were unable to discover any correlation between haemoglobin levels and the presence or absence of the sickle-cell trait in Uganda Africans. And Humphreys (1952) could find no evidence that the sickle-cell trait was responsible for any morbidity in Nigerian soldiers.

However, during the course of field work undertaken in Africa in 1949 I was led to question the view that the sickle-cell trait is neutral from the point of view of natural selection and to reconsider the possibility that it is associated with a selective advantage. I noted then that the incidence of the sickle-cell trait was higher in regions where malaria was prevalent than elsewhere. The figures presented by Lehmann and Raper (1949) for the frequency of the sickle-cell trait in different parts of Uganda lent some support to this view, as did the published reports from elsewhere. Thus the trait is fairly common in parts of Italy and Greece, but rare in other European countries ; in Greece the trait attains its highest frequencies in areas which are conspicuously malarious (Choremis et al., 1951).

Relation between Malaria and Sickle-cell Trait

Other reports appeared suggesting more directly that there might be a relationship between malaria and the sickle-cell trait. Beet (1946) had observed that in a group of 102 sicklers from the Balovale district of Northern Rhodesia only 10 (9.8%) had blood slides showing malaria parasites, whereas in a comparable group of 491 non-sicklers 75 (15.3%) had malaria parasites. The difference in incidence of malaria in the two groups is statistically highly significant ($x^2 = 19.349$ for 1 d.f.)* ; hence Beet's figures imply strongly that malaria is less frequent among individuals with the sickle-cell trait than among those without the trait. The difference in malarial susceptibility between sicklers and others seemed to be most pronounced at the time of the year when malaria transmission was lowest.

Later, in the Fort Jameson district of Northern Rhodesia, Beet (1947) found that the same difference was present, although it was much less pronounced. Of 1,019 non-sicklers, 312 (30.6%) had blood slides with malaria parasites, whereas of 149 sicklers 42 (28.2%) showed malaria parasites. This difference is not statistically significant. However, among the sicklers from Fort Jameson enlarged spleens were less common than among non-sicklers. In a series of 569 individuals there were 87 with the sickle-cell trait ; 24 (27.9%) of these had palpable spleens, as compared with 188 (39.0%) with splenomegaly out of 482 non-sicklers. This difference is again statistically significant ($x^2 = 4.11$ for 1 d.f.). Beet concluded that Africans with the sickle-cell trait were probably liable to recurrent attacks of thrombosis, with resultant shrinkage of the spleen.

Brain (1952a), also working in Rhodesia, confirmed Beet's observation that the spleen is palpable in a much lower proportion of sicklers than of non-sicklers ; he went on to suggest that the finding might be explained by diminished susceptibility to malaria on the part of the sicklers. Moreover, Brain (1952b) compared the proportion of hospitalized cases in groups of African mine-workers with and without the sickle-cell trait. He found that the sicklers actually spent less time in hospital, on an average, than did the control group of non-sicklers. The incidence of malaria and pyrexias of unknown origin was much lower in the group with sickle cells.

It became imperative, then, to ascertain by more direct methods of investigation whether sickle cells can afford some degree of protection against malarial infection, thereby conferring a selective advantage on possessors of the sickle-cell trait in regions where malaria is hyperendemic. An opportunity to do this came during the course of a visit to East Africa in 1953.

Incidence of Malarial Parasitaemia in African Children with and without the Sickle-cell Trait

The observations of Beet and of Brain on differences in parasite rates and spleen rates are open to criticism because the populations were heterogeneous and were drawn from relatively wide areas. It was decided, therefore, to carry out similar tests on a relatively small circumscribed community, where all those under observation belong to a single tribe. Children were chosen rather than adults as subjects for the observations so as to minimize the effect of acquired immunity to malaria. The recorded incidence of parasitaemia in a group of 290 Ganda children, aged 5 months to 5 years, from the area surrounding Kampala (excluding the non-malarious township) is presented in Table I. The presence of sickling was demonstrated by chemical reduction of blood with isotonic sodium metabisulphite (Daland and Castle, 1948). Fresh reducing solutions were made up daily.

*These and other statistics in this paper are my own, using available figures.

TABLE I

	With Parasitaemia	Without Parasitaemia	Total
Sicklers	12 (27·9%)	31 (72·1%)	43
Non-sicklers ..	113 (45·7%)	134 (53·3%)	247

It is apparent that the incidence of parasitaemia is lower in the sickle-cell group than in the group without sickle cells. The difference is statistically significant ($\chi^2 = 5.1$ for 1 d.f.). In order to test as many families as possible only one child was taken from each family. There is no reason to suppose that these groups are not comparable, apart from the presence or absence of the sickle-cell trait.

The parasite density in the two groups also differed : of 12 sicklers with malaria, 8 (66.7%) had only slight parasitaemia (group 1 on an arbitrary rating), while 4 (33.3%) had a moderate parasitaemia (group 2). Of the 113 non-sicklers with malaria, 34% had slight parasitaemia (group 1), the parasite density in the remainder being moderate or severe (group 2 or 3).

It may be noted, incidentally, that of the four cases in the sickle-cell group with moderate parasitaemia three had *P. malariae*, even though this species is much less common than *P. falciparum* around Kampala. It seems possible from these and other observations that the protection afforded by the sickle-cell trait is more effective against *P. falciparum* than against other species of plasmodia, but much further work is necessary to decide the point.

These results suggest that African children with the sickle-cell trait have malaria less frequently or for shorter periods, and perhaps also less severely, than children without the trait. Further evidence regarding the protective action of the sickle-cell trait could be obtained only by direct observation on the course of artificially induced malarial infection in volunteers.

Progress of Malarial Infection in Adult Africans with and without the Sickle-cell Trait

Fifteen Luo with the trait and 15 Luo without the trait were accepted for this investigation. All the volunteers were adult males who had been away from a malarious environment for at least 18 months. The two groups were

of a similar age and appeared to be strictly comparable apart from the presence or absence of the sickle-cell trait. Two strains of *P. falciparum* were used—one originally isolated in Malaya and one from near Mombasa, Kenya ; in Table II these are marked with the subscripts 1 and 2 respectively. The infection was introduced by subinoculation with 15 ml. of blood containing a large number of trophozoites (B in the table) or by biting with heavily infected *Anopheles gambiae* (M in the table). At least 3 out of the 10 mosquitoes applied had bitten each individual, and the presence of sporozoites was confirmed by dissection of the mosquitoes.

The cases were followed for 40 days. Parasite counts for each case were made by comparison with the number of leucocytes in 200 oil-immersion fields of thick films, the absolute leucocyte counts being checked at intervals. The abbreviated results of these counts are shown in Table II. In the few cases in which parasitaemia was pronounced and the symptoms were relatively severe the progress of the disease was arrested. At the end of the period of observation in every case a prolonged course of antimalarial chemotherapy was given.

Discussion

It is apparent that the infection has become established in 14 cases without the sickle-cell trait. The parasitaemia is relatively light, however, when compared with that observed in non-immune populations—for example, the Africans described by Thomas *et al.* (1953). This is to be expected : the Luo come from a part of the country where malaria is hyperendemic, and they have acquired a considerable immunity to malarial infection in childhood. This factor makes the interpretation of the observations rather more difficult ; however, it could not be avoided, since all the East African tribes who have high sickling rates come from malarious areas, and the acquired immunity should operate with equal force in the groups with and without sickle-cells. The acquired immunity was actually an advantage, since the symptoms were mild and the chances of complication very slight.

In the group with sickle cells, on the other hand, the malaria parasites have obviously had great difficulty in establishing themselves, in spite of repeated artificial infec-

TABLE II

No.	Mode of Infection and Strain	Day after Infection																
		8	10	12	14	16	18	20	22	24	26	28	30	32	34	36	38	40
Luo With No Sickle-cells																		
1	M₂ B₁	0·03	—	0·07	2·5	5·0	2·5	5·0	1·2	0·4	0·02	0·01	—	—	0·1	0·01	0·01	ST
2	M₂ B₁	—	—	—	—	—	—	—	0·03	0·13	0·41	—	—	—	—	0·03	—	ST
3	M₂ B₁	—	—	—	—	—	—	—	0·1	0·02	0·20	5·0	2·5	1·25	1·67	0·2	5·0	2·0 S
4	M₂ B₁	—	—	—	—	0·02	0·02	0·5	0·83	0·12	0·2	1·0	0·83	0·25	0·17	—	—	ST
5	M₂ B₂	—	—	—	—	0·05	1·0	1·67	0·25	0·05	0·07	0·25	1·2	1·0	0·03	—	—	ST
6	B₁	0·02	5·0	10·0	10·0	1·0	0·1	0·01	ST	—	—	—	—	—	—	—	—	—
7	B₂	—	—	—	—	15·0	50·0	ST	—	—	—	—	—	—	—	—	—	—
8	B₃	—	—	0·13	5·0	1·67	0·33	—	—	ST	—	—	—	—	—	—	—	—
9	B₁	—	—	—	—	5·0	—	0·1	0·5	2·5	—	1·0	0·1	2·5	10·0	5·0	0·5	ST
10	B₂	—	—	—	—	—	—	0·05	0·05	—	—	0·67	—	0·1	0·05	5·0	5·0	ST
11	B₂	—	0·05	0·3	—	—	—	0·2	ST	—	—	—	—	—	—	—	—	—
12	B₂	—	—	0·3	0·3	0·3	0·1	0·3	ST	—	—	—	—	—	—	—	—	—
13	B₂	—	—	—	—	—	—	—	—	—	—	—	—	—	—	ST	—	—
14	B₂	2·0	1·7	2·0	60·0	5·0	0·6	ST	—	—	—	—	—	—	—	—	—	—
15	B₂	0·05	0·3	—	0·4	0·1	0·3	ST	—	—	—	—	—	—	—	—	—	—
Luo With Sickle-cell Trait																		
1	M₁ B₂	—	—	—	—	—	—	—	—	—	—	—	—	—	—	—	—	ST
2	M₁ B₂	—	—	—	—	—	—	—	—	—	—	—	—	—	—	—	—	,,
3	M₁ B₁	—	—	—	—	—	—	—	—	—	—	—	—	—	—	—	—	,,
4	M₁ B₁	—	—	—	—	—	—	—	—	—	—	—	—	—	—	—	—	,,
5	M₁ B₁	—	—	—	—	—	—	—	—	—	—	—	—	—	—	—	—	,,
6	M₁ B₁	—	—	—	—	—	—	—	—	—	—	—	—	—	—	—	—	,,
7	M₁ B₁	—	—	—	—	—	—	—	—	—	—	—	—	—	—	5·0	0·5	,,
8	M₁ B₁	—	—	—	—	—	—	—	—	—	—	—	—	—	—	—	—	,,
9	M₁ B₁	0·7	—	—	—	—	—	—	—	—	—	0·03	0·1	0·03	0·03	—	—	,,
10	M₁ B₁	—	—	—	—	—	—	—	—	—	—	—	—	—	—	—	—	,,
11	B₂ M₂	—	—	—	—	—	—	—	—	—	—	—	—	—	—	—	—	,,
12	B₂ M₂	—	—	—	—	—	—	—	—	—	—	—	—	—	—	—	—	,,
13	B₂ M₂	—	—	—	—	—	—	—	—	—	—	—	—	—	—	—	—	,,
14	B₂ M₂	—	—	—	—	—	—	—	—	—	—	—	—	—	—	—	—	,,
15	B₂ M₂	—	—	—	—	—	—	—	—	—	—	—	—	—	—	—	—	,,

Figures represent parasite counts in hundreds per mm.³ of blood. ST = Stopped by chemotherapy.

tion. Only two of the cases show parasites, and the parasite counts in these cases are comparatively low. The striking difference in the progress of malarial infection in the two groups is taken as evidence that the abnormal erythrocytes in individuals with the sickle-cell trait are less easily parasitized than are normal erythrocytes.

It can therefore be concluded that individuals with the sickle-cell trait will, in all probability, suffer from malaria less often and less severely than those without the trait. Hence in areas where malaria is hyperendemic children having the trait will tend to survive, while some children without the trait are eliminated before they acquire a solid immunity to malarial infection. The protection against malaria might also increase the fertility of possessors of the trait. The proportion of individuals with sickle cells in any population, then, will be the result of a balance between two factors ; the severity of malaria, which will tend to increase the frequency of the gene, and the rate of elimination of the sickle-cell genes in individuals dying of sickle-cell anaemia. Or, genetically speaking, this is a balanced polymorphism where the heterozygote has an advantage over either homozygote.

The incidence of the trait in East Africa has recently been investigated in detail (Allison, 1954), and found to vary in accordance with the above hypothesis. High frequencies are observed among the tribes living in regions where malaria is hyperendemic (for example, around Lake Victoria and in the Eastern Coastal Belt), whereas low frequencies occur consistently in the malaria-free or epidemic zones (for example, the Kigezi district of Uganda ; the Kenya Highlands ; and the Kilimanjaro, Mount Meru, and Usumbara regions of Tanganyika). This difference is often independent of ethnic and linguistic grouping : thus, the incidence of the sickle-cell trait among Bantu-speaking tribes ranges from 0 (among the Kamba, Chagga, etc.) to 40% (among the Amba, Simbiti, etc.). The world distribution of the sickle-cell trait is also in accordance with the view presented here that malarial endemicity is a very important factor in determining the frequency of the sickle-cell trait. The genetical and anthropological implications of this view are evident.

The fact that sickle cells should be less easily parasitized by plasmodia than are normal erythrocytes is presumably attributable to their haemoglobin component, although there may be other differences, not yet observed, between the two cell-types. Sickle-cell haemoglobin is unlike normal adult haemoglobin in important physico-chemical properties, notably in the relative insolubility of the sickle-cell haemoglobin when reduced (Perutz and Mitchison, 1950). The malaria parasite is able to metabolize haemoglobin very completely in the intact red cell, the haematin pigment remaining as a by-product of haemoglobin breakdown (Fairley and Bromfield, 1934 ; Moulder and Evans, 1946). That plasmodia are greatly affected by relatively small differences in their environment is suggested by their remarkable species specificity. Thus the difficulty of establishing an infection in monkeys with human malaria parasites, and vice versa, is generally recognized.

How far species differences in the haemoglobins themselves, known from immunological and other studies, are responsible for the species specificity of plasmodia it is impossible to say. However, the physico-chemical differences between human adult haemoglobin and monkey haemoglobin appear to be less pronounced than the differences between either type and sickle-cell haemoglobin. It is clear that the natural resistance to malaria among individuals with the sickle-cell trait is relative, not absolute. This is perhaps attributable to differences in the expressivity of the sickle-cell gene, which may be responsible for the production of from nearly 50% to only a very small amount of sickle-cell haemoglobin (Wells and Itano, 1951 ; Singer and Fisher, 1953). Moreover, the sickle-cell haemoglobin may not be evenly distributed in the cell population : most observers recognize that there are cases in which only some of the red cells are sickled even after prolonged reduction. However, even a relative resistance to malaria may be

enough to help those with the sickle-cell trait through the dangerous years of early childhood, during which an active immunity to the disease is developed.

The above observations focus attention upon the importance of haemoglobin to plasmodia in the erythrocytic phase. Hence it is worth considering whether erythrocytes containing other specialized or abnormal types of haemoglobin might be resistant to malaria also. Thus, human foetal haemoglobin differs from human adult haemoglobin in many properties. Red cells containing foetal haemoglobin continue to circulate in the newborn for the first three months of life, after which they are quite rapidly replaced by cells containing normal adult haemoglobin. It has long been known that the newborn has a considerable degree of resistance to malarial infection : Garnham (1949), for instance, found that in the Kavirondo district of Kenya at the end of the second month of life only 10% of babies were infected ; after this age the percentage affected rises rapidly, until by the ninth month practically all children have the disease. The correspondence between the appearance of cells containing normal adult-type haemoglobin and malarial susceptibility is illustrated in the Chart. The correspondence may of course be fortuitous, but it is striking enough to merit further investigation, even though other factors, such as a milk diet deficient in p-aminobenzoic acid (Maegraith et al., 1952 ; Hawking, 1953) and immunity acquired from the mother (Hackett, 1941) may play a part in the natural resistance of the newborn to malaria.

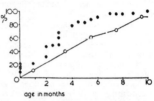

The apparent relationship between the appearance of adult-type haemoglobin (dots) and malarial infection (circles) in the newborn. Each dot represents a test on a single individual, using an alkali denaturation technique (Allison, unpublished observations) ; the circles represent the percentage of Luo children showing malaria parasites (Garnham, 1949).

Finally, it is possible that the explanation offered above for the maintenance of the sickle-cell trait may also apply to thalassaemia. The problems presented by the two diseases are very similar ; many homozygotes, and possibly some heterozygotes, are known to die of thalassaemia, and yet the condition remains remarkably common in Italy and Greece, where as many as 10% of the individuals in certain areas are affected (Bianco et al., 1952). Greek and Italian authors have commented that cases of thalassaemia usually come from districts severely afflicted with malaria (Choremis et al., 1951). Perhaps those who are heterozygous for the thalassaemia gene suffer less from malaria than their compatriots : the fertility of the heterozygotes appears to be greater (Bianco et al., 1952). Selective advantage of the heterozygote with the sickle-cell gene, and possibly the heterozygote with the thalassaemia gene also, would explain why such high gene frequencies can be attained in the case of these conditions while other genetically transmitted abnormalities of the blood cells remain uncommon, not very much above the estimated mutation rate—for example, hereditary spherocytosis (Race, 1942).

Summary

A study has been made of the relationship between the sickle-cell trait and subtertian malarial infection. It has been found that in indigenous East Africans the sickle-cell trait affords a considerable degree of protection against subtertian malaria. The incidence of parasitaemia in 43 Ganda children with the sickle-cell trait was significantly lower than in a comparable group of 247 children without the trait. An infection with P. falciparum was established in 14 out of 15

Africans without the sickle-cell trait, whereas in a comparable group of 15 Africans with the trait only 2 developed parasites.

It is concluded that the abnormal erythrocytes of individuals with the sickle-cell trait are less easily parasitized by *P. falciparum* than are normal erythrocytes. Hence those who are heterozygous for the sickle-cell gene will have a selective advantage in regions where malaria is hyperendemic. This fact may explain why the sickle-cell gene remains common in these areas in spite of the elimination of genes in patients dying of sickle-cell anaemia. The implications of these observations in other branches of haematology are discussed.

This investigation was made possible by a grant received from the Colonial Office at the recommendation of the Colonial Medical Research Committee. Acknowledgment is made to the Directors of Medical Services, Kenya and Uganda, to Dr. E. J. Foley, Nairobi, to the staff of the Mulago Hospital, Kampala, and to the volunteers for their kind co-operation. To Dr. G. I. Robertson, Nairobi, a special debt of gratitude is due for constant help and advice. Drs. A. E. Mourant, R. G. Macfarlane, and A. H. T. Robb-Smith read the manuscript and made valuable comments and suggestions.

REFERENCES

Allison, A. C. (1954). *Trans. roy. Soc. trop. Med. Hyg.* In press.
Beet, E. A. (1946). *E. Afr. med. J.*, **23**, 75.
—— (1947). Ibid., **24**, 212.
Bianco, I., Montalenti, G., Silvestroni, E., and Siniscalo, M. (1952). *Ann. Eugen. Lond.*, **16**, 299.
Brain, P. (1952a). *British Medical Journal*, **2**, 880.
—— (1952b). *S. Afr. med. J.*, **26**, 925.
Choremis, C., Zervos, N., Constantinides, V., and Zannos, Leda (1951). *Lancet*, **1**, 1147.
—— Ikin, Elizabeth W., Lehmann, H., Mourant, A. E., and Zannos, Leda (1953). Ibid., **2**, 909.
Daland, G. A., and Castle, W. B. (1948). *J. Lab. clin. Med.*, **33**, 1082.
Edington, G. M. (1953). *British Medical Journal*, **2**, 957.
Fairley, Sir N. H., and Bromfield, R. J. (1934). *Trans. roy. Soc. trop. Med. Hyg.*, **28**, 141.
Foy, H., Kondi, A., and Brass, W. (1951). *E. Afr. med. J.*, **28**, 1.
Garnham, P. C. C. (1949). *Ann. trop. Med. Parasit.*, **43**, 47.
Hackett, L. W. (1941). *Publ. Amer. Ass. Advanc. Sci.*, **15**, 148.
Haldane, J. B. S. (1947). *Ann. Eugen., Lond.*, **13**, 262.
Hawking, F. (1953). *British Medical Journal*, **1**, 1201.
Henderson, A. B., and Thornell, H. E. (1946). *J. Lab. clin. Med.*, **31**, 769.
Humphreys, J. (1952). *J. trop. Med. Hyg.*, **55**, 166.
Lambotte-Legrand, J., and Lambotte-Legrand, C. (1951). *Institut Royal Colonial Belge, Sect. d. Sci. Nat. et Med.*, Memoires, Tome XIX. fasc. 7, p. 98.
Lehmann, H., and Cutbush, Marie (1952). *British Medical Journal*, **1**, 404.
—— and Milne, A. H. (1949). *E. Afr. med. J.*, **26**, 247.
—— and Raper, A. B. (1949). *Nature, Lond.*, **164**, 494.
Maegraith, B. G., Deegan, T., and Jones, E. S. (1952). *British Medical Journal*, **2**, 1382.
Moulder, J. W., and Evans, E. A. (1946). *J. biol. Chem.*, **164**, 145.
Neel, J. V. (1950). *Cold Spr. Harb. Symp. quant. Biol.*, **15**, 141.
—— (1952). *Blood*, **7**, 467.
Pauling, L., Itano, H. A., Singer, S. J., and Wells, I. C. (1949). *Science*, **110**, 543.
Perutz, M. F., and Mitchison, J. M. (1950). *Nature, Lond.*, **166**, 677.
Race, R. R. (1942). *Ann. Eugen., Lond.*, **11**, 365.
Singer, K., and Fisher, B. (1953). *Blood*, **8**, 270.
Thomas, A. T. G., Robertson, G. I., and Davey, D. G. (1953). *Trans. roy. Soc. trop. Med. Hyg.*, **47**, 338.
Vandepitte, J. M., and Louis, L. A. (1953). *Lancet*, **2**, 806.
Wells, I. C., and Itano, H. A. (1951). *J. biol. Chem.*, **188**, 65.

26

Reprinted from *Proc. Roy. Soc. London* B **164**:298–310 (1966)

Enzyme polymorphisms in man

By H. HARRIS*

*Department of Biochemistry and M.R.C. Human Biochemical Genetics
Research Unit, King's College, Strand, London*

There are a large number of different enzymes synthesized in the human organism, and many of these probably contain more than one structurally distinct polypeptide chain. If current theories about genes and proteins are correct we must suppose that the primary structure of each of these different polypeptides is determined by a separate gene locus, and that there are probably also other loci which are specifically concerned with regulating the rate of synthesis of particular polypeptides or groups of polypeptides. Furthermore, we may expect that genetical diversity in a human population will to a considerable extent be reflected in enzymic diversity. That is to say, in differences between individuals either in the qualitative characteristics of the enzymes they synthesize, or in differences in rates of synthesis.

The work I am going to discuss was largely aimed at trying to get some idea of the extent and character of such genetically determined enzyme diversity among what may be regarded as normal individuals. When my colleagues and I started on this line of work about three years ago the information available about this aspect of the subject was very limited. It had of course been recognized for quite a long time that there are many rare metabolic disorders, the so-called 'inborn errors of metabolism', which are due to genetically determined deficiencies of specific enzymes (Harris 1963). These conditions can in general be attributed to mutant genes which result either in the synthesis of an abnormal enzyme protein with defective catalytic properties, or in a gross reduction in rate of synthesis of a specific enzyme protein. By and large such genes appear to be relatively uncommon and have frequencies of between 0·01 and 0·001 in the general population. Heterozygotes often show a partial enzyme deficiency though they are usually in other respects quite healthy. A few cases are also known where a specific enzyme deficiency occurs quite commonly in certain populations. The most extensively studied example of this is glucose 6-phosphate dehydrogenase deficiency, and it seems likely that in this particular case the relatively high incidence in certain populations is attributable to a specific selective advantage which the deficiency may confer in situations where endemic malaria is an important selective agent (Motulsky 1964).

Virtually all these enzyme deficiencies have been identified in the first instance because of some more or less striking clinical or metabolic disturbance of which they were the cause. They therefore represent a highly selected group of mutants,

* Present address: The Galton Laboratory, University College London.

and cannot be expected *per se* to provide us with any clear picture of how extensively genetically determined enzyme variation which does not result in overt pathological manifestations may occur in the general population. Nor can they provide us with any precise indication of what the general character of such 'concealed' variation might be. Whether it is, for example, mainly a matter of minor quantitative differences in rates of synthesis, attributable to genes at so-called 'regulator' or 'operator' loci, or whether qualitative differences involving enzyme structure are an important feature.

In attempting to tackle this rather general problem we have adopted a quite empirical and perhaps somewhat simple-minded approach. Our idea was to see whether, if we examined a series of arbitrarily chosen enzymes in normal individuals in sufficient detail, we would find genetically determined differences, and if so whether such differences were common or rare, and whether they were peculiar to one class of enzyme rather than another.

Because we would need to examine the selected enzymes in quite a large number of different people, and because we wished to carry out family studies on any enzyme differences that turned up, we were in the first instance largely forced to confine our attention to enzymes present in blood. We had, of course, also to make some decision about the kind of techniques we would utilize in looking for such differences. A wide variety of methods suitable for examining the many different properties of enzyme proteins are available, and it would have been impractical to attempt to utilize more than a few of these. In practice we have mainly relied in the first instance on the technique of starch gel electrophoresis. This is known to be capable, if one gets the conditions right, of detecting quite subtle differences in molecular charge and molecular size. It is however not designed to pick up other sorts of molecular differences, and it is also not very sensitive in the detection of small quantitative differences. Thus we could expect to detect at best only a proportion of all possible forms of enzyme variation.

Despite these limitations, we have found, during the course of examining in varying degrees of detail some ten arbitrarily chosen enzymes, three quite striking examples of genetically determined polymorphism.

RED CELL ACID PHOSPHATASE

The first enzyme we selected for study on this arbitrary basis was red cell acid phosphatase (Hopkinson, Spencer & Harris 1963, 1964). This enzyme is known to differ both in its pattern of substrate specificity and in its inhibition characteristics from the acid phosphatase present in other tissues, and it is thought to occur only in the erythrocyte. Its precise function, however, is not known.

A fairly simple method was developed for detecting the enzyme after electrophoresis in starch gel. The surface of the gel is incubated in a reaction mixture containing phenolphthalein diphosphate at pH 6·0, so that at any site of acid phosphatase activity free phenolphthalein is liberated. This can then be detected by making the surface of the gel alkaline, so that the sites of enzyme activity appear as bright red zones.

When haemolysates from a series of normal individuals were examined using this procedure, it was found that every sample showed more than one zone of enzyme activity. We were evidently dealing with what is now generally called a set of isoenzymes. Furthermore, there were clear-cut person-to-person differences in the number, the mobilities, and the relative activities of these isoenzyme components (figure 65). Five distinct phenotypes were soon identified. They are now referred to as A, BA, B, CA and CB, and in the British population occur with frequencies of about 0·13, 0·43, 0·36, 0·03 and 0·05 respectively. The initial family studies showed that these phenotypes are genetically determined and led to the hypothesis that three allelic genes (P^a, P^b and P^c) at an autosomal locus are involved (phenotypes A and B being produced by the homozygous genotypes $P^a P^a$ and $P^b P^b$ respectively, and phenotypes, BA, CA and CB by the heterozygous

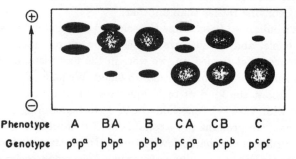

FIGURE 65. Diagram of isoenzyme components seen in the various red cell acid phosphatase phenotypes after electrophoresis at pH 6·0.

genotypes $P^a P^b$, $P^a P^c$ and $P^b P^c$). The hypothesis predicted the occurrence of a sixth phenotype corresponding to the genotype $P^c P^c$. Gene frequency considerations indicated that this would be fairly uncommon (about 1 in 625 of the general population), and indeed a few examples of what is probably this phenotype have now been observed (Lai, Nevo & Steinberg 1964).

With the five common phenotypes fifteen different mating types are possible and the segregation pattern in most of these has now been studied (table 7). These quite extensive family data have proved to be fully consistent with the hypothesis and the findings have also been confirmed by several other groups working with different populations. There seems little doubt therefore that these acid phosphatase variations reflect a polymorphism involving at least three alleles, and a variety of studies on the properties of the isoenzymes in the different phenotypes makes it appear reasonably certain that these alleles determine the synthesis of structurally different forms of the enzyme.

A particularly interesting feature of the polymorphism is that the qualitative differences between the phenotypes are reflected quantitatively by differences in the levels of enzyme activity (Spencer, Hopkinson & Harris 1964a). Levels of total acid phosphatase activity were determined by a standard method in a series of haemolysates from individuals of the different phenotypes, using p-nitrophenyl phosphate as substrate. Although there was considerable variation in activity between individuals of any one phenotype, nevertheless quite marked differences between the mean values for different phenotypes could be demonstrated (table 8). Using these values one may examine the question as to whether the quantitative

TABLE 7. SEGREGATION OF RED CELL ACID PHOSPHATASE
PHENOTYPES IN 216 FAMILIES

| parents | number of matings | children | | | | | | total |
		A	BA	B	CA	CB	C	
A × A	4	8	—	—	—	—	—	8
A × BA	25	31	25	—	—	—	—	56
A × B	11	—	20	—	—	—	—	20
A × CA	4	3	—	—	2	—	—	5
A × CB	5	—	2	—	5	—	—	7
BA × BA	51	24	52	21	—	—	—	97
BA × B	50	—	52	44	—	—	—	96
BA × CA	7	6	3	—	3	2	—	14
BA × CB	8	—	5	5	2	6	—	18
B × B	24	—	—	58	—	—	—	58
B × CA	9	—	7	—	—	16	—	23
B × CB	16	—	—	23	—	16	—	39
CA × CA	—	—	—	—	—	—	—	—
CA × CB	1	—	1	—	1	—	—	2
CB × CB	1	—	—	—	—	1	1	2
totals	216	72	167	151	13	41	1	445

TABLE 8. MEANS AND STANDARD DEVIATIONS OF RED CELL ACID
PHOSPHATASE ACTIVITY IN INDIVIDUALS OF KNOWN PHENOTYPES

The activity is expressed as μM p-nitrophenol liberated in $\frac{1}{2}$ h at 37 °C/g haemoglobin.

phenotype	number of individuals	mean activity	standard deviation
A	33	122·4	16·8
BA	124	153·9	17·3
B	81	188·3	19·5
CA	11	183·8	19·8
CB	26	212·3	23·1

effects of the three postulated alleles are additive in a simple way or not. If they are additive one could expect the following relationships to be true:

$$(a)\quad \tfrac{1}{2}\overline{A}+\tfrac{1}{2}\overline{B} = \overline{BA},$$
$$(b)\quad \overline{CA}-\tfrac{1}{2}\overline{A} = \overline{CB}-\tfrac{1}{2}\overline{B},$$

where $\overline{A}$, $\overline{BA}$, $\overline{B}$, etc., are the mean values for the various phenotypes. It will be seen that the results support the idea of simple additivity rather well ($\tfrac{1}{2}\overline{A}+\tfrac{1}{2}\overline{B}$ = 155·35, $\overline{BA}$ = 153·9, $\overline{CA}-\tfrac{1}{2}\overline{A}$ = 122·6 and $\overline{CB}-\tfrac{1}{2}\overline{B}$ = 118·15). Estimates from these data of the average activity attributable to each allele are:

$$P^a \rightarrow \; 60\cdot7 \pm 1\cdot1 \text{ units,}$$
$$P^b \rightarrow \; 93\cdot7 \pm 1\cdot0 \text{ units,}$$
$$P^c \rightarrow 120\cdot3 \pm 3\cdot7 \text{ units.}$$

Somewhat unexpectedly the three values turn out to be very close to the simple ratio 2:3:4, and it is tempting to think that this may have some special significance in terms of enzyme structure.

It is of interest to note that if one determines red cell acid phosphatase activities in a series of randomly selected individuals one obtains a continuous unimodal distribution which, in fact, is not dissimilar in form to the distributions usually obtained when other enzymes in man are examined quantitatively in randomly selected populations. In particular the variance when related to the mean is of the same order of magnitude as is found with many other enzymes. In the present case however it is clear that the overall distribution represents a summation of a

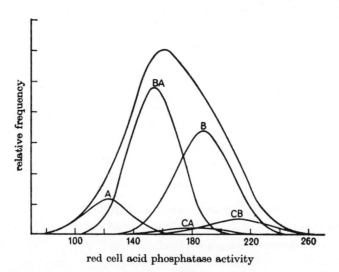

red cell acid phosphatase activity

FIGURE 66. Distribution of red cell acid phosphatase activities in the general population (top line) and in the separate phenotypes. The curves are constructed from the values given in table 8 and from the relative frequencies of the phenotypes observed in a randomly selected population.

series of separate but overlapping distributions corresponding to each of the qualitatively different phenotypes (figure 66). Furthermore, the genetical component of the variance of the overall distribution can be largely if not entirely attributed simply to the effects of the three alleles. It is not unreasonable to suppose that the genetical component of other examples of continuous variation in enzyme levels may have a similar simple underlying basis.

PHOSPHOGLUCOMUTASE

One of the main problems in studying enzymes by starch gel electrophoresis is the development of sensitive and specific methods for the detection of the zones of enzyme activity. The general approach has been to utilize substrates which will yield coloured products or products capable of reacting rapidly with some chemical included in the reaction mixture to give a coloured compound. Phenolphthalein diphosphate, for example, proved to be a useful substrate for red cell acid phosphatase, and naphthyl phosphates have been widely utilized for the study of other phosphatases. However, for the majority of enzymes this approach is not feasible because of their very restricted range of substrate specificity.

One way round this difficulty is to utilize other enzymes in the reaction mixture so as to build up a sequence of reactions culminating in the formation of some detectable substance. Such reaction mixtures are usually complex and often include six or more different interacting components whose relative concentrations require careful adjustment. However, the general method is proving extremely valuable and has opened up the possibility of examining many previously inaccessible enzymes. Our first successful application of this idea was with phosphoglucomutase, and it led to the discovery of another example of enzyme polymorphism (Spencer, Hopkinson & Harris 1964 *b*).

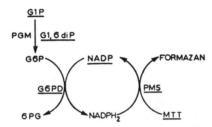

FIGURE 67. Sequence of reactions in the detection of phosphoglucomutase (PGM) after starch gel electrophoresis. The underlined components are contained in the reaction mixture. [Key: G1P, glucose 1-phosphate; G6P, glucose 6-phosphate; G1,6diP, glucose 1,6-diphosphate; 6PG, 6-phosphogluconate; G6PD, glucose 6-phosphate dehydrogenase; NADP and NADPH$_2$, oxidized and reduced nicotinamide adenine dinucleotide phosphate; PMS, phenazine methosulphate; MTT, tetrazolium salt.]

Phosphoglucomutase catalyses the reversible transfer of phosphate between glucose 1-phosphate and glucose 6-phosphate, and it has an important role in carbohydrate metabolism. Its detection following starch gel electrophoresis was accomplished via the sequence of reactions shown in figure 67. The underlined components are included in the reaction mixture and the sites of phosphoglucomutase activity are located by the deposition of a blue-coloured formazan formed by the reduction of the tetrazolium salt *MTT*.

When haemolysates from different individuals are subjected to starch gel electrophoresis and this reaction system is applied, one obtains the rather complex isoenzyme patterns shown in figure 68. At least seven different zones of activity (*a–g*) may be detected, and three quite distinct types of pattern can be identified in different individuals. These are referred to as PGM 1, PGM 2-1 and PGM 2. The phenotypes differ in the occurrence of components a, b, c and d; a and c being present in PGM 1 and PGM 2-1, but not PGM 2, while b and d are present in PGM 2-1 and PGM 2 but not PGM 1. Components e, f and g are present in all the three phenotypes.

In the British population the incidence of the three phenotypes has been found to be PGM 1 0·58, PGM 2-1 0·36 and PGM 2 0·06. Studies on the segregation of these phenotypes in more than 150 different families involving all the possible mating types make it clear that two autosomal alleles (PGM^1 and PGM^2) determine these differences. Phenotypes PGM 1 and PGM 2 represent the homozygotes $PGM^1\,PGM^1$ and $PGM^2\,PGM^2$, and phenotypes PGM 2-1 the heterozygote $PGM^1\,PGM^2$.

This suggests that the isoenzyme components a and c determined by PGM^1 may be molecular alternatives of components b and d determined by PGM^2. Possibly these isoenzymes contain a common polypeptide chain, and the difference between the two homozygous phenotypes depends on a small structural difference in this, which involves perhaps a single amino acid substitution. If this is so then one would presume that this polypeptide chain is not present in the isoenzyme components e, f and g, as they appear to be uninfluenced by this gene substitution, and

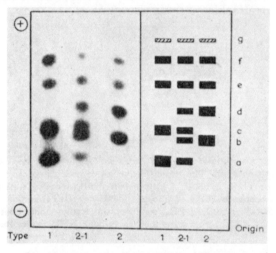

FIGURE 68. Photograph and diagram of phosphoglucomutase isoenzyme patterns obtained by starch gel electrophoresis at pH 7·4.

their structures are presumably therefore determined by other loci. Some support for this idea has recently been obtained by the discovery of an uncommon variant involving components e and f but not affecting a, b, c or d. This variant was found to segregate independently from phenotypes 1, 2-1 and 2, and the family study indicated that it was determined at a separate and not closely linked locus. It is also possible that other structurally distinct polypeptide chains may be contained in these isoenzyme components. These might, for instance, account for the mobility differences between isoenzymes a and c, or between e and f, and would if present imply the existence of further loci involved in the determination of this enzyme. No doubt structural studies on the isolated isoenzyme components will enable these questions to be resolved.

Unlike red cell acid phosphatase, phosphoglucomutase occurs in many different tissues, and it was therefore of some importance to see whether the isoenzyme components and the polymorphism found in erythrocytes also occurred elsewhere. It has in fact been possible to demonstrate that this is the case. The tissues studied included liver, kidney, muscle, brain, skin and placenta. The phosphoglucomutase isoenzymes have also been demonstrated in tissue culture cells grown *in vitro*. The tissue cultures were started from small skin biopsies from different individuals and were kept going for up to ten passages for more than three months. The cells were harvested at different times and the phosphoglucomutase examined. In each case the PGM phenotype found was the same as that originally observed in the red cells of the donor whose skin was used to start the culture.

ADENYLATE KINASE

The third polymorphism discovered in the screening programme of arbitrarily selected enzymes involves the enzyme adenylate kinase (Fildes & Harris 1965). This catalyses the reversible reaction

$$2\,ADP \rightleftharpoons ATP + AMP$$

and two procedures for detecting the enzyme after starch gel electrophoresis have been developed. Both of these require complex and different multienzyme reaction mixtures, but reveal the same pattern of zones of activity. It has been found that the enzyme as it occurs in erythrocytes, and also in skeletal muscle, includes several distinct isoenzyme components, and so far two discrete phenotypes have been recognized. One of these occurs in about 1 in 10 of the general population, and family studies indicate that the individuals showing it are heterozygous for two autosomal alleles. This work is still in its preliminary stages and it has not yet been possible to determine whether all or only some of the isoenzymes present are involved in the polymorphism.

PLACENTAL ALKALINE PHOSPHATASE

The various enzymes studied in this screening programme were selected because among other reasons they were known to occur in the erythrocyte, and this is obviously convenient if one wishes to carry out population and family investigations. However, the erythrocyte is a rather specialized cell type, and even though it is often possible to demonstrate that many red cell enzymes occur in essentially the same form in other tissues, it might be considered that a survey restricted to one cell type could present a somewhat biased picture of human enzyme variation in general. Analogous studies on enzymes localized to other tissues are for obvious reasons very much more difficult to pursue, and so far have not been carried out in any systematic way. However, it has been possible to investigate in some detail one particular example of a polymorphism involving what can be regarded as an organ-specific enzyme, and the results illustrate something of the possibilities and problems which such enzymes may present.

The enzyme is an alkaline phosphatase present in quite large amounts in the human placenta. It appears to be peculiar to this organ and to be different from the alkaline phosphatases present in other tissues such as liver, kidney and bone.

Following earlier work by Boyer (1961), it has now been possible to demonstrate that placentae may be classified into at least six distinct phenotypes according to the electrophoretic behaviour of the alkaline phosphatase they contain (Robson & Harris 1965). The electrophoretic patterns are illustrated in figure 69, and one may note that complete discrimination of the six phenotypes requires electrophoresis at two different pH's. The six phenotypes are referred to as S, F, I, SF, SI and FI. In types S, F and I most of the alkaline phosphatase activity is present in a single rapidly moving component, which however has a different mobility in each type. In the I phenotype, the characteristic component has a mobility very close to that of the F phenotype at pH 8·6, but is indistinguishable from that of the S phenotype at pH 6·0. In phenotypes SF, SI and FI, three such components are found, two of them in each case having mobilities similar to the component present in S, F or I,

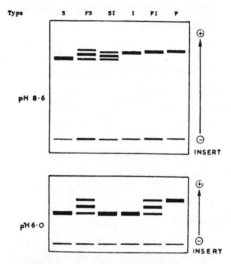

FIGURE 69. Placental alkaline phosphatase patterns obtained by electrophoresis at pH 8·6 and pH 6·0.

while the third has an intermediate mobility. There are reasons for thinking that this third component in these three phenotypes may represent a 'hybrid' enzyme containing polypeptide chains characteristic of the two other components present. In each of the six phenotypes at least one other component which migrates only very slowly may be seen. These slow components do in fact exhibit slight differences in mobility in the different phenotypes, and these can be shown to correlate with the more striking differences observed in the major and more rapidly moving components.

On the basis of the phenotypic patterns and their relative frequencies, it is possible to construct a simple genetical hypothesis which will account for these variations. This suggests that three autosomal allelic genes are concerned, phenotypes F, I and S representing the three homozygous genotypes, and phenotypes SF, SI and FI the corresponding heterozygotes. From the observed incidence of the six phenotypes in the British population, one may readily obtain values for the frequencies of the three postulated genes. They are 0·27, 0·09 and 0·64, and using these values one finds a very good agreement between the observed incidence of the phenotypes and those expected assuming a Hardy–Weinberg equilibrium (table 9)

It is of course obvious that family studies of the ordinary kind are impracticable in the case of a characteristic peculiar to the placenta. However it occurred to us

TABLE 9. OBSERVED AND EXPECTED NUMBERS OF PLACENTAL ALKALINE PHOS-
PHATASE TYPES IN A POPULATION SAMPLE ASSUMING A HARDY–WEINBERG
EQUILIBRIUM

$(p = 0{\cdot}27, q = 0{\cdot}09 \text{ and } r = 0{\cdot}64)$

placental alkaline phosphatase type	expected incidence		expected numbers in population sample	observed numbers in population sample
S	r^2	0·410	135·9	141
SF	$2pr$	0·346	114·7	111
F	p^2	0·073	24·2	28
SI	$2qr$	0·115	38·2	32
FI	$2pq$	0·049	16·1	15
I	q^2	0·008	2·7	5
totals	$(p+q+r)^2$	1·001	331·8	332

TABLE 10. PLACENTAL ALKALINE PHOSPHATASE PHENOTYPES IN
130 DIZYGOTIC TWIN PAIRS

dizygotic twins		expected incidence assuming three alleles with frequencies p, q and r	expected incidence $p = 0{\cdot}27$ $q = 0{\cdot}09$ $r = 0{\cdot}64$	observed incidence
like pairs				
S	S	$\frac{1}{4}r^2(1+r)^2$	35·80	39
SF	SF	$\frac{1}{2}pr[pr+(1+p)(1+r)]$	25·34	27
F	F	$\frac{1}{4}p^2(1+p)^2$	3·82	6
SI	SI	$\frac{1}{2}qr[qr+(1+q)(1+r)]$	6·90	6
I	I	$\frac{1}{4}q^2(1+q)^2$	0·31	1
FI	FI	$\frac{1}{2}pq[pq+(1+p)(1+q)]$	2·24	3
			74·41	82
unlike pairs				
S	SF	$pr^2(1+r)$	23·58	26
S	F	$\frac{1}{2}p^2q^2$	1·95	0
S	SI	$qr^2(1+r)$	7·85	9
S	I	$\frac{1}{2}q^2r^2$	0·21	0
S	FI	pqr^2	1·30	1
SF	F	$p^2r(1+p)$	7·70	3
SF	SI	$pqr(1+2r)$	4·60	4
SF	I	pq^2r	0·18	0
SF	FI	$pqr(1+2p)$	3·12	1
F	SI	p^2qr	0·55	2
F	I	$\frac{1}{2}p^2q^2$	0·03	0
F	FI	$p^2q(1+p)$	1·09	0
SI	I	$q^2r(1+q)$	0·73	1
SI	FI	$pqr(1+2q)$	2·39	1
I	FI	$pq^2(1+q)$	0·31	0
			55·59	48

that we might be able to test the hypothesis by studying a series of pairs of placentae from dizygotic twins, because such twin pairs can be regarded as pairs of sibs. We were fortunate in being able to obtain such material from an extensive investigation of twin births which is being carried out in the Birmingham area under the general direction of Dr John Edwards. So far we have been able to examine the alkaline phosphatase types in 260 placentae from 130 dizygotic twin pairs. The findings are summarized in table 10. In 82 pairs the alkaline phosphatase phenotypes were the same, and in 48 pairs they were different. This result excludes the possibility that these placental phenotypes are determined by the maternal genotypes, because if this were so none of the pairs should have shown any differences.

If the phenotypes depend on the foetal genotype then one may test the hypothesis by calculating the expected incidence of the different sorts of sib pair using the gene frequencies previously obtained. This is shown in table 10, and it will be seen that there is quite good agreement between the numbers of the different sorts of twin pair observed, and those expected according to the hypothesis.

Thus the evidence strongly suggests that these placental alkaline phosphatase phenotypes are determined by the foetal genotype and that at least three autosomal alleles are concerned. The biological significance of the polymorphism is still quite obscure, but it might well be of importance in problems concerned with maternal–foetal interaction. It may also perhaps be worth considering other enzyme polymorphisms from this point of view. The phosphoglucomutase polymorphism, for example, can also be readily demonstrated in the placenta and here again it has been shown that the placental phenotype is determined by the foetal and not the maternal genotype.

DISCUSSION

Although this work is still in its early stages, an interesting and perhaps in some ways an unexpected picture of enzyme variation in human populations is beginning to emerge.

In the course of examining some ten arbitrarily chosen enzymes, in none of which we had any particular reason to expect any degree of variation, and not all of which have been examined in great detail or by perhaps the most suitable methods, we have come across three quite striking examples of enzyme polymorphism. Although one can hardly draw firm numerical conclusions from such a small series, it seems likely, unless we have been excessively lucky in our choice of enzymes, that polymorphism to a similar degree may be a fairly common phenomenon among the very large number of enzymes that occur in the human organism.

Some idea of how extensive this diversity might be can be obtained by considering together the various enzymes which have been shown to exhibit some degree of polymorphism in our own population. Relevant data on seven such enzymes are given in table 11. For the present purpose only those variations where two or more allelic genes have been found to have frequencies greater than 0·01 have been included. In the case of one enzyme, serum cholinesterase, variation at two different loci fall into this category, so that eight loci are represented in all. Of these at least six can be regarded as 'structural' loci since the variation produced appears to involve qualitative differences. In the other two cases (serum cholinesterase E_2 and acetyl transferase) only quantitative differences in enzyme level have so far been identified, but it is possible that these may also reflect structural differences in the enzyme protein present.

TABLE 11. ENZYME POLYMORPHISM IN THE ENGLISH POPULATION

enzyme	number of alleles with frequency greater than 0·01	frequency of commonest phenotype	probability of two randomly selected individuals being of the same phenotype	reference
red cell acid phosphatase	3	0·43	0·34	Hopkinson, Spencer & Harris (1963)
phosphogluco-mutase	2	0·58	0·47	Spencer, Hopkinson & Harris (1964b)
placental alkaline phosphatase	3	0·41	0·31	Boyer (1961) Robson & Harris (1965)
acetyl transferase	2	0·50	0·50	Price Evans & White (1964)
adenylate kinase	2	0·90	0·82	Fildes & Harris (1965)
serum cholinesterase				
locus E_1	2	0·96	0·92	Kalow & Staron (1957)
locus E_2	2	0·90	0·82	Harris, Hopkinson, Robson & Whittaker (1963)
6-phosphogluconate dehydrogenase	2	0·96	0·92	Fildes & Parr (1963)
combined	—	0·037	0·014	—

Each of these variations occurs independently of the others so that quite a large number of different phenotypic combinations may be found in the general population. Indeed, the commonest of these will occur in less than 4 % of people and the probability that two randomly selected individuals would be found to have the same combination of phenotypes is less than 1 in 70. Thus just taking into account this very limited series of examples, quite a high degree of individual differentiation in enzymic make-up is demonstrable, and it is of interest that most of this is probably attributable to variation in enzyme structure.

These different polymorphisms pose a variety of intriguing problems both in biochemistry and in genetics. One would like to know, for example, what is the precise nature of the structural differences between the variant forms of a given enzyme, and whether these are reflected in kinetic differences and in differences in functional activity. It is notable that several of these enzymes apparently occur in multiple molecular forms even in homozygous individuals. The recognition of these so-called isoenzyme systems is a fairly new development in enzymology and the further investigation of these particular examples, and of their variant forms, may well help to throw light on the general biological significance of this phenomenon.

One would also like to know why these different enzyme phenotypes occur with the particular frequencies that we observe, and why, as is the case, for example, with red cell acid phosphatase and placental alkaline phosphatase, the gene frequencies may vary quite widely from one population to another. Presumably selective differences are important here but at present we have virtually no idea what these might be. However, one may reasonably hope that, if the metabolic and functional differences which presumably derive from the various enzyme differences can be elucidated, this may provide us with some indication of what selective factors may be important.

REFERENCES (Harris)

Boyer, S. H. 1961 Alkaline phosphatase in human sera and placentae. *Science*, **134**, 1002.

Fildes, R. A. & Harris, H. 1965 Genetically determined variation of adenylate kinase in man. (In preparation.)

Fildes, R. A. & Parr, C. W. 1963 Human red cell phosphogluconate dehydrogenases. *Nature, Lond.*, **200**, 890.

Harris, H. 1963 The 'inborn errors' today. In *Garrod's Inborn errors of metabolism*. Oxford University)Press.

Harris, H., Hopkinson, D. A., Robson, E. B. & Whittaker, M. 1963 Genetical studies on a new variant of serum cholinesterase detected by electrophoresis. *Ann. Hum. Genet., Lond.* **26**, 359.

Hopkinson, D. A., Spencer, N. & Harris, H. 1963 Red cell acid phosphatase variants: a new human polymorphism. *Nature, Lond.* **199**, 969.

Hopkinson, D. A., Spencer, N. & Harris, H. 1964 Genetical studies on human red cell acid phosphatase. *Amer. J. Hum. Genet.* **16**, 141.

Kalow, W. & Staron, N. 1957 On the distribution and inheritance of atypical forms of human serum cholinesterase as indicated by dibucaine numbers. *Canad. J. Biochem. Physiol.* **35**, 1305.

Lai, L., Nevo, S. & Steinberg, A. G. 1964 Acid phosphatases of human red cells: predicted phenotype conforms to a genetic hypothesis. *Science*, **145**, 1187.

Motulsky, A. 1964 Hereditary red cell traits and malaria. *Amer. J. Trop. Med. Hyg.* **13**, 147.

Price Evans, D. A. & White, T. A. 1964 Human acetylation polymorphism. *J. Lab. Clin. Med.* **63**, 394.

Robson, E. B. & Harris, H. 1965 Genetics of the alkaline phosphatase polymorphism in the human placenta. *Nature, Lond.* **207**, 1257.

Spencer, N., Hopkinson, D. A. & Harris, H. 1964a Quantitative differences and gene dosage in the human red cell acid phosphatase polymorphism. *Nature, Lond.* **201**, 299.

Spencer, N., Hopkinson, D. A. & Harris, H. 1964b Phosphoglucomutase polymorphism in man. *Nature, Lond.* **204**, 742.

EPILOGUE:
THE LEGACY OF SUCCESS

Christian religious traditions tell us that envy with its implied in-equity began in the Garden of Eden. Its origin, however, undoubtedly transcends organized religions and is embodied in the hierarchy of status and perquisites that are integral to all cultures. Efforts to institutionalize social stratification must have begun early, but we can only conjecture when. Equally obscure is the origin of notions of genetic superiority or inferiority. Certainly by the nineteenth century a pseudo-scientism had developed to perpetuate, through the notion of racial, ethnic, or group inequality, position, privilege, and perquisite.

It took the twentieth century, however, to translate these concepts into national polity, as in the case of Nazi Germany, and the venomous, self-serving diatribes so blatantly set forth in Madison Grant's *The Passing of the Great Race.* Unfortunately, many of the early names associated with human genetics in the United States and in Europe directly or inadvertently contributed to the support of these concepts. As a consequence, there developed in the years immediately prior to and following World War II a reaction to these excesses—a reaction that led almost to a rejection of any willingness to examine the social implications involved in the existence of genetic variability. Matters have changed, however, and a new social awareness exists that attempts to eschew the implication of inferiority and superiority in genetic variability and to encourage responsible but individual judgments. New diagnostic techniques offer heretofore unavailable alternatives in individual decision making. But they have also brought their overtones, in the form of obligatory screening for genetic variability, of a new racism. Some of this changing perspective, and willingness to reexamine issues that have been so painful in the recent past, must be ascribed to the prospect of genetic

339

engineering, cloning, and those other seemingly science-fiction alternatives that appear to be just over the horizon.

Amniocentesis, the transabdominal or vaginal aspiration of amniotic fluid, coupled with an increased number of prenatal diagnostic procedures based largely upon cell culture, and a liberalized attitude toward abortion, both on the part of the general public and the law, promise revolutionary changes in reproductive decision making. Today, under ideal circumstances, it is possible to diagnose with certainty from cell cultures several dozen inherited biochemical diseases or conditions and virtually all chromosomal abnormalities. Tay-Sachs disease, a rapidly progressive fatal neurological disorder of infancy that results from the accumulation of a ganglioside in cells of the nervous system, illustrates well the changes these developments have occasioned. Previously, one could merely state to the parents of such a child that the disease was a simply inherited, recessive trait and that subsequent children would run one chance in four of developing the disorder. We now know that Tay-Sachs disease stems from the deficiency of an enzyme, hexosaminidase A, and that if this enzyme is found in fetal cells, the fetus is free of Tay-Sachs disease. Parents need not, therefore, live in apprehension throughout gestation and for some period thereafter; if the fetus is unhappily found to have Tay-Sachs disease, an abortion can be considered. Most of the diseases that can be recognized prenatally are, like Tay-Sachs disease, rare, but some may be frequent in certain groups of individuals. Sickle cell anemia is such a disease, but as yet no fully satisfactory prenatal diagnostic procedure exists. Relatively small technical advances should, however, add this disease and possibly another dozen disorders to the list soon.

These new alternatives pose ethical, psychological, and sociomedical issues that neither medicine nor genetics has previously confronted. Given, for example, that it is possible to diagnose with certainty the genotype of an infant with respect to a particular genetic locus prior to birth and that such a diagnosis reveals an abnormality, abortion of the pregnancy becomes a possible course of conduct. However, is it in the "best" interest of the persons involved and of society, or is it merely a further manifestation of the decline in moral values that some perceive? Who defends the fetus's rights if rights exist? If abnormal fetuses are to be aborted, what should be the attitude towards carriers of the gene that results in an abnormality when in double dose? At what point in time should the scientific evidence be viewed as sufficiently compelling to entertain a decision as irrevocable as abortion? Had the presence of an ancillary Y chromosome been viewed as sufficient evidence to consider an abortion when the association between this chromosomal abnormality and antisocial behavior was first reported (Casey et al. 1966; see also Jacobs, Price, and Court Brown 1968), would society rue its support? The evidence of such an association seems much less clear now. Efforts to identify whose interests are involved and determine a "best" strategy have prompted a lively, often stimulating, occasionally acrimonious debate among geneticists, lawyers, physicians, ethicists, theologians, and just plain citizens. No consensus has been forthcoming nor does one seem likely. Indeed, at this juncture in time some of the issues involved seem insuperable.

Possibly one of the most rewarding resultants of this debate has been an increasing interest in genetic counseling, whom it reaches, how the "message" is projected, and the support that it may or may not provide to decision making. Should genetic counseling be directive—that is, should it seek to imprint upon the counseled a specific message—or should it be nondirective—that is, provide information but encourage independent decisions? How is its effectiveness to be evaluated—on the basis of the counselee's capacity to recapitulate a specific set of information or upon the ultimate impact of counseling upon decision making? Should it be programmed—that is, can films, readings, and the like substitute in large measure or in part for personal interaction—or should it be counselor structured—that is, should the thrust of the counselor-counselee interaction be determined largely by the counselor? Should counseling be restricted to physicians or otherwise specially trained biologists or can there be a role for the priest, the informed friend, and the like in this enterprise? Are there certain "global strategies" that at least in the short run might have a more immediate impact upon man's burden of inherited disease and disability than disease-specific counseling? For example, would a general encouragement to all prospective parents to have their children in the years of reproductive optimality—that is, between the ages of twenty and thirty-five (for the mother)—have as profound an effect (certainly a more cost-effective one) than the discouragement of specific reproductive events with the stigmatization that this almost automatically implies? What is the proper role of the law in this process? Should it be restricted to encouraging the availability of information from which informed decisions can be made, or should it have a regulatory role? Most thoughtful people would probably not prefer the latter. The earlier decades of this century saw a plethora of eugenic legislation that frequently equated homelessness, deafness, or blindness (irrespective of its origin) with either social irresponsibility, as best, or social depravity, at worst. How can society protect itself from ill-informed judgments on hereditary disease that stem from polar points of view? Can, indeed should, un-fettered research occur in a democracy, if the majority views, for whatever reason, the prospective or actual results inadmissible as bases for social action? Illustrative of this is the petulant argumentation that has surrounded issues of the relationship between certain tests of intellectual performance (often ill-advisably equated to measures of intelligence) and different ethnic or racial origins. Possibly a more simplistic formulation of these issues is, "How can science best serve the society that nurtures it?" It seems to us unlikely that societally imposed constraints upon the nature of the information that is sought achieves this end, but possibly priorities in research would.

Most societies are no longer content to accept the practice of science for its own sake. They demand of the scientist a share in the responsibilities for the abuses that scientific developments can spawn and a participation in the decisions these developments may foist upon us all as citizens. A fitting conclusion to this selection of papers, or so it seems to us, would be a series of articles that attempt to support the change in social awareness that pervades contemporary human genetics and to which we have alluded in the previous paragraphs. Limitations of space preclude such a selection; moreover, Carl Bajema's *Eugenics Then and Now* preempts some of the need. However, we cannot close this volume without some suggestions. The latter can, at best, project a small moment of this self-inspection and social awareness. We make no effort to defend the specific choices, but do feel compelled to share with the harried reader our three candidates for a "short course."

L. C. Dunn's presidential address to the American Society of Human Genetics in 1962, "Cross Currents in the History of Human Genetics," captures the issues and the tenor of the first several decades of this century admirably well. More comprehensive treatments of this era of social awareness in genetic circles are to be found in three books, namely, Mark H. Haller's *Eugenics: Hereditarian Attitudes in American Thought*, Donald K. Pickens' *Eugenics and the Progressives*, and Kenneth M. Ludmerer's *Genetics and American Society*. The reader interested in a more personal, eloquently stated and intellectually provocative response to these excesses would do well to read J. B. S. Haldane's *Heredity and Politics*. A measured modern presentation of the eugenical point of view that shifts the emphasis from the negative to the positive is Frederick Osborn's *The Future of Human Heredity*. Allen Chase's *The Legacy of Malthus* describes some of the costs of "scientific racism."

Joshua Lederberg's "Experimental Genetics in Human Evolution" is a highly readable account of some of the scientific portents of the future, such as cloning, surrogate motherhood, and genetic engineering. Others have written on these issues (see, for example, Sinsheimer 1969), but few with the flair of this pioneer in molecular biology.

Herman J. Muller's "Means and Aims in Human Genetic Betterment" is but one of the numerous papers written on this general subject by this now deceased Nobel laureate. As early as 1935 Muller (1935) advocated an elitist approach to human betterment. He was confident that prevailing trends would lead to a world so burdened by inherited disease and disability that eventually the halt would be obliged to nurse the lame. He was one of the first, if not the first, public advocate of sperm banking and espoused an activist attitude toward steps to better man's genetic lot. He encouraged these steps to be voluntary in nature, rather than compulsive, but his overall attitudes toward these issues did not set him much apart from earlier eugenicists, and through much of his life he was often critical of medicine's failure to encourage through genetic counseling what he saw to be more responsible parenthood.

If the late 1940s and 1950s saw a timorous attitude toward socially controversial issues on the parts of many human geneticists, the last decade has certainly not. We are literally inundated with opinions and suggestions. Book follows on book. Lipkin and Rowley (1974), we are told, explore the complex and highly controversial problem, *Genetic Responsibility: On Choosing our Children's Genes*, and in Milunsky and Annas (1976) "the new genetics" confronts the law, or so the advertisements claim. Too frequently the casts in these confrontations remain the same, and issues that are, indeed, important have begun to take on aspects of a sideshow. An important exception is the National Academy of Science's *Genetic Screening: Programs, Principles and Research* (1975). Here, a complex, often controversial issue is explored with probity and without fanfare. All of these efforts to guide and inform warrant perusal. Each has its provocative moments.

REFERENCES

Bajema, C. 1976. *Eugenics Then and Now.* Dowden, Hutchinson & Ross, Stroudsburg, Pa.

Casey, M. D., C. E. Blank, D. R. K. Street, L. J. Segall, J. H. McDougall, P. J. McGrath, and J. L. Skinner. 1966. YY chromosomes and antisocial behavior. *Lancet* ii:859–960.

Chase, A. 1977. *The Legacy of Malthus.* A. A. Knopf, New York.

Dunn, L. C. 1962. Cross currents in the history of human genetics. *Am. J. Hum. Genet.* 14:1–13.

Grant, M. 1921. *The Passing of the Great Race.* Charles Scribner's Sons, New York.

Haldane, J. B. S. 1938. *Heredity and Politics.* W. W. Norton and Co., New York.

Haller, M. H. 1963. *Eugenics: Hereditarian Attitudes in American Thought.* Rutgers University Press, New Brunswick, N.J.

Jacobs, P. A., W. H. Price, and W. M. Court Brown. 1968. Chromosome studies on men in a maximum security hospital. *Ann. Hum. Genet.* 31:339–47.

Lederberg, J. 1966. Experimental genetics and human evolution. *Am. Nat.* 100: 519–31.

Lipkin, M., Jr., and P. T. Rowley. 1974. *Genetic Responsibility: On Choosing our Children's Genes.* Plenum Press, New York.

Ludmerer, K. M. 1972. *Genetics and American Society.* Johns Hopkins University Press, Baltimore.

Milunsky, A., and G. J. Annas. 1976. *Genetics and the Law.* Plenum Press, New York.

Muller, H. J. 1935. *Out of the Night: A Biologist View of the Future.* Vanguard Press, New York.

Muller, H. J. 1965. Means and aims in human genetic betterment, Chapter 5. In T. M. Sonneborn (ed.), *The Control of Human Heredity and Evolution.* Macmillan, New York, pp. 100–22.

National Academy of Sciences. 1975. *Genetic Screening: Programs, Principles and Research.* National Academy of Sciences, Washington, D.C.

Osborn, F. 1968. *The Future of Human Heredity.* Weybright and Telley, New York.

Pickens, D. K. 1968. *Eugenics and the Progressives.* Vanderbilt University Press, Nashville, Tenn.

Sinsheimer, R. L. 1969. The prospect for designed genetic change. *Am. Sci.* **57:** 134–42.

AUTHOR CITATION INDEX

SUBJECT INDEX

Abnormalities, lethal chromosome, 159
Abortion, spontaneous, 159
Absence central incisors, 71
Acatalasemia, 200
Accipiter novae-hollandiae, 301
Acetyl transferase, 337
Achondroplasia, 139
Achromatopsia, 137
Acid
  aklapton, 9
  glycosuric, 12
  homogentisic, 9, 11, 12
  uroleucic, 9
Adalia bipunctata L., 301
Adaptation, 301
Adentia, 70
Adenylate kinase, 33, 337
Admixture, 318
Agglutinins
  immune, 39
  iso, 39, 317
Agglutination, 5, 34, 35, 39
  immune, 39
Albinism, 12, 213
Albuminuria, 88
Alkaline phosphatase, 334
  placental, 333, 335, 337, 338
Alkaptonuria, 4, 9, 15, 62, 152, 214, 215, 217, 222
Allelomorph, 4, 15, 39, 56, 59, 71, 74, 136, 298, 303, 305
Alopecia congenita, 128
Amauris niavius L., 300
Amauris niavius niavius, 298
Amauris psyttalea damoclides, 298
Amniocentesis, 340
Amylo-1,6-glucosidase, 243
Anencephaly, 113
Anemia
  pernicious, 63
  sickle-cell, 237, 239, 321, 340
Anidrosis, 70

Ankylosing spondylitis, 149
Anopheles gambiae, 323
Argynnis paphia L., 299
Ascertainment, 9, 94, 103
Atrophy
  Werdning-Hoffmann, 310
  peroneal, 87
  progressive muscular, 86, 90
Autolysis, 147
Autosome, supernumerary, 148

Beagle, The, 273
Biston betularia L., 294
Blood group antigens, 137
Blood groups, 5, 39, 42
  ABO, 5, 39, 52, 53, 274, 275
  Duffy, 149
  MN, 5, 40, 43, 52, 53, 274, 305
  Rh, 5, 274
Boarmia punctinalis, 294, 296
Brachydactyly, 4, 5, 16
5-Bromodeoxyuridine, 151
Buffy coat, 147

Cancer, mammary, 269
Cell function, 150
Centaurea nigra, 309
Centromere, 149
Character gradients, 299
Characteristics
  discontinuous, 113
  quasi-continuous, 113
Chondrodystrophy, 139
Choreoathetosis, 207
Chromatid exchange, 147
Chromosomal localization, 178
Chromosome
  B group trisomy, 162, 172
  heteropyknotic, 169
  sex, 58
  trisomy, 160, 163, 166, 168, 171
  trisomy of number 3, 163

355

About the Editors

WILLIAM J. SCHULL is Professor of Population Genetics and Director of the Center for Demographic and Population Genetics, University of Texas Health Science Center at Houston. He obtained his Ph.D at Ohio State University in 1949, has published extensively in human genetics, and is particularly well known for his studies (in association with James V. Neel) of the genetic effects of exposure to the atomic bombings of Hiroshima and Nagasaki. He is a past president of the American Society of Human Genetics and a frequent consultant to national and international agencies.

RANAJIT CHAKRABORTY received all of his post-school training at the Indian Statistical Institute, Calcutta, where he received his Ph.D. degree in Biostatistics with guidance from Professor C. R. Rao, F.R.S. in 1971. Author of some eighty articles, his current research interests range from theoretical molecular population genetics to medico-legal issues of human genetics and genetic epidemiology. He is currently an Assistant Professor of Population Genetics with the Center for Demographic and Population Genetics, University of Texas Health Science Center at Houston. Concurrently, he has an appointment in the School of Public Health of the Health Science Center where he teaches courses in statistical inference and statistical models in genetics.

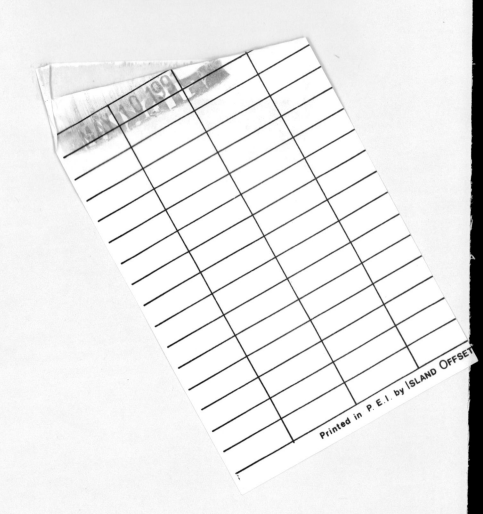

MAY 10 199

Printed in P.E.I. by ISLAND OFFSET